Smart Hybrid AC/DC Microgrids

Smart Hybrid AC/DC Microgrids

Power Management, Energy Management, and Power Quality Control

Yunwei Ryan Li
Farzam Nejabatkhah
Hao Tian

This edition first published 2023

Registered Offices
John Wiley & Sons Ltd, The Atrium, Southern Gate, Chichester, West Sussex, PO19 8SQ, UK

Editorial Office
The Atrium, Southern Gate, Chichester, West Sussex, PO19 8SQ, UK

For details of our global editorial offices, customer services, and more information about Wiley products visit us at www.wiley.com.

Wiley also publishes its books in a variety of electronic formats and by print-on-demand. Some content that appears in standard print versions of this book may not be available in other formats.

Library of Congress Cataloging-in-Publication Data is Applied for:

Hardback ISBN: 9781119598374

Cover Design: Wiley
Cover Images: © metamorworks/Shutterstock, KatieDobies/Getty Images

Set in 9.5/12.5pt STIXTwoText by Straive, Chennai, India
Printed and bound by CPI Group (UK) Ltd, Croydon, CR0 4YY

C9781119598374_130822

Contents

Author Biographies

Yunwei Ryan Li received his Bachelor degree from Tianjin University, China, in 2002, and his Ph.D. degree from Nanyang Technological University, Singapore, in 2006. In 2005, Dr. Li was a Visiting Scholar with Aalborg University, Denmark. From 2006 to 2007, he was a Postdoctoral Research Fellow at Toronto Metropolitan University (previously known as Ryerson University), Canada. In 2007, he also worked at Rockwell Automation Canada before he joined University of Alberta, Canada in the same year. Since then, Dr. Li has been with University of Alberta. Dr. Li is among the first few researchers on microgrids, since early 2000, and his research on microgrids, power quality, and interfacing converters has led to over 350 publications. He is recognized as a Highly Cited Researcher by the Clarivate Analytics.

Farzam Nejabatkhah received his Bachelor (Hons) and Master (Hons) degrees in electrical engineering from the University of Tabriz, Tabriz, Iran, in 2009 and 2011, respectively, and his Ph.D. degree in electrical engineering from the University of Alberta, Edmonton, Canada, in 2017. He was a Postdoctoral Fellow from 2017 to 2019 and an instructor in 2019 at the University of Alberta. Since September 2019, he has joined CYME International T&D, EATON, in Montreal, Canada. In 2021, he was also an instructor at Polytechnique Montreal University. His research interests include smart grids, hybrid AC/DC microgrids, power converters, and cyber-physical systems.

Hao Tian received his Bachelor and Master degrees in electrical engineering from Shandong University, Jinan, China, in 2011 and 2014, respectively, and his Ph.D. degree in energy systems from University of Alberta, Edmonton, Canada, in 2019. Since 2019, he has been working as a Postdoctoral Research Fellow at University of Alberta. His research interests include power quality control of hybrid AC–DC microgrid and multilevel converter topology.

Author Biographies

[illegible]

[illegible]

[illegible]

Preface

Smart grids are becoming as the next generation power systems, which encompass interconnected microgrids with high penetration of renewable generations and energy storage. Hybrid AC/DC microgrids are the most likely future microgrid structure since DC subgrids feature high efficiency, better power quality, better current carrying capacity, and faster response; DC subgrids can also be easily interfaced to the century-long AC utility grid.

During the past decade, smart hybrid AC/DC microgrids have received much attention and experienced significant development. However, as an emerging concept for the future grid structure, their technical details have not been understood very well. For example, with the high percentage of renewable energy and energy storage as well as the wide adoption of interfacing power electronics, there are great challenges in power management among the different elements in a microgrid within a very short time (e.g. a fundamental cycle). Therefore, topics such as power management, power converter control, integration of renewable generation and energy storage, cybersecurity, communication technologies, and power quality are critical for the sound operation of a microgrid. This book thoroughly covers the above topics for smart hybrid AC/DC microgrids and presents effective solutions.

This book contains 10 chapters that cover the basics as well as advanced materials in smart hybrid AC/DC microgrids. The 10 chapters are organized into three main parts: Part 1 Smart Hybrid AC/DC Microgrids, which includes a panoramic introduction to hybrid AC/DC microgrids and different microgrid structures. Renewable based generation and energy storage with their interfacing power converters as well as microgrid communications are presented. Part 2 Power Management Systems (PMSs) and Energy Management Systems (EMSs), which provides thorough discussions on hybrid microgrid planning and operation including interfacing power electronics converter control, instantaneous power management, and microgrid energy management systems. Part 3 Power Quality Issues and Control in Smart Hybrid Microgrids, covers more detailed power quality issues in microgrids as well as their control strategies. Additional opportunities to improve the microgrid power quality with smart interfacing power electronic converters are also addressed.

This book is designed to be suitable for senior undergraduate and graduate students as well as academic researchers and industry engineers in the areas of renewable energy, smart grids, microgrids, and power electronics. The book covers a comprehensive range of topics related to smart AC/DC microgrids and includes a good number of references in each chapter. Acknowledging that this is a rapidly developing field, any comments regarding further improvement to the book are always welcomed by the authors.

The authors wish to express their deep gratitude to their families for support, encouragement, and patience, which was essential to the completion of this book during the challenging pandemic that started in 2020. The authors would also like to thank the members of the ELITE (Electronics and Intelligent) Grid Research Lab at the University of Albert, whose contributions to chapter proofreading, figures, and many helpful comments are much appreciated.

Part I

Smart Hybrid AC/DC Microgrids

1

Smart Hybrid AC/DC Microgrids

Structures and Technical Challenges

1.1 Introduction to Microgrids

1.1.1 Concept of Microgrids

"Microgrids" became jargon in the electrical engineering field at the beginning of the twenty-first century. After nearly two decades of development, the core of this concept keeps expanding and growing along with the development of many other fields, such as power electronics and smart grids. In general, a microgrid refers to a less complex form of an electrical grid, consisting of power generation, energy storage, and consumption as well as essential interfaces. Its functions, on the other hand, entail many more differences than conventional grids [1], e.g. (i) it can work in grid-connected or standalone operation modes; (ii) To the grid, it operates as a self-controlled entity; (iii) it normally features an advanced control strategy to optimally regulate the intermittence from renewable energies, providing high reliability and high power quality; (iv) it is typically located near the users as well as the power generators in a distributed manner, providing high flexibility and cost-effectiveness.

Another important concept closely related to microgrids is distributed generation (DG). DG mainly refers to power generation with distributed forms, differing from the traditional centralized power plant. DG technologies can use sources such as: (i) renewable energy resources such as wind, photovoltaic, micro-hydro, biomass, geothermal, ocean wave, and tides; (ii) clean alternative energy generation technologies such as fuel cells and microturbines; (iii) traditional fossil fuel and rotational machine technologies, such as diesel generators. Due to several benefits of these sources, such as cleanness and simple technologies, compounded with increasing demand for electrical energy and the exhaustible nature of fossil fuels, renewable and clean-energy-based DGs play an essential role in microgrids. Generally speaking, the microgrid is a key concept to broadly adopt DGs into the conventional electrical grid.

Smart Hybrid AC/DC Microgrids: Power Management, Energy Management, and Power Quality Control, First Edition. Yunwei Ryan Li, Farzam Nejabatkhah, and Hao Tian.

1.1.2 Development of Microgrids

The affix "micro" in "microgrid" indicates one iconic nature of this technique, which is its scale compared to the utility grid. However, the traditional grid used to be much smaller when the first power plant was constructed in the 1880s – the Manhattan Pearl Street Station. In terms of scale, it is indeed micro, and can essentially fall into the generalized category of microgrids. It was also operated as the very early combined heat and power (CHP) demonstration where steam was used to heat nearby buildings as well as power the generators.

During the dawn of the electrical grid, Thomas Edison's direct current (DC) grid configuration showed superior performance when supporting power at a short distance. By 1886, Edison's firm had installed 58 DC "microgrids." Things quickly changed after Nikola Tesla, with the Westinghouse company, patented an electric motor in 1888. It exploited the rotating field invented by Galileo Ferraris, showing the promising potential of the alternative current (AC) generator. Further enabled by AC transformer technologies, high voltage AC transmission with high efficiency became possible. In 1891, an experiment regarding such an AC-based transmission technique took place in Germany, where a 175 km long, 15 kV transmission line was implemented [2]. The success of this experiment soon gained commercial attention, resulting in the monopoly of AC-type utility grids until now.

During this early stage of the electrical system, power quality issues like harmonic voltages and currents also gained their engineering-perspective investigation rather than pure mathematical problems. The word "harmonic" firstly appeared in electrical research in 1894 by *Houston* and *Kennelly*'s work entitled "The Harmonics of Alternating Current." The active compensation concept came later during the 1920s [3]: an AC-machine-based compensator was introduced by Boucherot and Kapp. It can adjust the reactive power produced by the machine which shares the similar methodology of modern static compensation equipment.

Alongside the rapid development of a centralized AC electrical system, electricity generation for remote areas (e.g. small islands, isolated mountain settlements, etc.) was challenging based on the traditional grid infrastructure with remote fuel-based power plants and long distance transmission. For those areas, small-scale AC off-grid systems or standalone-only microgrids provided electrical power utilizing techniques such as wind-diesel combinations in the early twentieth century, and even up until now. On the other hand, DC power systems, including DC microgrids, still exist and found their application in systems such as telecommunication systems.

During the last century, worldwide electrical grids experienced significant growth, driven by the everlasting demand for electricity generation. In 1924, the first event of the World Energy Congress was held in London. The concerns regarding limited sources of fossil fuels and dramatically increased energy demand embarked energy experts on exploring alternatives. Solar energy was described as a promising candidate in F. M. Jaeger's article published in *Science* in 1929 [4]. More detailed discussions of alternative energy forms covering water, wind, solar, and nuclear (at that time it was called atomic) were provided in C.C. Furnas's article published in *Science* in 1941 [5]. Similar discussions are scattered in historical publications but rarely conveyed into market driving forces toward sustainable energy eco-systems until the first energy crisis in the twentieth century. The 1973 Arab oil embargo, a turning point for the United States energy strategy, resulted in a chain reaction that soon spread out worldwide. One of the eventual reactions was the establishment of the International Energy Agency (IEA). Born from the oil security crisis, the IEA has evolved through the years, pursuing the enhancement of the reliability, affordability, and sustainability of energy. Another important point of progress in history was the 1992 Energy Policy Act in the United States, further strengthening the cost-competitiveness of renewable energy technologies.

In addition to utility-scale regulation, small scale distributed power generation was also taken care of by national policies, e.g. through the 1978 Public Utilities Regulatory Policy Act, the United States became the first country to establish fixed power buy-back rates (i.e. independent producers are allowed to connect to the grid and sell power). The rapid growth of electricity demand keeps pushing the electrical grids to their design limits. During the 1980s–1990s, the economic value of DG started to be recognized as a good complement to the monopoly of the traditional grid. In addition, DGs can support critical electrical needs in rural areas that are difficult to be covered by the centralized grid infrastructure.

At the end of the twentieth century, distributed-resources-based systems received dedicated research attention, which eventually spawned into the concept of modern microgrids, where power electronics serve as vital interfaces bridging renewable energy generation and the load and grid. In 1999, the United States microgrid research development and demonstration program was established under the Consortium for Electric Reliability Technology Solutions (CERTS). The 2005 Energy Policy Act was more energy legislation that was of great significance not only in the United States but also worldwide. It covers a wide scope of renewable energy forms, emphasizes research and development, and promotes the study of advanced energy technologies such as DG, integrated thermal systems, reliability of energy production, etc.

The following years witnessed intense research of microgrids. The trajectory of microgrid technology is shifting from technology demonstration pilot projects to commercial projects, which have grown into a multi-billion-dollar market. In addition to pure electrical power generation, microgrids with CHP applications brought significant opportunities by optimally regulating multiple energy forms for local customers to achieve much better overall efficiency. This is particularly true considering the much higher efficiency of transmitting electricity over a relatively long distance and the flexibility of DG locations. The concept of "district heating" presented in 1950 is a typical precedent that promoted the combination of thermal/electric stations to generate all the heat and power for a town [6]. In recent years, the philosophy of integration has been further extended to clusters of microgrids for a broader scope of energy generation, forming the virtual power plant (VPP) concept, which is not restricted to physical locations and can include assets connected to any part of the grid.

Moving forward to the third decade of the twenty-first century, a number of countries have announced pledges to achieve net-zero emissions in the future, e.g., IEA 2021 report "Net Zero by 2050" [7]. This is when microgrids as well as their larger interconnected systems will play key roles to better integrate renewable-based DGs with higher reliability, lower cost, and easier accessibility. The challenges are huge but there has been promising progress in recent years. Considerable research efforts have been dedicated to smarter operation for microgrids, e.g. multi-function optimization, fast and reliable power regulation, comprehensive power quality management, advanced communication, etc. Moving forward, microgrids also serve as one of the key enabling techniques for next-generation power systems, i.e. smart grids. These smart grids encompass interconnected microgrids, especially at the distribution level where DGs are increasingly used.

1.1.3 Features of Modern Microgrids

With many years of development of microgrids and the enabling technologies in power electronics, communications, and control, a modern microgrid includes a physical electrical system with renewable and non-renewable-based DGs, energy storage systems (ESSs), and various loads, as well as the communication and control systems as shown in Figure 1.1.

As illustrated, the microgrid has a higher-level control system that monitors and coordinates the physical components through communication systems. Control system information acquisition and command execution are mainly realized by the actuators, such as sensors, relays, and, most importantly, the interfacing converters (IFCs). IFCs are interfaces between the microgrid network and renewable energy, ESSs, loads, or another microgrid network, performing power conversions required for the interconnections. The modern microgrids are expected to have some distinctive features, as shown in the next sections.

High Percentage of Renewable Energy and Energy Storage

Carbon emission and pollution from fossil-fuel-based power generation have been considered as the major challenges confronted by human beings. However,

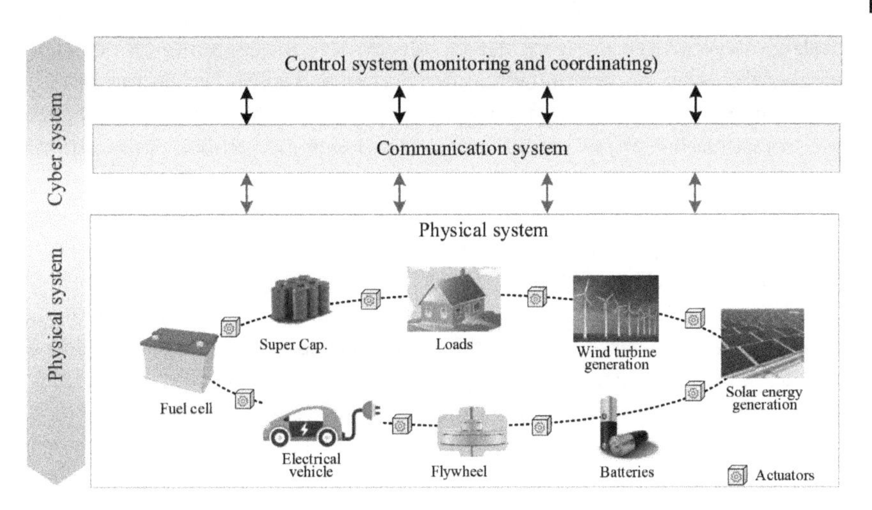

Figure 1.1 Block diagram of modern microgrids.

renewable-based electricity generation, such as the generation from solar and wind, generally suffers intermittencies and lacks complete control. Microgrid technology provides a solution to adopt more renewable-energy-based power generation without degrading the reliability and power quality of the grid through the integration of ESSs, controllable loads, and the corresponding coordination control.

High system efficiency. As mentioned earlier, the flexibility of providing CHP systems can significantly improve the energy efficiency of microgrids. In addition, the adoption of DGs closer to the loads can effectively reduce the losses in the traditional transmission and distribution systems. Moreover, much renewable-based DG such as photovoltaic (PV), fuel cell, ESSs, and modern loads (such as LED lighting) are based on DC technologies, where the many AC–DC conversion processes through IFCs for connecting them to the traditional AC buses can create additional losses. The suitable microgrid configuration where both AC and DC buses exist to interface different types of generation, storage, and load technologies can further improve the system efficiency.

Resiliency. Microgrids are designed to work in both grid-connected or standalone operation modes. This enables the autonomous operation of microgrids when the utility grid suffers from blackouts or major disturbances. In this case, power systems with microgrids are more resilient than traditional ones. Moreover, with more sensors and actuators, such as the controllable and flexible IFCs, microgrids have enhanced the capability for power control and management, leading to sound system operation and higher reliability.

Intelligent. With the development of information and communication technologies, their role in modern microgrid operation and control is becoming more important than ever. Modern microgrids have the capability of system status monitoring, intelligent power and energy management, operation optimization, and outage control.

Superior power quality. With the increasing penetration of interfacing power converters in microgrids, they can be properly controlled to optimize the network power quality in addition to their power management targets. This is a promising idea since most IFCs are not always operating at full rating due to the intermittent nature of renewable-based DGs. Therefore, their available rating can be used in a smart way to support microgrids. Such microgrids will benefit from fewer harmonics, better power factors, lower unbalance, and well-regulated voltage amplitude and frequency.

1.2 Smart Hybrid Microgrid Configurations

According to the type (AC or DC) of buses or feeders to integrate the generation or loads, microgrids can be classified into AC microgrids, DC microgrids, and hybrid AC/DC microgrids. In AC-coupled microgrids, only AC buses or feeders are available and all sources and loads are connected to the AC buses. In a DC-coupled microgrid, all sources and loads are connected to DC buses, and the DC microgrid is interfaced to the main grid through an AC/DC IFC. In hybrid AC/DC microgrids, both AC and DC buses are available, and the microgrid generation sources and loads are connected to the respective buses to minimize the voltage conversion process or optimize the system operation.

1.2.1 AC-coupled Hybrid Microgrid

A simple example of an AC-coupled hybrid microgrid is shown in Figure 1.2. As illustrated, only AC buses are available in an AC-coupled microgrid, and various DGs and ESSs are connected to the AC buses through their IFCs. The ESSs need bidirectional converters to provide the bidirectional power flow capability. In this configuration, both AC and DC loads are also connected to the AC bus where the DC load will require an DC/AC IFC for such integration. This AC-coupled structure is commonly used when dominant generation sources in the microgrid produce grid-level AC voltages directly (such as from diesel generators) or indirectly through interfacing power converters.

In such an AC-coupled system, the control strategy and power management scheme are mainly focused on power generation/consumption balance and AC subgrid voltage/frequency control, especially in standalone operation mode. The

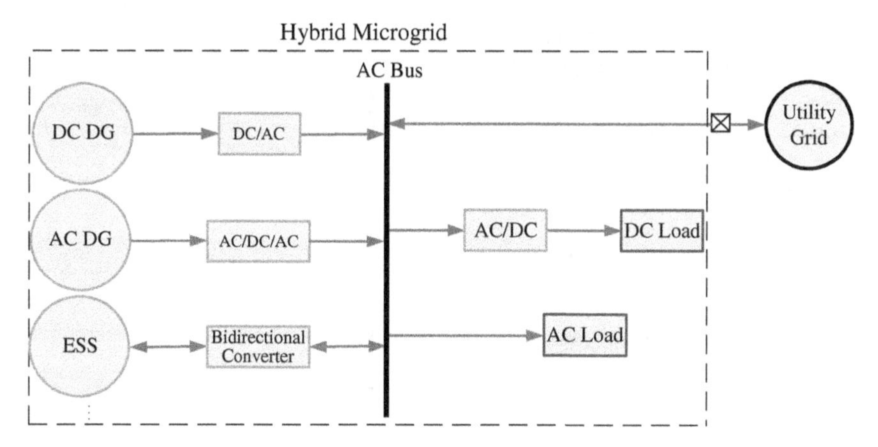

Figure 1.2 A typical example of an AC-coupled hybrid microgrid.

AC-coupled microgrid has been the dominant structure in the past due to its simple structure and simple control and power management scheme.

In some AC-coupled microgrids, instead of using IFCs for each DG or ESS, several power conversion stages can be replaced by multiple-port converters, which combine different power sources in a single power converter. Moreover, in some systems, high-frequency (higher than the power frequency) AC coupling can be adopted for the microgrid, where the microgrid then requires an AC/AC IFC to be connected to the main grid at power frequency.

1.2.2 DC-coupled Hybrid Microgrid

Figure 1.3 shows a DC-coupled hybrid microgrid, where only DC buses are available for integrating the DGs, ESSs, and loads. The AC-based source and loads will then require IFCs to be connected to the common DC bus. This DC-coupled configuration is typically adopted when DC power sources (e.g. PV or battery systems) are the major power generation units in the microgrid. Note that in this structure, all the DGs and ESSs are connected to the DC bus. In this DC-coupled microgrid, a variable frequency AC load such as adjustable speed motors can be connected to the DC bus with a DC/AC converter. In this case, the traditional front-end AC/DC grid side rectifier for AC bus connection can be removed, which brings obvious benefits in control, power quality, and efficiency. In this system, the microgrid DC/AC IFCs provide bidirectional power flow between the DC bus and AC bus. Depending on the power exchange requirement between DC and AC buses, parallel IFCs are typically used with increased rating and reliability.

The DC-coupled microgrid features a simple structure and does not need any frequency and phase angle related synchronisation when integrating different DGs.

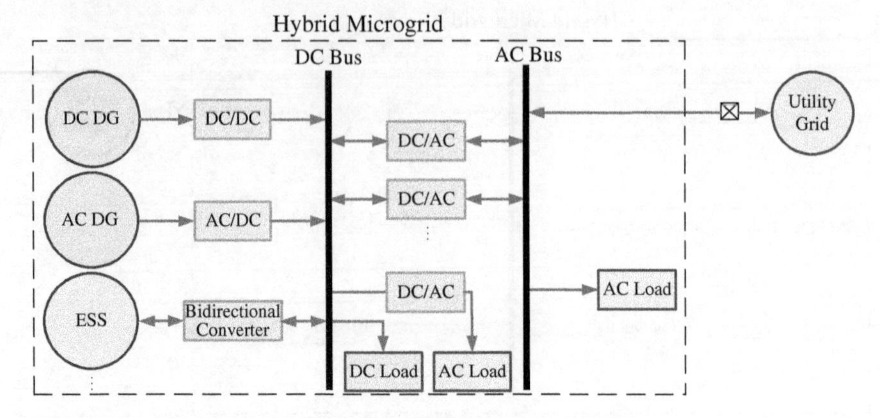

Figure 1.3 A typical example of a DC-coupled hybrid microgrid.

The control and power management of parallel microgrid IFCs, and their AC terminal voltage synchronization (with each other or with the grid in grid-connected mode) can present some challenges. Moreover, both DC and AC voltage control and subsystem power management are necessary for a DC-coupled system. In some DC-coupled hybrid microgrids, ESSs are connected to the DC bus directly without converters.

Similar to an AC-coupled hybrid microgrid, in DC-coupled hybrid microgrids, multiple-port power converters can be used to connect different input power sources to a common DC link in a unified structure.

1.2.3 AC-DC-Coupled Hybrid Microgrid

The structure of an AC-DC-coupled hybrid microgrid is shown in Figure 1.4. As seen, both DC and AC buses are available in such a system to integrate the DGs, ESSs, and loads. The AC and DC buses (or subgrids) are linked by IFCs. Different from the DC-coupled system, the AC-DC-coupled hybrid microgrid has DGs and ESSs on the AC subgrid too, which requires more coordination for the voltage and power control between the DC and AC subgrids. On the other hand, similar to the DC-coupled microgrid, parallel IFCs are desired to link AC and DC subgrids with increased capacity and reliability. In general, this structure is considered if major power sources include both DC and AC powers. This structure improves overall efficiency and reduces the system cost with a reduced number of power converters by connecting sources and loads to the AC and DC subgrids with minimized power conversion requirements.

Considering these benefits, AC-DC-coupled hybrid microgrids will be the most promising microgrid structures in the future.

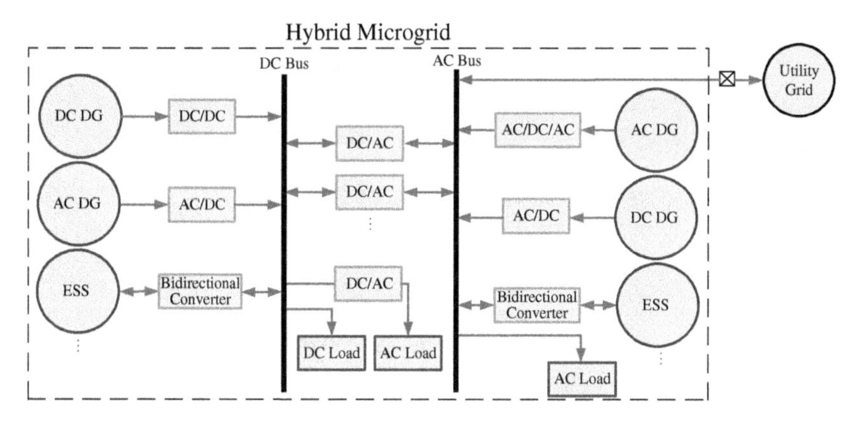

Figure 1.4 A typical example of an AC-DC-coupled hybrid microgrid.

1.2.4 Examples of Hybrid Microgrids

A hybrid microgrid can exist in many different forms, such as a community, campus, institutions, commercial center, or even a microgrid in the sky or ocean like more electric aircraft (MEA) or electrified ships. Considering the rapidly increasing online activities and demand of data centers as well as the wide acceptance of electric vehicles, two examples of hybrid microgrids, data centers, and electric vehicle charging stations, are briefly presented here.

Based on an EPRI report, data centers will consume 20% of electricity in the United States by 2030, and power quality issues are a significant concern in such systems (in the United States, low power quality and unreliable power supply of data centers can result in millions of dollar losses annually). In general, data center structure can be AC-coupled or DC-coupled, which are shown in Figure 1.5. The traditional configuration of the data center is an AC-couple structure, where the AC/DC and DC/DC voltage conversion happens right before the server load. In recent years, the research and implementation of DC data centers (400 V DC distribution) have seen increasing demand. This DC-based structure effectively reduces the current and losses in the system. They also have better performance compared to AC architecture in terms of reliability, efficiency, and power quality.

The charging stations of EVs can have AC or DC structures, which are shown in Figure 1.6. For commercial EV charging stations, level 2 (single/three-phase AC charger, around 20 kW) and level 3 (typically DC-based fast charger greater than 20 kW) chargers are popular. In the AC-coupled structure, all fast DC chargers are connected to a common AC-bus through AC/DC converters, while in a

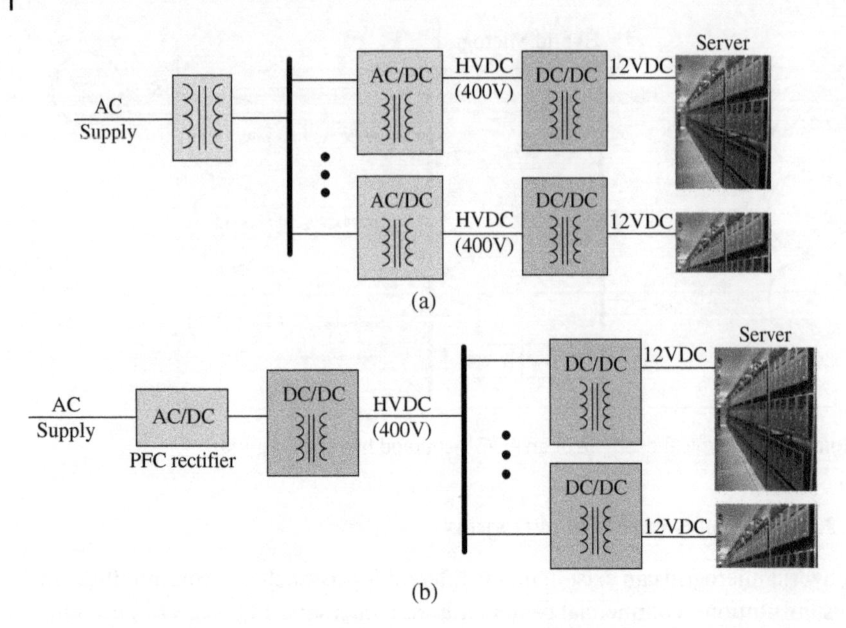

Figure 1.5 Examples of a data center: (a) AC-coupled structure, and (b) DC-coupled structure.

DC-coupled structure they are connected to a common DC bus. Considering the DC voltage requirement for EV batteries, the possibility of integrating renewable energy sources, higher efficiency, and better power quality, DC structures are more promising for charging stations.

Moreover, EV charging stations with level 2 or 3 chargers can easily consume power in the MW level, introducing significant peak power and stress to the station's electrical system (with potentially costly upgrades required). One solution for this is to also include local energy storage to reduce peak power demand. Again, the DC-coupled microgrid solution will ensure less AC/DC voltage conversion is required in such a charging station with battery energy storage.

As mentioned earlier, some standalone microgrids in the sky exist in the form of an MEA with a high-frequency AC bus. In Figure 1.7, an example of an MEA structure is shown. As can be seen, in this AC-coupled hybrid microgrid, the loads and power generators are connected to a common AC bus. The traditional loads such as starter, deicing, etc., are also powered by an electrical system to improve the overall system efficiency. The AC bus is a high-frequency bus with variable frequency

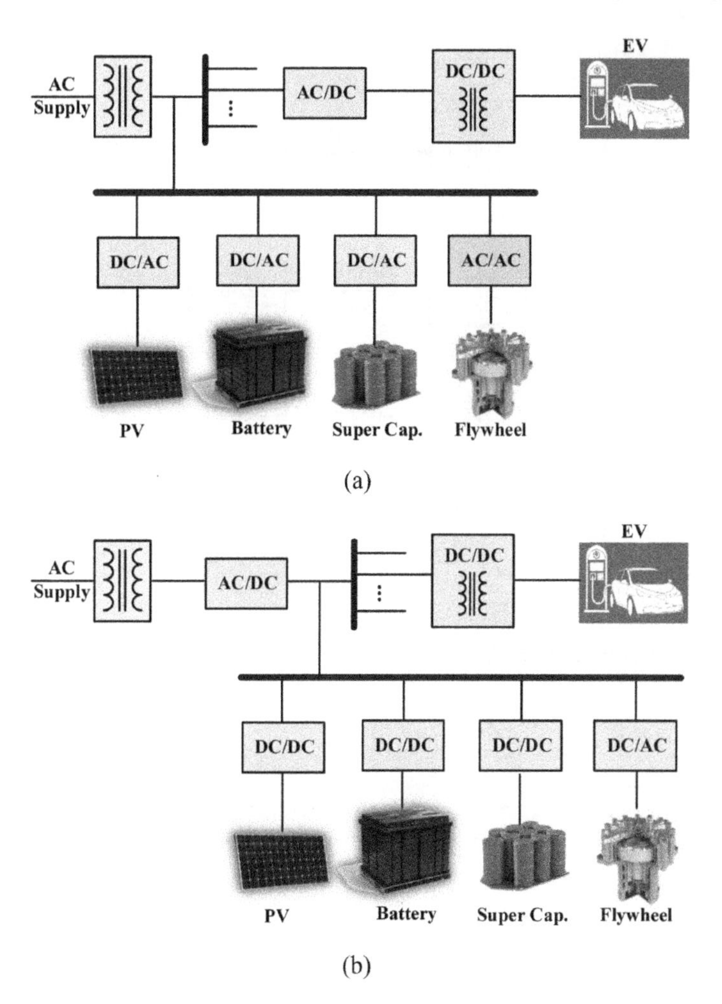

Figure 1.6 Examples of the electric vehicle charging station: (a) AC-coupled structure, and (b) DC-coupled structure.

in the range 350–800 Hz. Compared to the traditional 400 Hz fixed frequency, the variable frequency allows more efficient engine operation.

Similarly, electric ships can be considered as a standalone microgrid in the ocean. In an electric ship, the generators and loads are connected to the medium voltage DC bus on the ship to allow maximum controllability of the generators and loads to optimize the system operation with significant fuel savings.

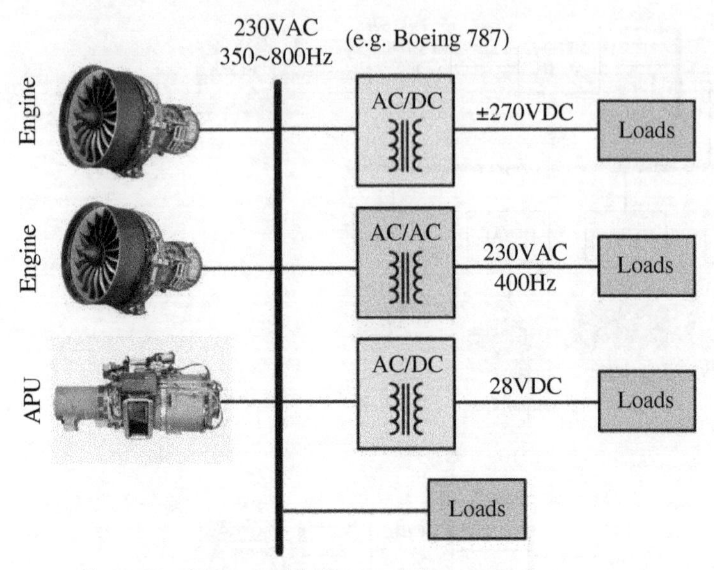

Figure 1.7 The more electric aircraft as a high-frequency AC-coupled microgrid.

1.3 Smart Hybrid Microgrid Operations

Different operation aspects or functions of a smart microgrid are executed at different time scales. For example, the primary control functions of IFCs (i.e. voltage and current controls) must be executed over a very short time (with control bandwidths of hundreds or thousands of hertz), especially when harmonics regulation is needed. The power and power quality management functions are also mostly waveform-based control and need to be implemented with a high control bandwidth. On the other hand, the energy management functions with long-term optimization objectives can be done over a relatively long time scale.

Figure 1.8 shows the basic and smart functions of smart microgrids with their expected time scales. The basic functions mainly focus on individual IFC operations while the smart functions improve smart microgrid stability, reliability, energy efficiency, and power quality. In the following, the key aspects of smart microgrid operation are briefly introduced.

1.3.1 Distributed Generation and Energy Storage Systems

Renewable and non-renewable energy-based DG systems are now widely adopted in microgrids. They can supply loads without long-distance power transmission,

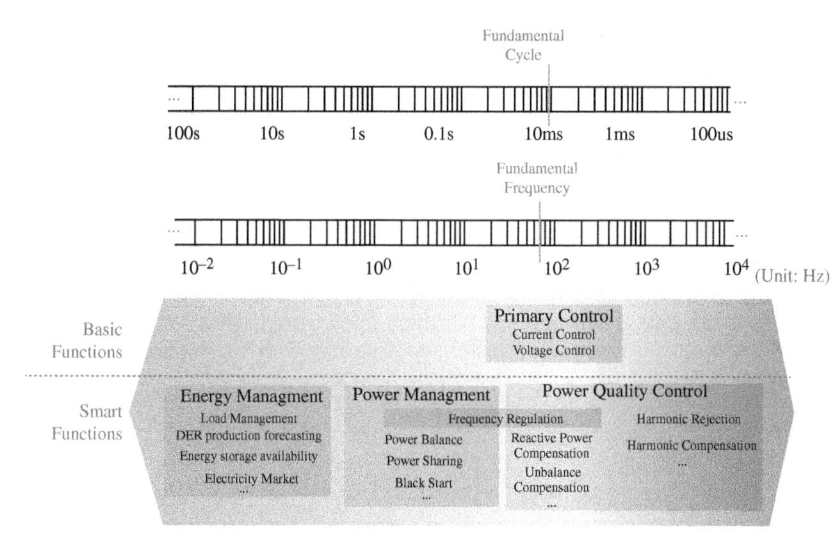

Figure 1.8 Functions of a smart microgrid.

reducing power losses compared to traditional power grids. However, despite the benefits of DG systems, challenges are also raised due to their integration. For example, the power flow within the microgrid is bidirectional, leading to complexity in operation control and system protection. Also, the output powers of DGs are usually not compatible with the grid; thus, appropriate power converters structures with high reliability, efficiency, controllability, and power quality are essential. As another challenge, the intermittence of renewable energy-based DG systems, such as wind or PV power systems, could lead to power fluctuations, power quality issues, or even stability problems, especially in high penetration levels.

Due to the intermittent nature of such renewable energy-based DGs and their low inertia, the load demands and their variations cannot be satisfied at all times, particularly in standalone operation. Therefore, ESSs are required, such as batteries, fuel cells, pumped-hydroelectric, flywheels, and supercapacitors. Some ESSs require proper power converters to be integrated into the microgrid. The ESSs should be appropriately sized to deal with the DG output variations and load demand variations. They should also be placed in a proper location, where different objectives such as improving voltage profile, reducing the system power loss, or reducing the occurrence and levels of abnormal conditions (overloads and over/under voltages) can be targeted. Furthermore, ESSs should be appropriately controlled and coordinated with DG systems to ensure system stability in steady-state and transition operation modes of microgrids.

1.3.2 Smart Interfacing Converters

As mentioned, a key component in a modern microgrid is smart IFCs, which enable flexible power flow control and are the actuators of smart functions. Those IFCs connect renewable/non-renewable energy-based DGs, ESSs, and loads to the AC and DC subgrids of microgrids. The main smart functions of IFCs can be classified into three categories: (i) information-level functions, (ii) microgrid-level functions, and (iii) converter-level functions.

The information-level functions of smart IFCs are mainly focused on data communication through a cyber system. In smart microgrids, physical components, such as IFCs, are usually interconnected to cyber systems, and their operations are coupled to cyber system functionality. Smart IFCs can communicate with control centers and each other either by wireline (such as power line communication and low bandwidth communication) or wireless technologies (such as a ZigBee, WiFi, and cellular communication networks). The data communications of IFCs can help to control microgrid operations in steady-state and transient conditions properly. Also, smart IFCs can be controlled remotely. However, as evident, an efficient, reliable, and timely data flow is required, and any cyber incidents can have devastating effects on the microgrid's operation.

The microgrid-level functions of smart IFCs are used to realize power and energy management system objectives in a microgrid. In general, microgrid energy and power management strategies can be realized in hierarchical control. There are three control layers in hierarchical control: primary, secondary, and tertiary layers. The primary control layer contains voltage and current regulations. The objectives of the second layer include system frequency regulations and power quality compensations such as unbalanced voltage compensation, and harmonic compensation. In tertiary control, an optimization problem is usually run to achieve a global optimum operation point and determine the operating power of each power source. The short-term power management system objectives are realized in the primary and part of the secondary control layers, while the tertiary control and part of the secondary control contain the long-term energy management system.

The converter-level functions are focused on proper power conversion. The IFCs track their reference powers provided by the power management and energy management strategies, using current or voltage controls. In addition, microgrid power quality control using IFCs is also part of the converter-level functions.

1.3.3 Cyber Systems

In general, cyber networks are important for smart microgrids to coordinate, monitor, and manage distributed devices, such as power converters, circuit breakers, and meters. In detail, the cyber system provides sensor information to control units (distributed or central) and transfers the control signals to the physical components such as IFCs and relays. Although smart IFC operation in microgrids is generally autonomous, cyber network presence can improve their control performance. Also, the cyber network is critical for the entire microgrid optimal operations (i.e. energy management scheme) and restoration and black start after faults occurrence.

Since the cyber network operation can impact the physical system performance and functionality, the cyber system's high reliability and security are expected. Malfunctions of cyber systems, such as communication failure or cyber-attacks, could adversely affect the smart microgrids' operation and stability.

1.3.4 Power Management and Energy Management Systems

In microgrids, the terms "energy management" and "power management" are different considering control tasks and time scales. The long-term energy management schemes match the total power production to the demand. The energy management strategies use measured data from sensors and predictions data (e.g. renewable sources prediction) and consider microgrid operational requirements (e.g. appropriate level of power reserve capacity) to optimize the microgrid operation. Generally, the energy management system needs to coordinate the various devices, such as power generators, energy storage, and loads in the microgrid, or even coordinate multiple microgrids.

The objective of the short-term power management system is to control the instantaneous operational conditions toward specific desired parameters such as voltage, current, power, and frequency within a very short time (e.g. a fundamental cycle). In other words, the power management strategies include voltage and frequency regulations, and real-time power dispatching among the microgrid different power sources. The power management strategies should also provide a seamless and smooth transition during microgrids transition between grid-connected and standalone modes (with minimum voltage and frequency disturbances and deviations, and ensure instantaneous power balancing of generation and demand to prevent DGs overloading and circulating powers).

1.3.5 Power Quality

Power quality issues are becoming urgent for future microgrids. They affect the operation of devices in the microgrid, including power converters, protection devices, and loads, while such device malfunctions can lead to further power quality issues. In microgrids, the integration of unbalanced/non-linear loads and unbalanced distributed sources cause the most significant power quality issues. In addition to conventional methods to improve power quality, IFCs from DGs and ESSs can be appropriately controlled to help address such power quality challenges.

In general, the power quality issues in the DC subgrid of hybrid microgrids can be voltage variations and harmonics. On the other hand, voltage variations (frequency deviation and magnitude change), unbalances, and harmonics are major power quality concerns in the AC subgrid. It is also important to note that in hybrid AC/DC microgrids, power quality issues on the AC or DC side can transfer to the other side through the operation of power electronics DC/AC IFCs.

1.4 Outline of the Book

Considering the most pressing technical challenges, this book focuses on the technologies for smart functions of smart hybrid microgrids, including power management, energy management, and power quality control. The book is organized into three parts:

Part 1: Smart Hybrid AC/DC Microgrids

Part 1 of the book includes Chapters 1–3, which introduce smart microgrid fundamentals and background technologies. Specifically:

Chapter 1 presents the concept, histories, and features of microgrids. The configurations of hybrid microgrids and key operation challenges of microgrids are also discussed.

Chapter 2 introduces the basics of renewable-based DGs and energy storage technologies suitable for microgrids. The widely used renewable generation, such as PV and wind power systems, are presented, and their interfacing power electronics converters and control strategies are discussed. Some power converter-based ESSs, including batteries, flywheels, and superconducting magnets, are also discussed, and their coordination with the renewable generators is studied.

Chapter 3 introduces the fundamentals of information and communication networks. The basic concepts, technologies, protocols, and standards of communication systems are comprehensively introduced. The importance of cyber security and the corresponding standards are also discussed in this chapter.

Part 2: Power Management Systems (PMSs) and Energy Management Systems (EMSs)

Part 2 of the book focuses on IFC control, microgrid power management, and energy management. This part includes the following three chapters.

Chapter 4 presents the control scheme for IFCs in a smart microgrid. In this chapter, the popular IFC control structures are reviewed. In addition, some advanced IFC control concepts for smart microgrid applications are presented, including virtual impedance control, droop control, and virtual synchronous generator control.

Chapter 5 presents power management strategies, such as voltage and frequency regulation and real-time power dispatching, for AC-coupled, DC-coupled, and AC/DC coupled microgrids under different operation modes or during operation mode transitions. The black start of the microgrid is also discussed.

In **Chapter 6**, the energy management strategies for a microgrid under hierarchical control are discussed. The applications of artificial intelligence and the multi-agent control for microgrids are presented. Considering the dependence on the communication system, a detailed discussion of cyber security, consisting of types of attacks, consequences, and solutions, is included in this chapter.

Part 3: Power Quality Issues and Control in Smart Hybrid Microgrids

Part 3 of the book addresses the important power quality issues in smart microgrids. Power quality events, such as grid disturbance, voltage harmonics, and unbalance, can be great challenges for the proper operation of microgrids, where smart IFC control can help address those power quality concerns as important auxiliary functions. This part includes four chapters.

In **Chapter 7,** the various power quality issues, such as transients, harmonics, short- and long-term voltage variations, and momentary power supply outages, are overviewed. Existing solutions based on passive filters and the more flexible power electronic compensation devices are reviewed. Finally, hybrid AC/DC microgrid specific power quality issues and challenges, and their compensation strategies are discussed.

Chapter 8 focuses on the control of IFCs during short-term and severe grid disturbances. The microgrid operation under a grid disturbance can be either switched to islanding operation or remain connected for a disturbance ride-through operation according to different grid codes. The islanding detection techniques, ride-through control strategies, and a protection coordination study considering the ride-through control of DG and microgrids are presented

Chapter 9 presents the strategies for smart IFCs to compensate for unbalanced voltages in a microgrid. These strategies utilize the IFC extra power rating to address the voltage unbalance issues without installing dedicated compensation devices. The control schemes discussed in the chapter include three-phase IFCs and single-phase IFCs. Coordinating multiple IFCs to share the unbalance compensation tasks properly is also discussed.

In **Chapter 10**, IFC control schemes to compensate harmonics in microgrids are presented. Moreover, the harmonic control scheme for IFCs with low switching frequency and the compensation of low-order harmonics in the DC subgrid are also discussed. Finally, control strategies to coordinate multiple IFCs to perform harmonic compensation in a hybrid microgrid are presented.

References

1 Li, D.V. and Loh, P.C. (2004). Design, analysis, and real-time testing of a controller for multibus microgrid system. *IEEE Transactions on Power Electronics* 19 (5): 1195–1204.

2 Maloberti, F. and Anthony, C.D. (2016). *A Short History of Circuits and Systems: From Green, Mobile, Pervasive Networking to Big Data Computing*. Stylus Publishing.

3 Emanuel, A.E. (2000). Harmonics in the early years of electrical engineering: A brief review of events, people and documents. *Ninth International Conference on Harmonics and Quality of Power*, Orlando, FL, USA.

4 Jaeger, F.M. (1929). The present and future state of our natural resources. *Science* 69 (1791): 437–445.

5 Furnas, C.C. (1941). Future sources of power. *Science* 94 (2445): 425–428.

6 Smith, D.V. (1950). District heating. *Journal of the Royal Sanitary Institute* 70 (4): 406–420.

7 Net Zero by 2050: A Roadmap for the Global Energy Sector (2021). International Energy Agency (IEA), France.

2

Renewable Energy, Energy Storage, and Smart Interfacing Power Converters

2.1 Renewable-based Generation

The significantly growing power generation from renewables is driven by global carbon reduction policies and the maturing power conversion techniques. Further adoption of renewables in today's electric grids is vital to tackle climate change. This is particularly the case for solar and wind energy generation, which is forecasted to support more than half of the global power generation by 2050. In Figure 2.1, the global electricity generation mix, historical and forecast to 2050, is shown [1].

This section introduces photovoltaic (PV) power systems and wind power systems in detail.

2.1.1 Photovoltaic (PV) Power Systems

PV devices, or solar cells, were developed in 1954 at Bell Labs. PV power generation technology experienced a significant boost in 1974 when the first oil crisis occurred. In recent years, PV power penetration into power systems continues its rapid increase worldwide, serving as an important renewable energy source for a promising low-carbon society. Since 2000, the global PV system installed capacity has grown nearly 400-times, and it had reached about 633.7 GW in 2019. In 2019, China, the US, and India were the top three PV system installers by 30.1, 13.3, and 8.9 GW installed capacity. Even though the installation growth has been significant up to today, PV power generation still only accounted for 2.6% of global power generation as of 2019 [2].

The major components in a PV power system include PV panels and the power conversion system. As the core component, the power conversion system is made of power converters, which track the maximum power of the PV system and convert the PV output voltage from DC to DC or AC, depending on the interconnected

Smart Hybrid AC/DC Microgrids: Power Management, Energy Management, and Power Quality Control, First Edition. Yunwei Ryan Li, Farzam Nejabatkhah, and Hao Tian.

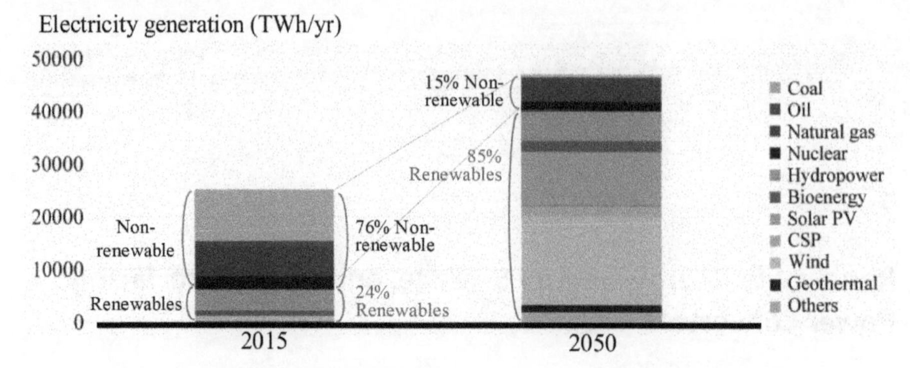

Figure 2.1 Global electricity generation mix, historical and forecast until 2050.

load or grid voltage requirements. Along with the massive production of solar generation components, the PV power production cost has dropped significantly in the last few decades and is expected to continue this trend.

Fundamental Features and Electrical Characteristics of PV

The electrical characteristic of a solar cell can be considered similar to the classical diode with a PN junction, as shown in Figure 2.2. The PV effect converts the photon energy into a voltage across the PN junction, which can typically be modeled as a current source paralleled with one diode and resistors, as demonstrated in Figure 2.3a.

The current of a PV cell based on the one-diode model is expressed as:

$$I_{\mathrm{pv}} = I_{\mathrm{ph}} - I_{\mathrm{D}} - \frac{V_{\mathrm{D}}}{R_{\mathrm{sh}}} \tag{2.1}$$

where I_{ph} is the photocurrent (which is proportional to the solar irradiation), I_{D} is the diode current, V_{D} is the diode voltage, and R_{sh} is the shunt resistance which represents the leakage current. In Figure 2.3, $\mathrm{R_s}$ is the PV cell internal resistance.

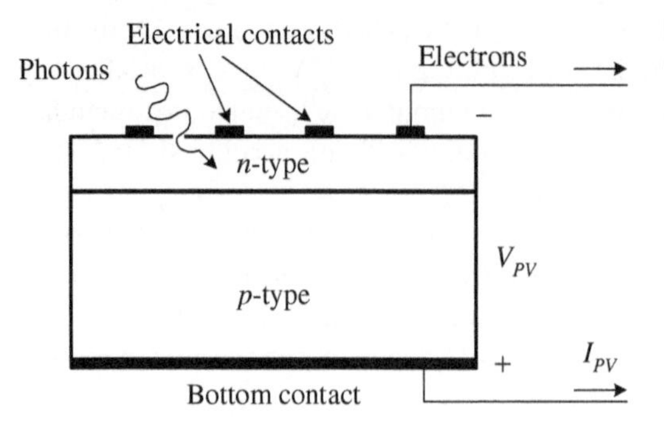

Figure 2.2 Photon energy converted into electrical energy in a PV cell.

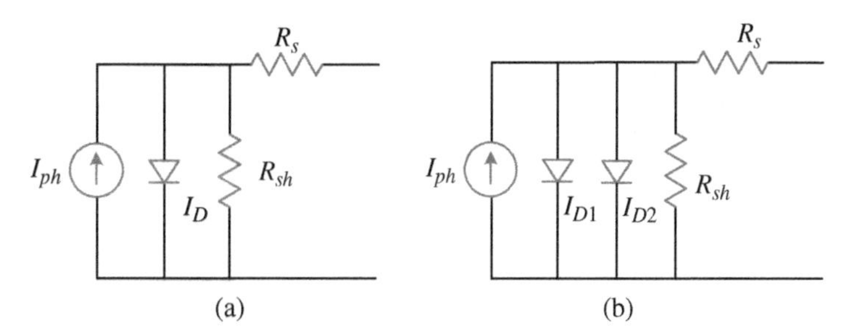

Figure 2.3 Equivalent solar cell electrical model.

The diode current can be expressed as follows:

$$I_{\mathrm{D}} = I_0 \left(\mathrm{e}^{\frac{V_{\mathrm{D}}}{V_{\mathrm{T}} A}} - 1 \right) \tag{2.2}$$

where I_0 is the cell saturation current, V_{T} is the thermal voltage of the PV cell, and variable A is the diode ideality factor and depends upon the material doping.

The PV cell can also be modeled using two diodes for increased accuracy, as demonstrated in Figure 2.3b. Similarly, the current of PV cell based on the two-diode model is expressed as:

$$I_{\mathrm{pv}} = I_{\mathrm{ph}} - I_{\mathrm{D1}} - I_{\mathrm{D2}} - \frac{V_{\mathrm{D}}}{R_{\mathrm{sh}}} \tag{2.3}$$

where I_{D1} and I_{D2} represent the two diode currents, respectively, and they can be expressed as follows:

$$\begin{aligned} I_{\mathrm{D1}} &= I_{01} \left(\mathrm{e}^{\frac{V_{\mathrm{D}}}{V_{\mathrm{T}} A_1}} - 1 \right) \\ I_{\mathrm{D2}} &= I_{02} \left(\mathrm{e}^{\frac{V_{\mathrm{D}}}{V_{\mathrm{T}} A_2}} - 1 \right) \end{aligned} \tag{2.4}$$

where I_{01} and I_{02} are the cell saturation currents, and the variables A_1 and A_2 are the diode ideality factors for diodes 1 and 2, respectively. Although the modeling accuracy is higher, the required parameters are increased significantly for the two-diode model.

The nonlinear feature of PV cells comes from their semiconductor nature. To further illustrate the electrical (I–V) characteristics of a PV cell, two curves, a "dark" curve and a "light" curve, are shown in Figure 2.4 (where I_{SC} is the PV cell short circuit current and V_{OC} is the PV cell open-circuit voltage). Without light, the PV cell is like a simple diode, described by the "dark" curve (a diode curve turned upside down). Without load, the PV cell short circuit current is directly proportional to the solar irradiation. Therefore, the "light" curve can be obtained by adding the "dark" curve with the short circuit current.

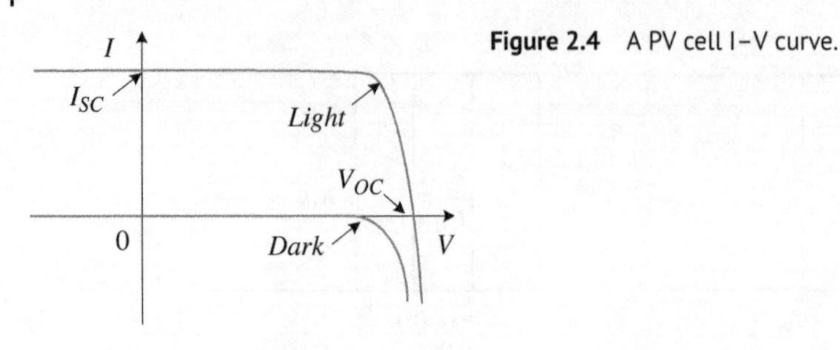

Figure 2.4 A PV cell I–V curve.

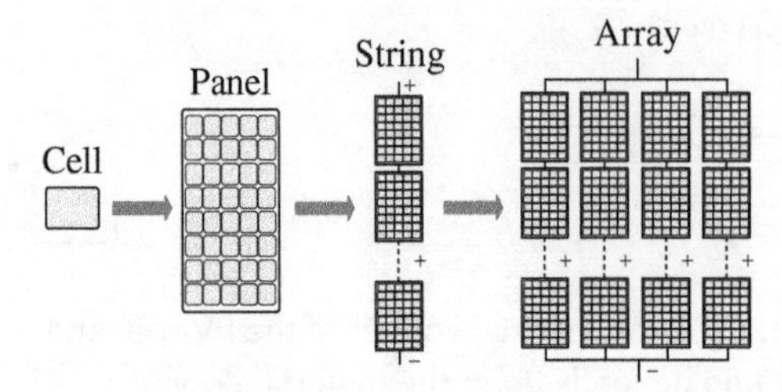

Figure 2.5 PV configuration from cell to array.

The basic building block for PV applications is a panel that consists of series-wired cells to increase the producing voltage and parallel-wired cells to increase the producing current. Multiple PV panels can be integrated into series to form a PV string. Then, several strings can be connected in parallel to form the PV array. Therefore, the practical PV generation system typically contains both series and parallel connected PV panels to obtain optimal performance. In Figure 2.5, PV configuration from cell to array is shown.

Maximum Power Point Tracking of PV Systems

In general, the PV panel's I–V (current–voltage) and P–V (power–voltage) curves are non-linear, as shown in Figure 2.6, and could be influenced by the rapid changing of environmental conditions, such as radiation intensity and temperature.

More energy will be collected if the PV panel is installed on a tracker with an actuator that follows the sun. A sun-tracking design can increase the energy yield by up to 40% over the year compared to no tracker design. The sun trackers

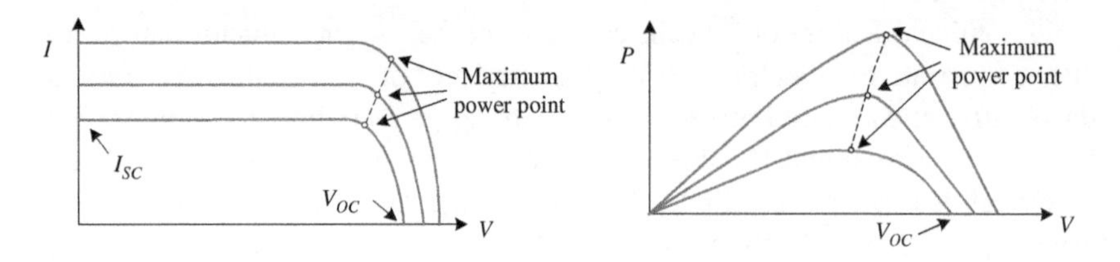

Figure 2.6 (a) I–V and (b) P–V characteristics of a PV panel at different environmental conditions.

drive the PV panel mechanically to face the sun to guarantee the maximum solar radiation input to the PV panel. Another way to guarantee the maximum power output from the panel is typically enabled by actively controlling the PV system to track the maximum power output, called maximum power point tracking (MPPT). These points are shown in Figure 2.6.

The PV system output power is quite sensitive to the ambience and I–V permutations. This means PV generation power could be dramatically lost within a small range of operation point offset. Therefore, to cope with the dynamics of PV characteristics, the MPPT is usually realized by fast-dynamic power electronics devices, e.g. DC/DC or DC/AC power converters. A general scheme of a PV system with an MPPT is shown in Figure 2.7.

There are different methods for MPPT control, which can be classified into three major groups:

- Direct methods that work under any meteorological conditions. Typical methods are Hill Climbing, Perturb and Observe (P&O), and Incremental Conductance (IncCond).
- Indirect methods require prior knowledge of the PV characteristics or are based on mathematical relationships without meeting all meteorological conditions. Typical methods are fractional short-circuit current, fractional open-circuit voltage, and pilot cell.
- Soft computing methods consist of more recent techniques with functions like prediction, data training, and multiobjective tracking. Typical approaches are Kalman filter, fuzzy logic control, neural network, partial swarm optimization, ant colony optimization, artificial bee colony, and bat algorithm.

Among the various approaches, the two most used MPPT methods are introduced in detail below: (i) the P&O method and (ii) the IncCond method. These two methods are compatible with any PV generation and are simple to implement on a digital controller and integrate into commercial inverters.

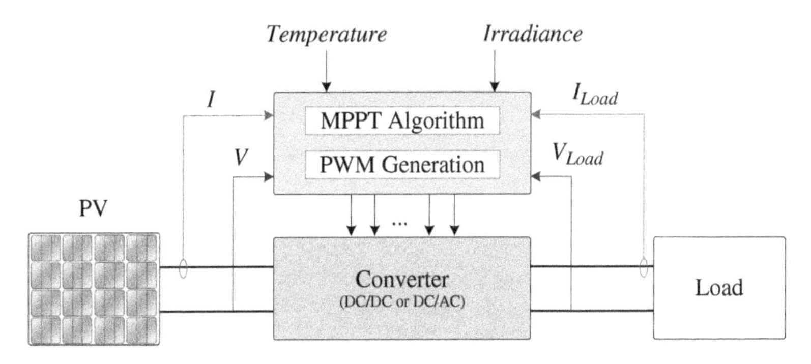

Figure 2.7 General scheme of a PV system with MPPT.

P&O Method The P&O method operates periodically by perturbing the operating voltage point and observing the power variation to deduct the change direction to give to the voltage reference. It is widely employed in practice due to its low cost, simplicity and ease of implementation.

The operating voltage V can be perturbed at every MPPT cycle (fixed or variable) with this algorithm. However, V will oscillate around the ideal operating voltage when the MPP is reached. Thus, the P&O method with fixed perturbation suffers an inherent tracking oscillations trade-off problem. In Figure 2.8, the P&O algorithm is shown. Starting at point A, the operating power gradually increases with a fixed value of 'offset'. Then, based on the P&O algorithm, it passes through point B until it reaches point C and will constantly oscillate between C and B. The oscillation at the MPP region is dependent on the value of step size. While the step size can be made relatively small as $\mathrm{d}P/\mathrm{d}V$ approaches zero and the panel operates in the proximity of the MPP. The adaptive P&O has a good tracking speed and accuracy at a steady state.

IncCond Method The IncCond method utilizes the fact that the slope of the PV power versus voltage (P–V) curve is zero at the MPP, which is expressed by the following equations:

$$\begin{aligned}
&\frac{\mathrm{d}P}{\mathrm{d}V} = 0, \text{if } \frac{\mathrm{d}I}{\mathrm{d}V} = -\frac{I}{V} \text{ (at MPP)}\\
&\frac{\mathrm{d}P}{\mathrm{d}V} > 0, \text{if } \frac{\mathrm{d}I}{\mathrm{d}V} > -\frac{I}{V} \text{ (left of MPP)}\\
&\frac{\mathrm{d}P}{\mathrm{d}V} < 0, \text{if } \frac{\mathrm{d}I}{\mathrm{d}V} < -\frac{I}{V} \text{ (right of MPP).}
\end{aligned} \tag{2.5}$$

Because of measurement errors and the quantification, the condition of the MPP is rarely achieved. To tolerate this, a small marginal error (ε) can be added to the

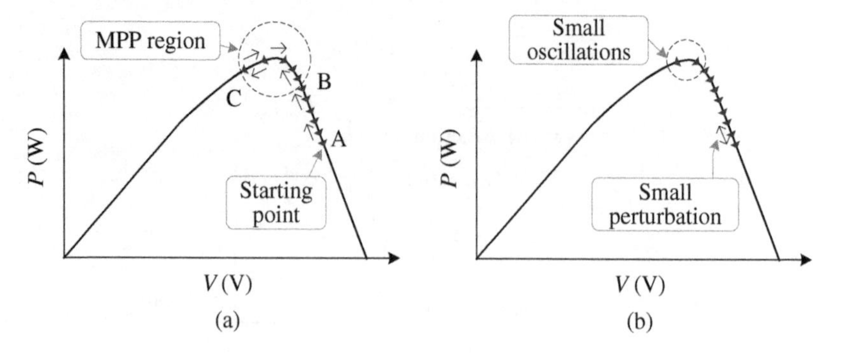

Figure 2.8 (a) The movement of the operating point in MPP tracking operation for large perturbation. (b) For small perturbation.

maximum power condition:

$$\left|\frac{dI}{dV} + \frac{I}{V}\right| \leq \varepsilon. \tag{2.6}$$

Compared to the P&O method, the IncCond method avoids the wrong direction of perturbation under rapidly varying conditions and improves the tracking accuracy and dynamic performance.

PV Power System

The PV power system can operate under grid-connected mode or standalone mode. The power conversion unit converts DC output from PV arrays to the DC or AC form for the load and/or grid in the grid-connected mode. An energy storage system (ESS) (such as a battery) is typically required in the standalone operation mode. In such systems, The PV array supplies power to the loads (DC loads or AC loads through DC/AC inverters) and charges the batteries with sunlight. Also, batteries provide power to the loads when there is no sunlight. The PV output can also be directly coupled to the load without a power conversion unit (e.g. PV-pump system).

In general, each PV system can have:

- different panel configurations
- different topologies for power converters
- different control strategies (MPPT, grid synchronization, and islanding detection are three main required control functions).

The three main configurations for a PV power system connected to an AC system depend on how the PV panels and converters are configured, as shown in Figure 2.9. Each of them has its specific MPPT strategy:

1. Central converters: In this configuration, usually used for PV power plants, PV strings are put in parallel and connected to a central DC/AC converter. Such converters are usually three-phase converters that range from tens of kW to MW. This configuration has high efficiency and low cost due to the small number of power converters. However, it has low reliability and does not have optimal MPPT.
2. String converters: Each PV string is connected to its own DC/AC converter in this configuration. The power ratings of string converters are traditionally 1.5–5 kW, usually used for residential applications. However, a recent trend is that larger capacity string converters (20–100 kW) are used in large PV farms. Each string has its MPPT function with a higher energy yield in this configuration. The strings can also have different orientations.
3. Micro-converters: Each PV panel has its inverter in this configuration, enabling optimal MPPT individually. This configuration has lower efficiency with tens to

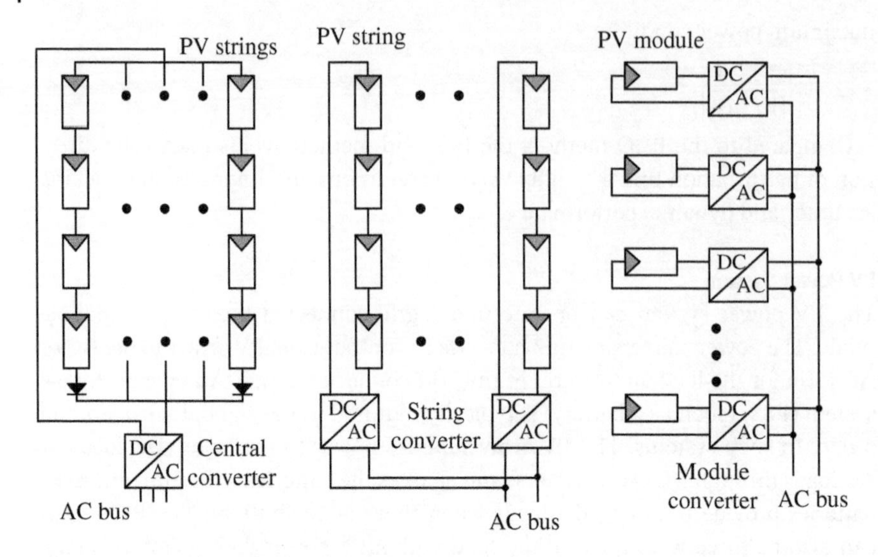

Figure 2.9 Different configurations of PV panels.

hundreds watt power rating due to the many micro-inverters used. For the same reason, this configuration has a higher cost per kW of PV generation. However, it can be used as a plug-in device with a simple installation process with lower maintenance.

Depending on the PV panel configurations, the DC/AC PV voltage conversion may or may not contain a DC/DC converter stage. Thus, the PV converter configurations can be classified into dual power stages configuration and single power stage configuration.

Dual Power Stage Configuration The dual power stage PV converter configuration includes a DC/DC converter and a DC/AC converter. The DC/DC converter performs the MPPT and DC voltage boost, and the DC/AC inverter performs the output voltage and current control. The dual power stage PV converters may contain a high-frequency (HF) isolation transformer (in the DC/DC converter stage in the form of an isolated DC/DC converter) or low frequency (LF) or power frequency transformer (in the AC side of the DC/AC converter stage). The HF transformer leads to a more compact solution and, therefore, lower weight and smaller size but complex design. The isolation transformers separate grounded DC systems from grounded AC systems, and it may not be necessary for ungrounded PV systems. On the other hand, the transformer-less inverters can suffer from large ground leakage current and injected DC current. Due to restricting grid requirements for DC current injection, an insolation transformer is mandatory in many countries. In Figure 2.10, the diagrams of dual power stage PV converters with the LF and HF insolation are shown.

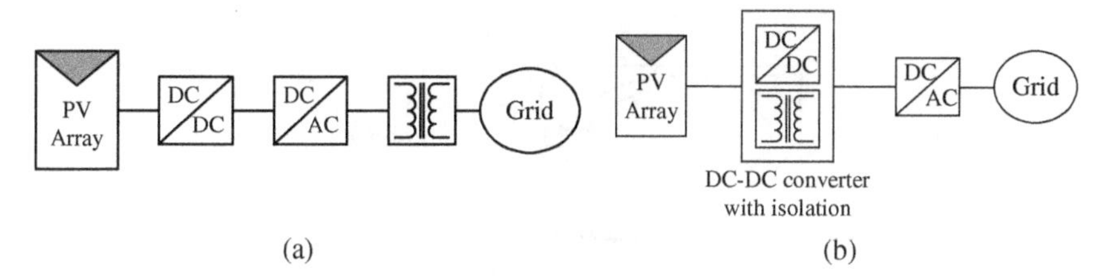

Figure 2.10 Dual power stage PV converters configuration: (a) low-frequency (LF) isolation and (b) high-frequency (HF) isolation.

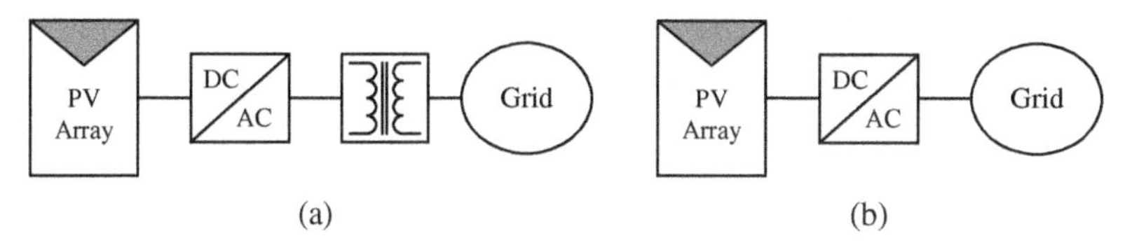

Figure 2.11 Single power stage PV converter configuration: (a) with an isolation transformer, and (b) without an isolation transformer.

Single Power Stage Configuration In a single power stage PV converter configuration without a DC/DC converter, the DC/AC inverter handles all tasks such as MPPT, and output voltage and current control. Such configuration may contain an isolation transformer. In this case, an LF isolation transformer is typically used since this configuration does not contain a DC/DC converter. Some single-stage DC/AC matrix converters with HF isolation are also suitable for this configuration if high power density is desired. In Figure 2.11, the block diagrams of single power stage PV converter configurations with and without isolation transformers are shown.

2.1.2 Wind Power Systems

Wind energy is a pollution-free, infinitely suitable form of energy. The wind energy system converts kinetic energy present in the wind into more useful energy forms such as electricity. The wind power system is now a popular and economical renewable energy source with high penetration into the power grid. Early wind–electric systems can be traced back as early as 1888 to the Brush wind turbine in Cleveland, Ohio. Then, starting from the 1920s, such technology was implemented speedily in isolated communities, e.g. farms of Denmark and France, and spreading later in the US. By 2020, more than 59 900 utility-scale wind turbines with a cumulative capacity of 107.4 GW had been installed in the US. The US had a 10% average annual increase in wind power installation between 2010 and 2020, while the global wind capacity annual increase was around 15% between 2009 and 2019, reaching 651 GW in 2019 [3]. Moreover, wind power system costs have declined drastically in recent years.

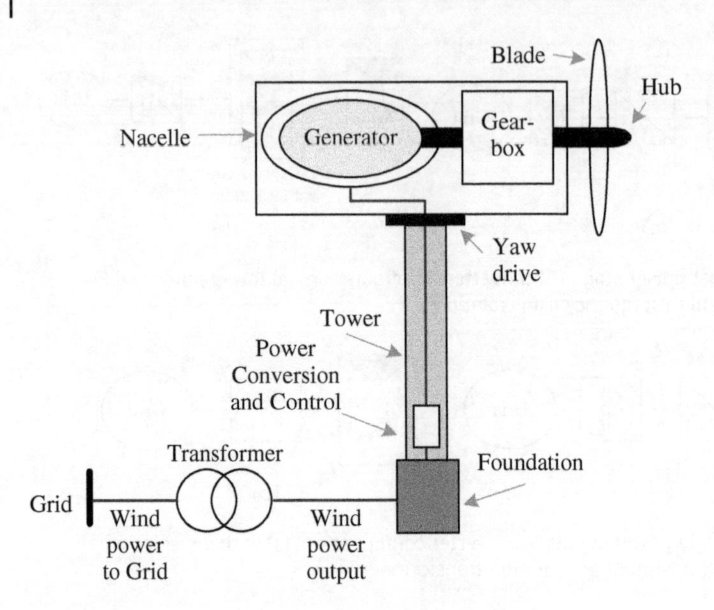

Figure 2.12 Main components of a wind power system.

A typical wind power system consists of a tower, rotor with blades, shaft with mechanical gear, electrical generator, and power converters. In Figure 2.12, a typical wind power system structure is shown.

Wind Turbines

Today, wind turbines range from a few kW (for standalone systems) to a few MW (for utility-scale). Based on the axis around which the turbine blades rotate, the wind turbines can be classified into horizontal axis wind turbines (HAWTs) and vertical axis wind turbines (VAWTs), as shown in Figure 2.13.

VAWTs do not need yaw control to keep them facing the wind, their nacelle can be located on the ground, and their structure is light and balanced with less stress on their blades. However, the turbine blades are close to the ground with low wind speed and have small start torque at low speed. Also, VAWTs cannot be easily controlled at high-speed winds (no pitch control). Therefore, such systems are typically used for small-rating energy harvesting purposes (e.g. street lighting).

Almost all utility-scale wind turbines are of the horizontal axis type. Such turbines can be classified into upwind and downwind HAWTs. The downwind turbines are naturally oriented to the wind direction (wind controls the yaw); however, the wind shadowing effects of the towers cause blades to flex. The upwind turbines are the most popular modern turbines, and they operate more smoothly and deliver more power. Such turbines require yaw control systems to face the wind. Most modern turbines typically have three blades.

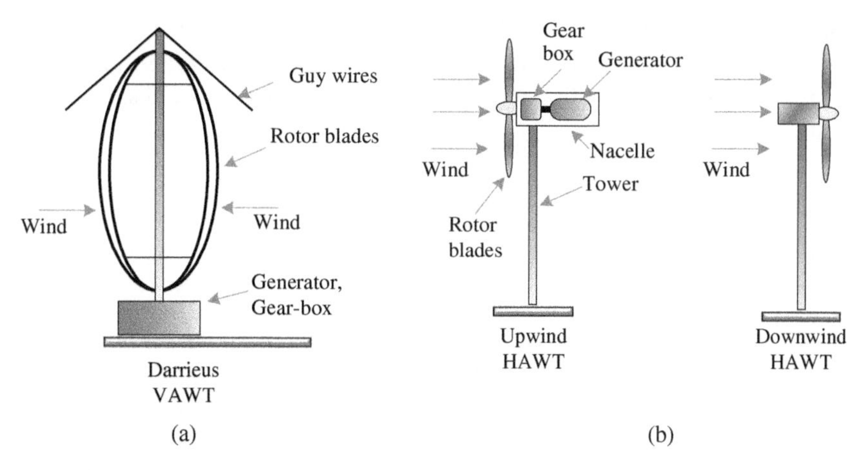

Figure 2.13 Wind turbine classification: (a) VAWT and (b) HAWT.

Fundamental Features and Characteristics of Wind Power

Power in the Wind The power carried by the wind is defined in (2.7), where it is evident that the power increases as the cube of the wind speed, where ρ is the density of air, A is the turbine blade swept area, and v is the air velocity at the turbine blades.

$$P_0 = \frac{1}{2}\,\rho A v^3 \tag{2.7}$$

In a wind turbine generator system, the power extracted from the wind into the wind power system can be calculated as:

$$P_0 = \frac{1}{2}\rho A v^3 \cdot C_\mathrm{P} \tag{2.8}$$

where C_P is the rotor (blade) efficiency, defined as:

$$C_\mathrm{P} = \frac{1}{2}(1+\lambda)(1-\lambda^2) \tag{2.9}$$

and

$$\lambda = \frac{v_\mathrm{d}}{v}. \tag{2.10}$$

In (2.10), v_d and v are downwind (behind the rotor blade) speed and upwind (undisturbed wind) speed. The maximum rotor (blade) efficiency that can be achieved in theory is:

$$\lambda = \frac{v_\mathrm{d}}{v} = \frac{1}{3} \Rightarrow C_{\mathrm{P\_max}} = 59.3\%. \tag{2.11}$$

In practice, for a modern wind turbine system, the maximum rotor efficiency is typically in the range of 40%.

Power Curves of a Wind Turbine As shown above, with the increase in wind velocity, the power delivered by the generator tends to rise by the cube of the wind speed (assume a constant rotor efficiency). There are three important wind speeds in a wind turbine:

- Cut-in wind speed (v_C): minimum speed to generate net power.
- Rated wind speed (v_R): at which the generator produces rated power.
- Cut-out or furling wind speed (v_F): at which the turbine system has to shut down.

In Figure 2.14, the idealized power curve of the wind turbine is shown. As seen from the figure, no power is generated at wind speeds below v_C. Also, at speeds between v_R and v_F, the output power equals the generator's rated power. Above v_F, the turbine is shut down. The wind shed technique is required when the speed is higher than v_R.

Turbine Speed Control One reason to control the turbine's speed is to shed wind in wind power systems. This mechanism helps to protect the rotor, generator, and power electronic equipment from overloading during high-gust winds. Also, when the generator is disconnected from the electrical load, accidentally or for a scheduled event, the rotor speed may run away, damaging the mechanical system if it is not appropriately controlled.

Another reason to control the turbine's speed is to capture more energy from the wind. It should be mentioned that the giant wind turbines should be of a variable-speed design, incorporating pitch control and power electronics, while the small wind turbines should have simple, low-cost power and speed control.

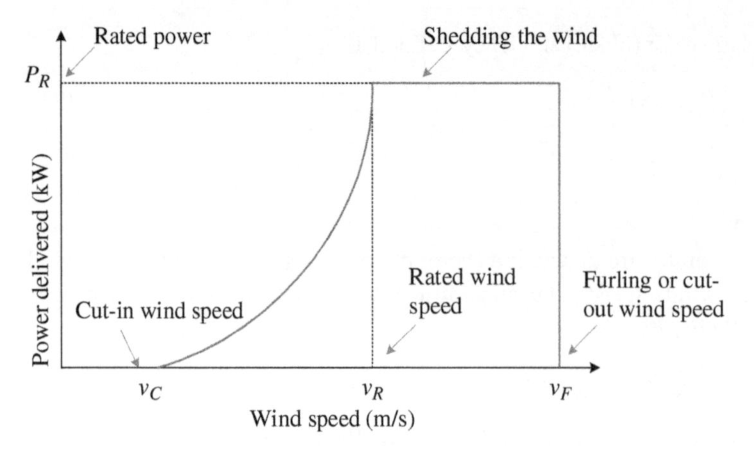

Figure 2.14 Idealized power curve of a wind turbine.

There are a few methods to control the turbine speed mechanically: pitch control, stall control, active stall control, and yaw and tilt control. Increasing the angle of attack (or pitch angle) on the wind turbine blade improves lifting forces; however, increasing the angle too much can result in a stall. The mechanical turbine speed controls can be briefly explained as follows:

- Pitch control: Adjusts the blade's angle of attack by rotating the blades about their axes. For example, when winds are high or if the generator output exceeds the limit, the pitch of the blades is adjusted to shed some wind.
- Stall control: With properly designed blades, stall automatically occurs when excessive winds result in a high angle of attack.
- Active stall control: The blades rotate just like in pitch control. However, the blades rotate in the opposite direction to increase the angle of attack to induce a stall.
- Yaw and tilt control: The rotor axis is shifted out of the wind direction when the wind speed exceeds the design limit. In yaw control, the nacelle is rotated to place the turbine's profile to the wind. In tilt control, the nacelle tilts back until the axis of rotation is perpendicular to the ground.

Maximum Power Point Tracking It is important to ensure the maximum amount of power is extracted from the wind through proper control of the wind turbine generator system by ensuring maximum possible rotor efficiency. To illustrate how maximum power in the wind turbine output can be tracked, the output power of the wind turbine versus turbine speed under different wind speeds is shown in Figure 2.15. From the figure, the rotor power versus the turbine speed curves for given wind speeds v_1, v_2, and v_3 are different, and maximum possible power happens at P_1, P_2, and P_3 for wind speeds v_1, v_2, and v_3, respectively. Therefore, the turbine should be operated at the peak power point at all wind speeds to extract the maximum possible energy.

In Figure 2.16, wind turbine power and rotor efficiency versus wind speed are shown. In the optimum constant C_p region, generator power is increased with increasing wind speed. In the power-limited region, constant power is generated

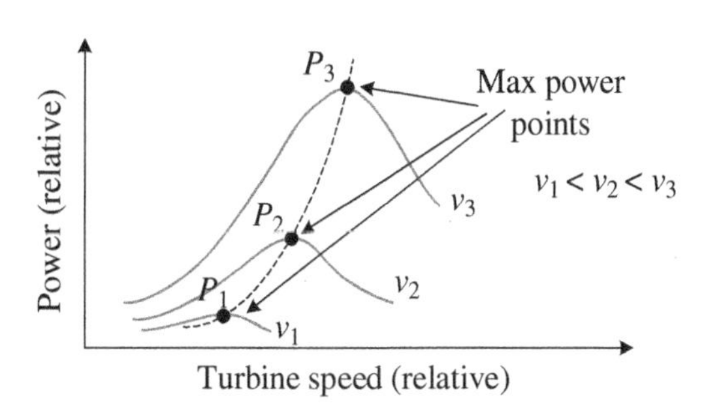

Figure 2.15 Wind turbine output power versus turbine speed under different wind speeds.

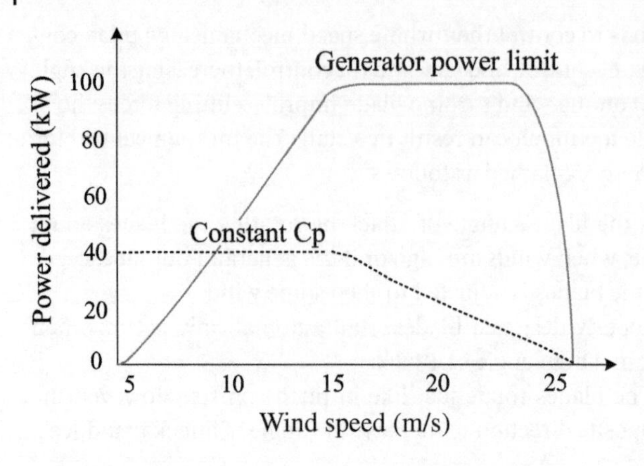

Figure 2.16 A typical wind turbine power and rotor efficiency versus wind speed.

even at higher winds by decreasing the rotor efficiency C_p. Finally, the generated power goes to zero in the power shut-off region.

Wind Turbine Generator Systems

Based on different combinations of the major wind power system components, there are four main configurations for wind turbine generator systems in the industry, Type-1 to Type-4, as shown in Figure 2.17. The Type-3 and Type-4 systems are leading the market in the industry, with Type-4 being adopted more in recent years.

Type-1 Wind Turbine System The wind turbine can be directly connected to the grid through an induction generator (IG), shown in Figure 2.17. This structure has a gearbox to transfer the rotation of the wind turbine to the IG. The IG is for relatively fixed speed operation, and the rotor speed variation range is very limited (almost constant) due to the limitation of slip range in the induction machine. A cage rotor induction machine is typically used in the Type-1 system to lower costs and maintenance requirements. This structure has a reactive power compensation (capacitor banks) required for standalone operation to provide excitation to the IG. Reactive power compensation is also usually adopted for grid-connected operation for voltage support. The Type-1 system also needs a starter circuit for inrush current reduction.

Type-2 Wind Turbine System The IGs can also operate in a small range of variable slips (e.g. 10%), giving a small range of generator speed variation for wind turbine speed control (e.g. 10%). Such generators are wound rotor IGs, in which the external circuit controls their total rotor resistance to control the slip. In Figure 2.17, a

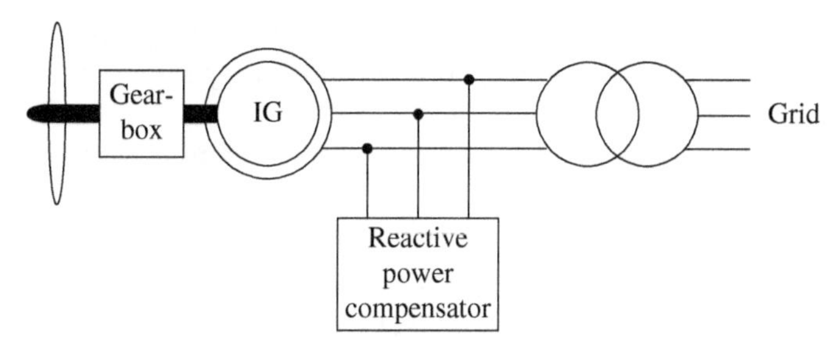

Type-1: Wind turbine direct induction generator-grid connection system.

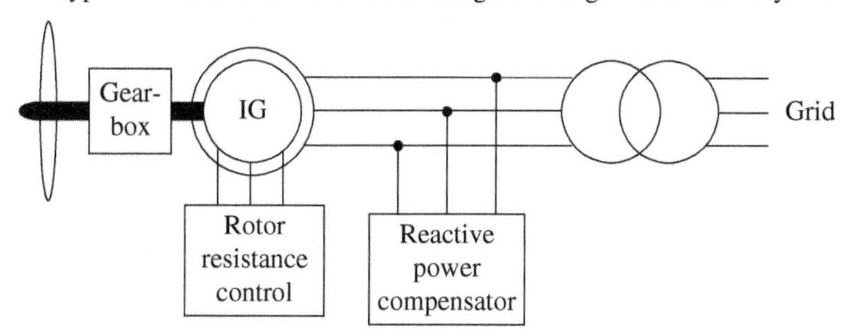

Type-2: Variable-slip induction generator wind turbine with wound rotor.

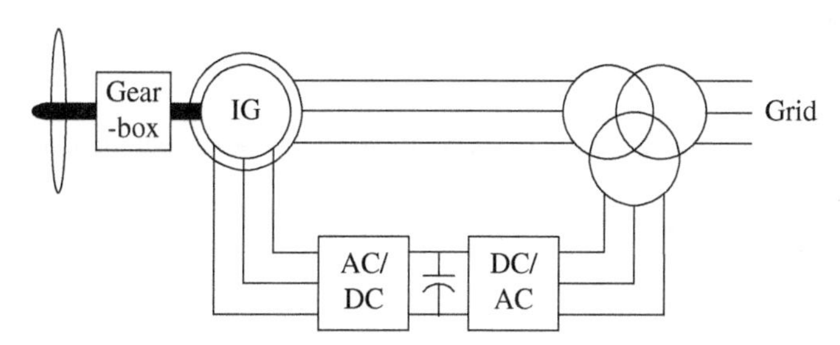

Type-3: Doubly fed induction generator wind turbine.

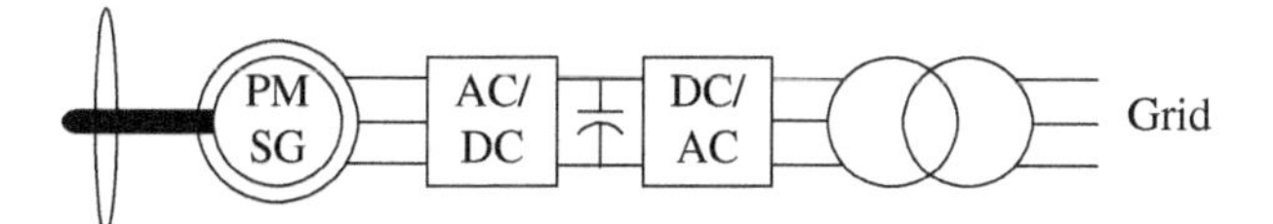

Type-4: Full-scale power converter drive system with a synchronous generator wind turbine.

Figure 2.17 Four types of wind turbine generator systems.

grid-connected variable-slip IG with a wound rotor is shown. Similarly, in the type 2 wind turbine system, reactive power support is provided by capacitor banks.

Type-3 Wind Turbine System The Type-3 system refers to the wind turbine generator system using the doubly fed induction generator (DFIG), shown in Figure 2.17. In this system, a voltage-source power converter feeds rotor-frequency power to the three-phase rotor. The induction machine is fed from both the stator and the rotor. The figure shows that the rotor voltage is supplied from the grid through the power converter, and only part of the power (20–30%) flows through the power converters. In this structure, rotor speed can be controlled by about 20–30%. Also, complete control of real and reactive powers (without capacitors) can be achieved by controlling the power converters.

Type-4 Wind Turbine System The Type-4 system refers to a variable speed system with a generator and a full-scale power converter as a direct drive that can provide full speed range (100%) control. The generator can be an IG or synchronous generation. Typically, a multi-pole permanent magnet synchronous generator (PMSG) is used for high power density and low rated speed. In Figure 2.17, a schematic of the full-scale power converter drive system of a PMSG is shown. The PMSG does not need brushes to be connected to the drive system. With multi-pole construction, it is possible to avoid gears in such a Type-4 system, resulting in reduced weight, size, and maintenance.

Typically commercialized wind turbines of Type-1 to Type-4 systems can be found in the power range of 3–8 MW. These products are manufactured by GE Energy (USA), Vestas (Denmark), Siemens (Germany), Enercon (Germany), Suzlon (India), Gamesa (Spain), Goldwind (China), etc.

Power Converters of Wind Generator-drive Systems From Figure 2.17, the Type-3 and Type-4 wind generators have power electronics-based drive systems. This section briefly introduces the power electronics converters in the drive systems. The Type-3 system typically utilizes back-to-back AC/DC and DC/AC pulse width modulation (PWM) converters to connect to the grid and regulate the doubly fed IG. Similarly, back-to-back AC/DC and DC/AC converter systems are used in the Type-4 system with full-scale speed control. A typical example of the back-to-back PWM converters in a Type-4 system using PMSM is shown in Figure 2.17. Also, it is possible to use diode rectifier on the machine side for a synchronous machine-based Type-4 system, and an example of such a converter system using a diode rectifier, DC/DC converter, and a grid side DC/AC PWM converter is shown in Figure 2.18.

In practice, voltage-source converters dominate the wind turbine market. Although other configurations have not been commercialized yet, they have

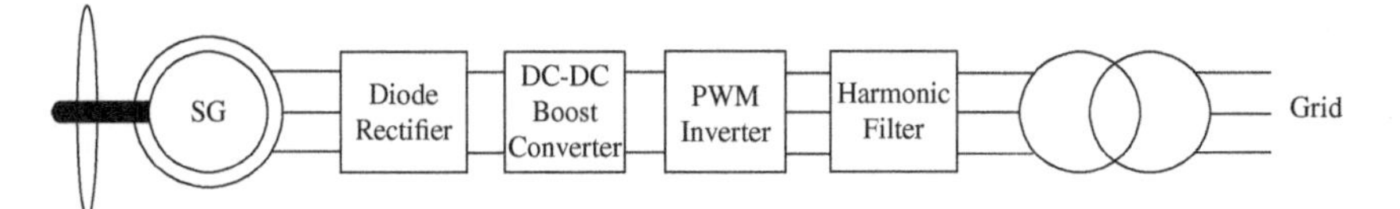

Figure 2.18 Variable-speed wind energy conversion system with DC/DC boost converter in Type-4 wind turbine system.

shown promising features for various applications. These include current-source configurations, modular-converter configurations, and matrix-converter configurations.

2.2 Energy Storage Systems

Electricity cannot be easily economically stored on a large scale. Thus, power generation is varied for conventional power plants to match the load demand. However, for wind or PV power systems with intermittent nature, the generated power cannot meet load demand all the time, particularly in standalone microgrid operation. Therefore, the ESS is key for microgrids or electric grids with a high percentage of renewables.

Based on the physical nature of different energy storage transformations, four categories of ESS can be summarized as:

- Electrical–chemical, including lead-acid battery, lithium-ion battery, and hydrogen energy storage.
- Electrical–mechanical, including pumped hydroelectric storage (PHS), compressed-air energy storage (CAES), and flywheels.
- Electrical–electrical, including supercapacitor and superconducting magnetic energy storage (SMES).
- Electrical–thermal, including molten salt energy storage.

Among these technologies, PHS and CAES are typically for utility-scale applications, and they can provide up to a gigawatt of power for tens of hours. However, their applications are geographically limited. By contrast, the batteries and flywheel technologies are usually not limited by geography; however, they have lower discharge power and shorter discharge time (usually seconds to six hours). The characteristics of widely used energy storage technologies are summarized in Table 2.1. Wisely implementing different ESSs can benefit power grids significantly. In practice, various functions have been essential to clean, efficient, and resilient smart grids, such as ancillary services (e.g. voltage and frequency support), energy arbitrage, and transmission and distribution system upgrade deferral. Furthermore, especially in microgrids, ESSs help increase the

Table 2.1 Characteristics of widely used energy storage technologies.

	Max power rating	Discharge time	Max cycles or lifetime	Max efficiency (%)
Compressed air	1000 MW	2–30 hours	20–40 years	70
Pumped hydro	3000 MW	4–16 hours	30–60 years	85
Moltern salt (thermal)	150 MW	hours	30 years	90
Li-ion battery	100 MW	1 min–8 hours	1000–10 000	95
Lead-acid battery	100 MW	1 min–8 hours	400–1500	90
Flow battery	100 MW	hours	12 000–14 000	85
Hydrogen	100 MW	minutes–week	5–30 years	45
Flywheel	20 MW	seconds–minutes	20 000–100 000	95

Source: Adapted from [4].

adoption of renewable-based generations by storing their excess energy and discharging when intermittent renewable generations are not available.

2.2.1 Battery Energy Storage System

Batteries have been one of the most common ESSs since the nineteenth century. A low-cost battery makes the most economic sense among the various energy storage technologies today. The working principles of a battery fundamentally rely on the electrical–chemical process: the charge mode converts electrical energy into chemical energy while it converts chemical energy into electric energy in discharge mode.

There are different types of batteries in industries, including lead–acid (Pb-acid), nickel–cadmium (NiCd), nickel–metal hydride (NiMH), lithium-ion (Li-ion), lithium-polymer (Li-poly), and zinc–air. For example, in the US, there is operational battery energy storage based on lead-acid, lithium-ion, sodium-based, nickel-based, and flow batteries.

The lead-acid battery was the most commonly used in the energy storage industry due to its maturity and high performance-over-cost ratio. However, lithium-ion batteries are becoming powerhouses in power systems due to their higher energy density and better performance. Furthermore, it is forecast that the lithium-ion battery price could drop 10-fold from 2010 (around $1000/kWh) to 2030 (below $100/kWh), so its penetration will continue to increase.

Fundamental Features

The complex electrochemical reactions inside the batteries are multidisciplinary, and their accurate modeling has always been an important and challenging problem in academia and industry. Three main categories of the battery models can be found in practice:

- physics-based electrochemical models
- electrical equivalent circuit models (ECM)
- data-driven models.

Physics-based Electrochemical Models The physics-based electrochemical models describe the physical and chemical phenomena inside the battery. They are vital for accurately simulating both the external and the internal characteristics of the battery with explicit physical meanings compared to other battery models. For example, the pseudo-two-dimensional model is one of the fundamental models established in the 1990s and applied to different Li-ion batteries.

Electrical Equivalent Circuit Models Compared to internal characteristics, the external characteristics of the battery are frequently more concerned by electrical engineers. They are usually described by the electrical ECM, using resistors, capacitors, and voltage sources to form a circuit network. A typical model consisting of n numbers of RC networks is named the n-RC model, shown in Figure 2.19. In the figure, U_{OC} indicates the open-circuit voltage of the battery, R_{i} represents the Ohmic internal resistance, and the RC network describes the dynamic characteristics of the battery. The general equations of an n-RC model can be expressed as in (2.12).

$$\begin{cases} \dot{U_{\mathrm{D1}}} = -\dfrac{U_{\mathrm{D1}}}{R_{\mathrm{D1}}C_{\mathrm{D1}}} + \dfrac{i_{\mathrm{L}}}{C_{\mathrm{D1}}} \\ \dot{U_{\mathrm{D2}}} = -\dfrac{U_{\mathrm{D2}}}{R_{\mathrm{D2}}C_{\mathrm{D2}}} + \dfrac{i_{\mathrm{L}}}{C_{\mathrm{D2}}} \\ \cdots \\ U_{\mathrm{t}} = U_{\mathrm{OC}} - U_{\mathrm{D1}} - U_{\mathrm{D2}} - \ldots - i_{\mathrm{L}}R_{\mathrm{i}} \end{cases} \tag{2.12}$$

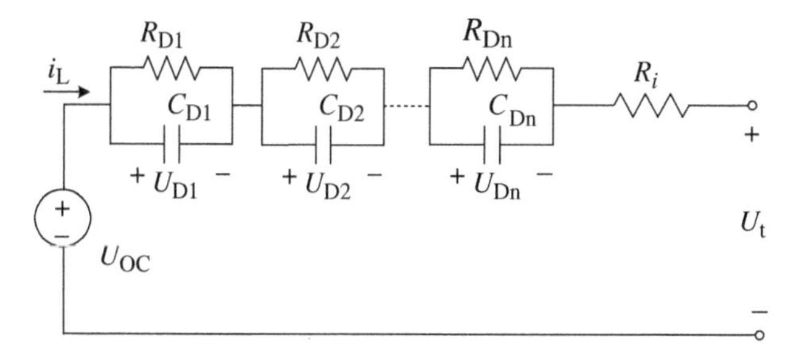

Figure 2.19 Diagram of an n-RC model of a battery.

Exceptional cases for the n-RC equivalent circuit model when $n = 0$, $n = 1$, and $n = 2$ have been extensively used for battery state estimation and management algorithms and are named as:

1. R_{int} model (zero-order)
2. Thevenin model (first-order)
3. dual polarization model (second-order).

Compared to the simplest R_{int} model, the added RC branch(es) usually serve(s) to describe the polarization characteristics during the charging and discharging of the battery. With a higher number of circuit orders, the accuracy increases as well. The battery models may also need to be discretized to blend in with the digital control purpose for battery management systems (BMSs).

Data-driven Models Data-driven modeling approaches are also popular and widely used in developing battery models, e.g. artificial intelligence (AI) algorithms including the neural network and support vector machine. These models can have good performance in nonlinear problems. On the other hand, the training datasets and training methods should be carefully dealt with since they could significantly influence the performance and accuracy of these methods.

State Estimation of Battery

The energy charged in a battery (C) is given by amp-hours (Ah). For example, a fully charged 12 V battery with a 10 hour and 200 Ah capacity could deliver 20 A for 10 hours. The battery capacity also determines the discharge rate of the battery, which is represented by $\frac{C}{hour}$. For example, a 200 Ah battery delivering 20 A is said to be discharging at C/10 rate (C/10 = 200 Ah/10 h = 20A).

Based on the battery capacity, the state of charge (SOC) of a battery can be defined as follows:

$$SOC = \frac{\text{Ah capacity remaining in the battery}}{\text{Rated Ah capacity}}. \tag{2.13}$$

From (2.13), the SOC is the charge level of a battery relative to its capacity. Due to the nonlinear nature of the electrochemical process, the accurate estimation of SOC is not straightforward. In practice, battery SOC estimation methods can be divided into different categories [5]:

- Filter-based methods: e.g. the Gaussian process-based filter approaches (this type utilizes the Kalman filter and its variants to help predict the system state and output) and the probability-based filter approaches (this type uses a particle filter and its variants to overcome limitations of the Kalman filter).
- Observer-based methods: mainly to deal with system linear/non-linear, time-varying, model uncertainties and environmental interference. Some typical

approaches include the Luenberger observer, the sliding mode observer, the proportional-integral observer, and the H-infinity observer.

- Data-driven based methods: consider the battery as a black box and learn the internal dynamics through extensive data, e.g. a neural network, fuzzy logic, genetic algorithm (GA), support vector machine.
- Other methods, including the open-circuit voltage look-up table method, impedance look-up table method, and ampere-hour integral method.

The state of health (SOH) is another critical indicator in batteries, which draws increasing attention due to the demand for long lifetime batteries, especially for electric and hybrid vehicle applications. Both the battery capacity fade and impedance increase are needed to be considered for SOH estimation, typically realized through experimental and adaptive methods:

- Experimental methods measure battery voltage, current, and temperature to monitor the battery behavior.
- Adaptive methods estimate the SOH through estimation techniques on battery circuit models with critical parameters (e.g., capacity decrease, internal resistance growth, etc.), for example, using Kalman Filters and an equivalent circuit model or the least square method with an electrochemical model. They usually has better adaptability to different battery types and chemistries.

Other useful indicators are also recommended to provide a more holographic picture of the battery, such as the state-of-energy, remaining discharge time, state-of-temperature, and state-of-balance.

Battery Charging

Battery charging strategies can be classified into three types:

1. Multiple charge rates: the battery is charged gently in three steps; bulk (fast) charge, taper charge, and trickle (float) charge. In the bulk (fast) charge, 80–90% of the drained capacity is deposited (typically charging with a constant current). In the taper charge, the charge rate is gradually cut back to top off the remaining capacity (typically with constant voltage charging). The last trickle (float) charge step is when the battery is fully charged to counter the battery (the battery slowly self-discharges even with no load).
2. Single-charge rate: the charger is on/off with only one charge rate. The charger is turned off when the battery is fully charged in this strategy. Also, when the battery voltage drops below a preset value, the charger is turned on.
3. Unregulated charging: there is no charger control in this strategy. This strategy can be used in a PV-battery power system, where the system is appropriately designed for safe operation. When the battery is fully charged, the array is fully shunted to the ground by a shorting switch. An isolation diode blocks the battery from powering the array at night.

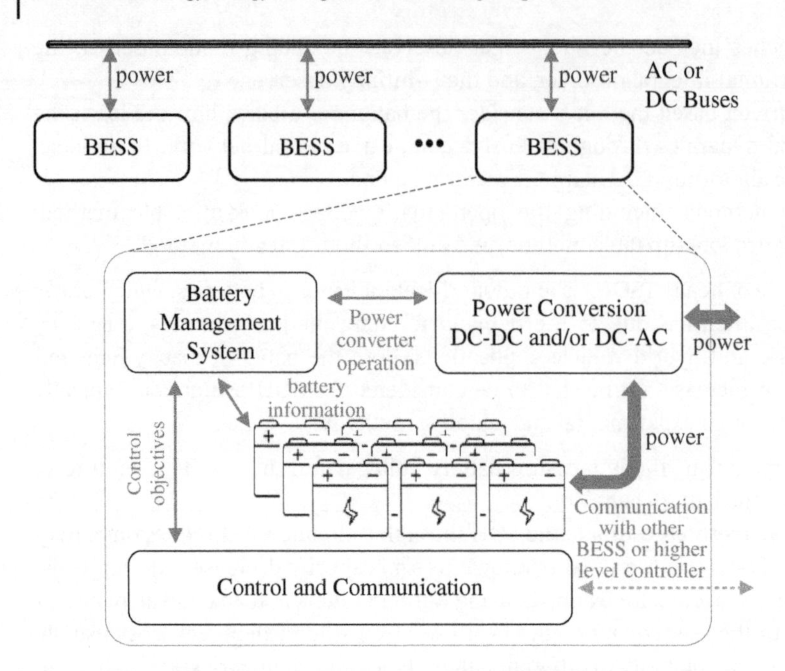

Figure 2.20 Battery energy storage system (BESS) components.

Battery Converters and Control

In smart microgrids, battery energy storage systems (BESSs) are used to smooth renewable-based generation in grid-connected mode. Also, the BESS maintains the bus voltage/frequency in the islanded mode of microgrids. The BESS contains a battery pack, a BMS for operation control, and a power conversion system, as shown in Figure 2.20. The BESS can be connected to the AC and DC subgrids in the hybrid AC/DC microgrids.

Power Conversion Systems of BESSs The BESS can generally be connected to the AC subgrids through bi-directional single-stage DC/AC converters or two-stage DC/DC and DC/AC converters. In addition, multiple parallel DC/DC modules plus a DC/AC converter are suitable for large-capacity BESS.

BESS can also be connected to the DC subgrid, which typically requires bi-directional DC/DC converters (e.g. single-stage or two-stage). Both parallel and/or series connections of multiple such converters can be adopted to increase power handling capability, e.g. interleaved DC/DC converters with increased current ratings and better quality and efficiency.

Controls of BESS Conventionally, power conversion systems in the BESS are controlled in current control mode in grid-connected operation and voltage control mode in islanded operation mode. Therefore, smooth switching between two control modes is necessary. Recently, advanced virtual synchronous generator (VSG) control of power conversion systems for BESSs has attracted much attention. This control strategy provides robust control for distributed BESSs. In such control, each BESS has grid-friendly interconnection as a synchronous generator, and seamless transition between grid-connected and islanded modes is one of the main advantages. In Chapter 4, detailed discussions about these control strategies are provided.

The BESS plays a similar role in DC subgrids/microgrids as in AC subgrids/microgrids; however, its controller is simpler. In DC subgrids/microgrids, the BESS can work on DC bus voltage control mode or output power/current control mode (when the DC bus voltage is regulated with other sources in DC networks). In both AC and DC microgrids, coordination among multiple BESSs in the network is required. One way is the SOC balancing control of distributed BESSs, where the droop coefficients of BESSs are adjusted by considering SOC at discharging.

2.2.2 Flywheel Energy Storage System

The flywheel ESS is mainly used for short-term power management rather than long-term control. It stores kinetic energy in rotating inertia. This energy can be converted from and to electricity with high efficiency. The round-trip conversion efficiency of a large flywheel system can approach 90%. The energy stored in a flywheel (limited by the mechanical stresses) is:

$$E = \frac{1}{2} J\omega^2 \tag{2.14}$$

where J is the system inertia, and ω is the flywheel rotation speed (rad/s). From (2.14), the energy stored in the system is proportional to the speed square. Thus, the SOC can be easily measured by speed. Also, deep discharge is not a concern for a flywheel. For example, reducing speed to one third can discharge about 90% of the energy in a flywheel system. The modern high-speed flywheel systems have a speed of up to 60 000 rpm, and therefore have a high energy storage capacity per unit of weight and volume.

The main flywheel system components are shown in Figure 2.21. The high-speed rotor is attached to the shaft via a strong hub. The bearings with a good lubrication system or magnetic suspension in high-speed rotors are used. In the flywheel system, an electromechanical energy converter is usually a machine that can work as a motor during charging (accelerating the rotor) and a generator while discharging the energy (decelerating the rotor). The power converter drives the motor (charging) and conditions the generator power (discharging). Control circuits for magnetic bearings and other functions are used in such systems.

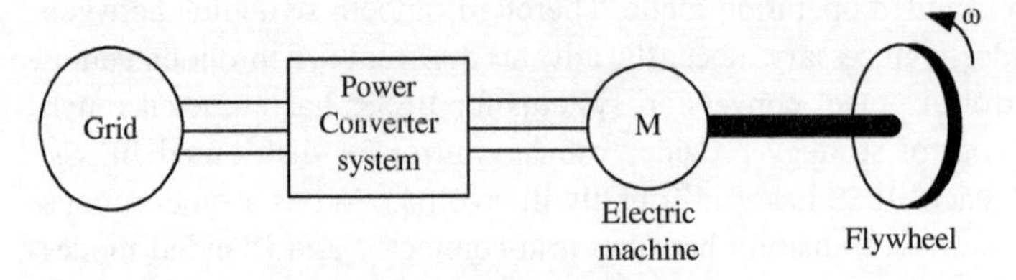

Figure 2.21 Flywheel system components.

2.2.3 Superconducting Magnet Energy Storage System

The energy stored in a superconductor coil or magnetic field is:

$$E = \frac{1}{2}LI^2 \tag{2.15}$$

where L is the coil's inductance (H) and I is the current flowing in the coil (A). From (2.15), the energy charging/discharging can be controlled by regulating the current. For dynamic current control, a voltage is required to be applied to the coil terminals. The relation between the coil current I and the voltage V is as follows:

$$V = RI + L\frac{di}{dt} \tag{2.16}$$

where R is the resistance of the coil (H). For steady-state current control (to store energy in steady-state where $\frac{di}{dt} = 0$), the voltage required to circulate the needed current is:

$$V = RI. \tag{2.17}$$

In a superconductor, the resistance of the coil is temperature dependent. In certain materials, the resistance abruptly drops to a precise zero at some critical temperature. Below this temperature, no voltage is required to circulate current in the coil, and the coil terminals can be shorted. However, the current continues to flow indefinitely in the shorted coil, with the corresponding energy also stored indefinitely. Thus, the coil has attained the superconducting state with zero resistance. The energy in the coil then "freezes." Usually, the abrupt loss of resistance happens at 0 K (Kelvin) = −273 °C (Celsius).

In Figure 2.22, the superconducting magnetic system components are shown. A refrigerator is used to keep the coil below the critical superconducting temperature (typically a few K). The converter system controls the voltage/current and DC power flow to charge/discharge the superconducting magnetic coil. It also provides a small voltage needed to overcome losses in the room temperature parts of the circuit components. This keeps a constant DC current flowing (frozen) in the superconducting coil.

The superconducting magnetic system has some superior characteristics. The round-trip efficiency of the charge–discharge cycle is high at 95%. The charge and

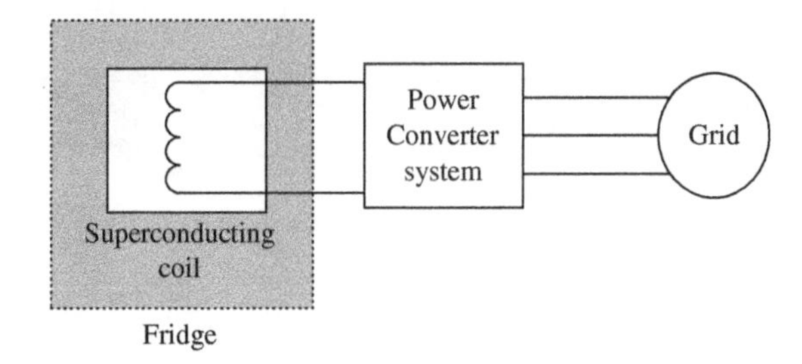

Figure 2.22 Superconducting magnetic system components.

discharge times can be extremely short, making it attractive for supplying high power for a short time if needed. Also, it has a much longer life, up to about 30 years. This system has no moving parts in the main system, except in the cryogenic refrigeration components. The main operating cost is to keep the coil below the critical superconducting temperature.

2.2.4 Hydrogen and Fuel Cell Energy Storage

The production of hydrogen (H_2) has been boosted since the oil embargo in 1973 and now is considered one of the promising energy storage types in addition to batteries, freewheel, and superconducting magnet systems in modern power systems.

Natural gas, oil, and coal-derived hydrogen is the lowest cost at large scales but has poor environmental credentials. With advanced carbon capture and storage, hydrogen from natural gas exhibits relatively low emissions. Biomass gasification may also offer large-scale centralized hydrogen production but higher capital cost. Hydrogen production from electrolysis has a higher cost but is more suitable for small-scale generation, given the modular nature of electrolyzers. Surplus electricity from the utility can power hydrogen production via water electrolysis. In addition to generating hydrogen through non-renewable ways, production methods by sustainable energy sources (e.g. PV, wind turbine, etc.) have emerged in recent decades and well suited for promoting global carbon reduction.

Such "green hydrogen" can be utilized as the common fuels fed into a fuel cell and fulfill the one-step transformation from chemical energy to electrical energy. The complete energy cycle is displayed in Figure 2.23, where the clean energy is generated and stored as hydrogen and recycled through a fuel cell technique, complementary to other sources in a coordinated way.

Electrolyzer Hydrogen Generation

The electrolyzer is a device that produces hydrogen and oxygen from water. The alkaline electrolyzer is a well-proven technology for various scales from less than 1 kW to large industrial electrolyzer plants over 100 MW. In addition, the cost of

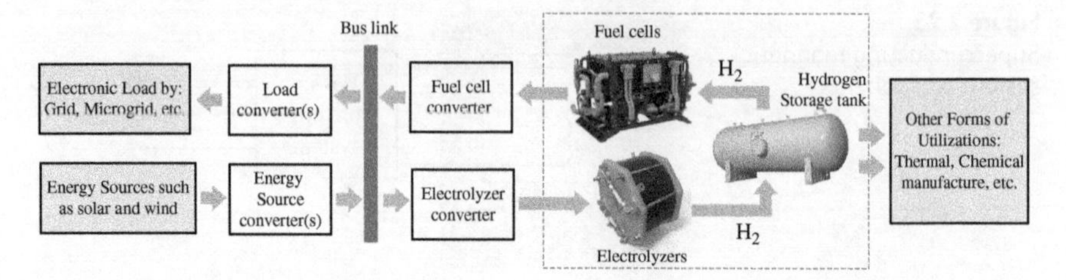

Figure 2.23 Generalized energy flow of a hydrogen-based storage system.

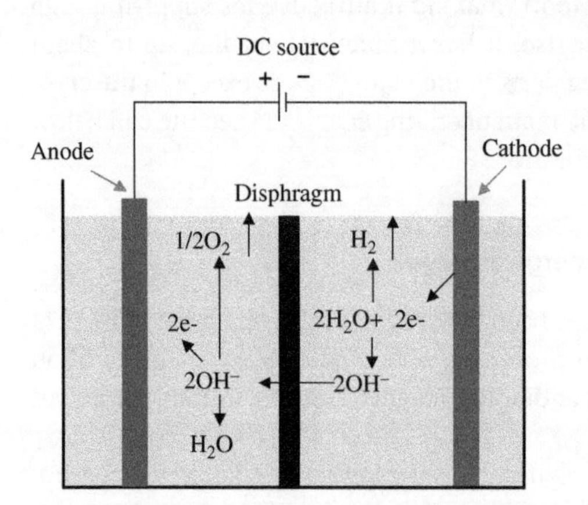

Figure 2.24 Schematic diagram of an alkaline electrolyzer.

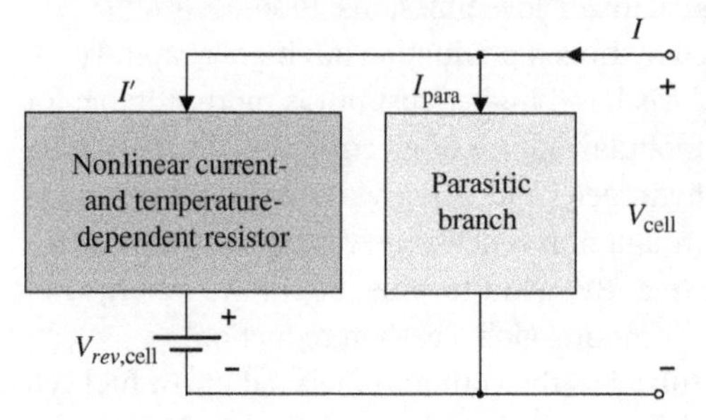

Figure 2.25 Electrolyzer equivalent circuit.

alkaline electrolyzers made in North America and Europe fell 40% between 2014 and 2019 [6]. An alkaline electrolyzer uses potassium hydroxide (KOH) as an electrolyte solution for transferring hydroxyl ions (OH−). Figures 2.24 and 2.25 show schematic diagrams of an alkaline electrolyzer and the electrolyzer equivalent circuit [7].

From Figure 2.25, the internal voltage for each cell can be written as:

$$V_{\text{cell}} = V_{\text{rev,cell}} + V_{\text{drop,cell}} \tag{2.18}$$

where $V_{\text{drop,cell}}$ is the voltage drop across the nonlinear current- and temperature-dependent resistor block. The cell reversible voltage $V_{\text{rev,cell}}$ is determined by the Gibbs free energy due to change in the electrolyzer's electrochemical process. The current can be expressed as:

$$I = I' + I_{\text{para}} \tag{2.19}$$

where $I'$is the current used to generate hydrogen (H_2), and I_{para} is the current due to parasitic current losses mainly determined by cell voltage. The current efficiency or Faraday efficiency is defined as:

$$n_F = \frac{I'}{I} = 1 - \frac{I_{\text{para}}}{I}. \tag{2.20}$$

According to Faraday's Law, the hydrogen production rate of an electrolyzer stack is determined by the number of cells, the current efficiency, and the power supply's input current. However, due to parasitic current losses, the actual hydrogen rate is lower than the theoretical value.

Fuel Cell

The fuel cell is an electrochemical device (similar to the battery technique but features unidirectional power conversion). By synergistic combination with hydrogen, the fuel cell can complement other power sources in practice. Moreover, as a potential alternative to batteries, the fuel cells can be implemented in stationary, portable, and transportation applications. With significant cost reductions in recent decades, several types of fuel cell have emerged, and this has led to increased market penetration. The following is a summary of different types of fuel cells:

- Polymer electrolyte membrane (PEM) fuel cell: leading candidates for use in hybrid electric vehicles with highest available efficiency of around 45%, power range from 30 W to 250 kW.
- Direct methanol fuel cell: also uses the same membrane as the PEM but offers the significant advantages of utilizing a liquid fuel, methanol (CH_3OH), instead of gaseous hydrogen.
- Phosphoric acid fuel cell: operating temperature is higher than that of the PEM fuel cell, close to 200 ° C; the electrochemical reactions are the same as in the PEM but with phosphoric acid as electrolyte.
- Alkaline fuel cell: developed for the Apollo and Space Shuttle programs, with high efficiency and reliability.
- Molten-carbonate fuel cell: operating in a very corrosive environment at a very high temperature, around 650 ° C; waste heat could reform internal fuel.
- Solid oxide fuel cell: with a solid ceramic electrolyte, unlike the liquids and solid polymers in other types of fuel cells; also operating at very high temperature, around 750–1000 °C, featuring internal fuel reforming capability.

A typical fuel cell consists of three components: a fuel electrode (anode), an oxidant electrode (cathode), and an electrolyte between the anode and cathode (see Figure 2.24). The electrochemical process can be described as follows:

1. The hydrogen is fed into the anode and should be oxidized by oxygen (from the cathode) to produce hydrogen ions and electrons.
2. The hydrogen ions migrate through the electrolyte while the electrons travel through the external circuits, forming the electrical current. The overall reaction can be expressed as follows:

$$H_2 + \frac{1}{2}O_2 = H_2O + W_{ele} + Q_{heat} \tag{2.21}$$

where H_2O is water, W_{ele} is electrical work, and Q_{heat} is heat.

The electrical losses of a fuel cell usually come from three ways:

- Activation losses: result from the energy required by the catalysts to initiate the reactions.
- Ohmic losses: result from current passing through the internal resistance posed by the electrolyte membrane, electrodes, and various interconnections in the cell.
- Mass-transport losses: result when hydrogen and oxygen gases have difficulty reaching the electrodes. This is especially true at the cathode if water is allowed to build up, clogging the catalyst.

The non-linear nature of the fuel cell voltage/current (often referred to as polarization curve in practice) results in a typical power curve shown in Figure 2.26. The maximum power point could be limited with specific operating voltage ranges, demanding optimized power tracking/management to maximize operational efficiency. Other factors (e.g. temperature, pressure, and humidity) can also contribute to the performance variation, typically addressed in a multi-physical sense.

The power conversion circuits are necessary to convert the generated electricity of a fuel cell into a grid-compatible form (see Figure 2.23), which often consists of a DC/DC converter and a DC/AC converter if AC power is desired (similar to

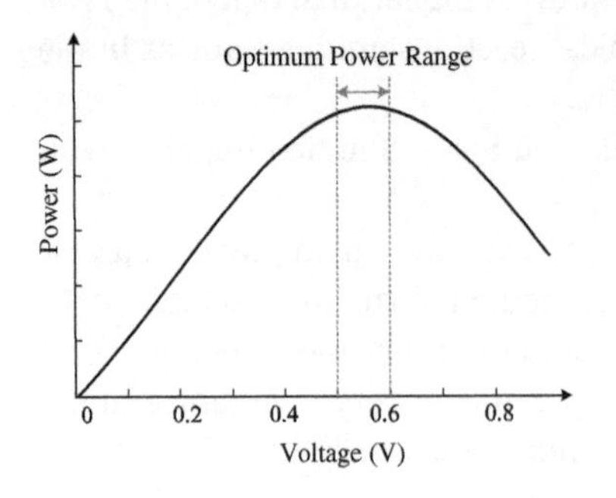

Figure 2.26 Typical polymer electrolyte membrane fuel cell power curve.

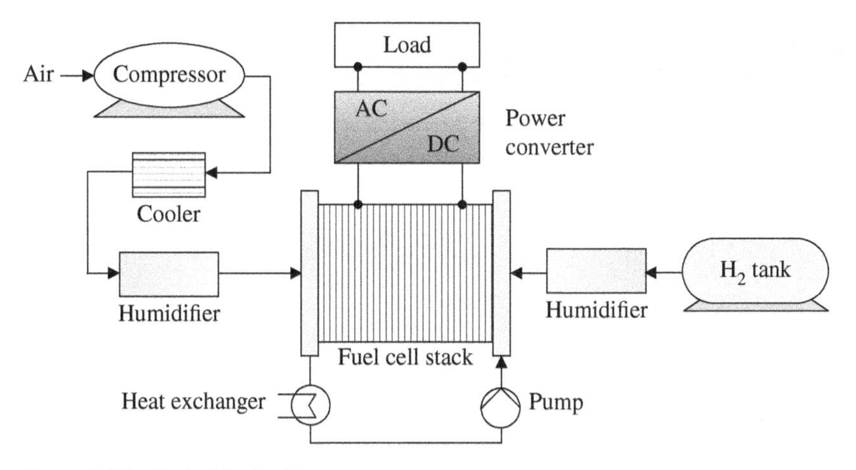

Figure 2.27 Typical fuel cell system structure.

the PV power converter system). If electrical isolation or a high voltage level is required, a transformer is usually integrated into the system, which is preferred to be put in the high-frequency section of the converter configuration for the potential size/material reduction. A general fuel cell power system is shown in Figure 2.27, where the hydrogen and air are compressed and humidified into the fuel cell stack, and the produced power is supplied to the load through a DC/AC power converter.

2.3 Integration of Renewable Energy and Energy Storage

2.3.1 Structure of Smart Interfacing Converters (IFCs)

Smart interfacing power converters (IFCs) are key for integrating distributed generation (DG) and energy storage system (ESS) into smart hybrid AC/DC microgrids and distribution power systems. In addition, they also link AC and DC subgrids in hybrid AC/DC networks.

DC/AC IFCs can be classified into single-stage and double-stage, as shown in Figure 2.28a. The single-stage structure has higher efficiency with a lower power conversion process; however, compromised control flexibility and limited operation range are shortcomings. A more flexible configuration would be the double-stage structure. In a double-stage structure, the DC/DC converter is mainly used to boost the DC-link voltage and control power flow (e.g. achieve a maximum power of PV and maximize fuel cell efficiency). The DC/DC converter can be isolated or non-isolated. Isolation is mandatory in some applications, such

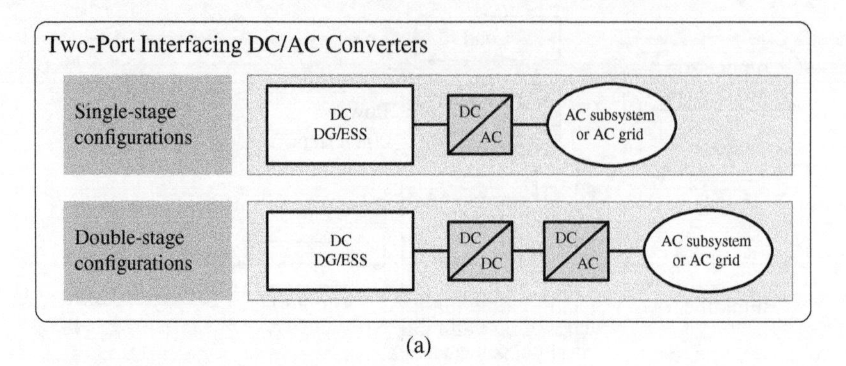

(a)

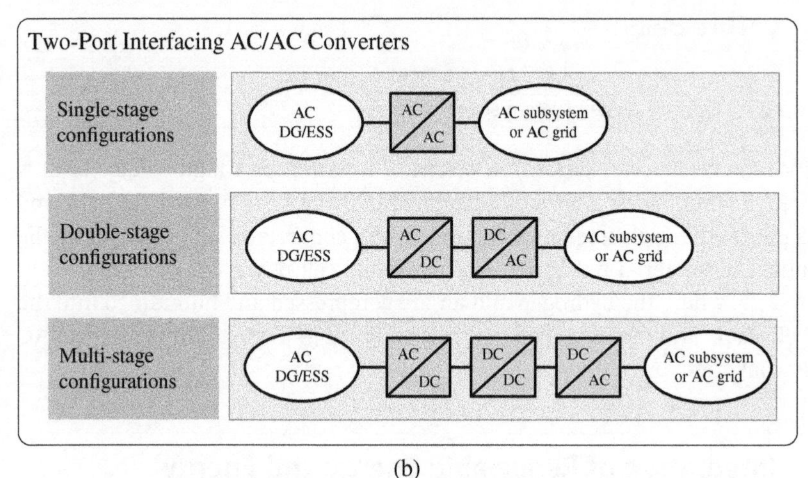

(b)

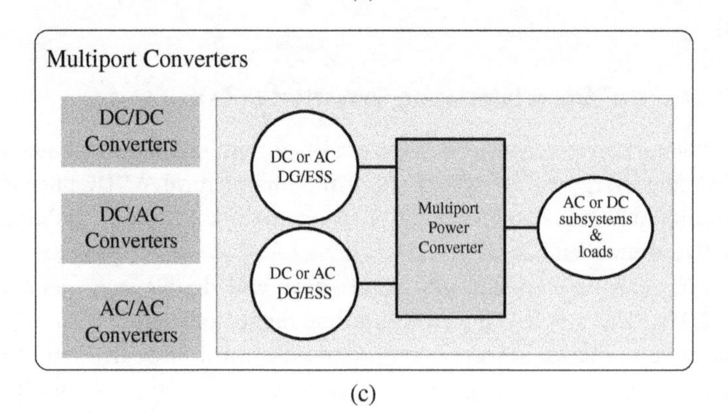

(c)

Figure 2.28 Interfacing power electronics converters (IFCs): (a) two-port DC/AC converters, (b) two-port AC/AC converters, and (c) multiport IFCs.

as EV charging and medical power supplies, where a high-frequency transformer (HFT) or medium-frequency transformer (MFT) can provide electrical isolation in the DC/DC converters, such as flyback, push-pull, and dual-active-bridge (DAB) converters, with lower weight and size. The DC/AC converter can be a voltage-source converter, current-source converter, or even a DC/AC matrix converter, allowing flexible selection according to different requirements. Depending on different control requirements in the second stage, the DC/AC converter can work with the current control method or voltage control method to meet the demand from the AC side.

In AC/AC IFCs, the power electronic interfaces can be classified as single-stage, double-stage, or multi-stage structures, as shown in Figure 2.28b. Matrix converters are considered promising candidates for the single-stage structure; however, the limited degree of freedom is a challenge due to the direct coupling of input and output currents. Therefore, the two-stage and multi-stage solutions are more popular in practice. A two-stage structure generally utilizes two DC/AC converters with connected DC sides, generally called a back-to-back structure. Utilizing a PWM converter on both sides allows bi-directional power flow. Also, when the unidirectional flow is needed, diodes can be used to build rectifiers to reduce system costs. Compared to single-stage and double-stage structures, a multi-stage structure generally has lower system efficiency due to the extra power conversions. However, introducing new functions with the added power stages can offset this. For instance, isolation can be added in the DC/DC stage in the multi-stage structure. Also, the two AC systems can work under different voltage levels when added stages can step-up or step-down voltages.

As mentioned above, the two-port DC/DC converters can be used in the two-port DC/AC and AC/AC converters structures. They are also used to interface DC renewable energy and energy storage to DC networks or interlink DC networks with different voltage levels. The two-port DC/DC converters have single-stage or more stages structures where isolated and non-isolated converters can be flexibly combined according to the requirements. Generally speaking, isolated converters can add isolation features with HFT or MFT and allow a higher voltage conversion ratio, which is extended by adjusting transformer turn ratios. On the other hand, non-isolated converters generally have a simpler structure and higher efficiency (transformers in isolated converters are widely considered efficiency bottlenecks).

Another classification of two-port DC/DC converters can be unidirectional and bi-directional. For example, unidirectional DC/DC converters interface DC distributed generations to DC networks while BESSs are connected to DC networks through bi-directional DC/DC converters.

With the development of power converter technologies, multiport IFCs are emerging. These converters can interlink various DGs and ESSs to the AC/DC subgrid or load, shown in Figure 2.28c. Such converters are generally derived from existing DC/AC, DC/DC, or AC/AC converter topologies, achieving a compact size and lower cost by duplexing some components. On the other hand, they usually have complex structures and control strategies. Multiport converters can be classified as electrically coupled and magnetically coupled structures. An example of an electrically coupled multiport converter is multilevel converters with multiple DC buses, such as neutral-point-clamped converters and cascaded H-bridge converters. The magnetically coupled structure generally relies on MFTs or HFTs. A typical way is to connect multiple converters to multi-winding HFTs and apply alternative voltages to the windings. The phase-angles can change the power flow direction among the multiple windings.

2.3.2 Operation and Coordination

Energy storage is an important part of microgrids with renewable energy sources. In particular, in the microgrid's standalone operation mode, the ESS must complement intermittent renewable energy sources to meet the demand. They also help avoid curtailment of renewable energy sources when the generation is higher than load consumption. The coordination between renewable generation and energy storage can be discussed in short-term and long-term time frames.

In the short-term time frame, the power management strategies coordinate renewable generations and energy storage operations. These strategies affect the instantaneous operational conditions toward certain desired parameters such as voltage, current, power, and frequency. Depending on the different structures of microgrids, energy storage has different responsibilities with respect to renewable generation. For example, in the AC-coupled hybrid microgrid, the energy storage helps renewable generations on power balancing within the microgrid and the AC bus voltage and frequency control. In the DC-coupled hybrid microgrids, the DC bus's energy storage and renewable generation coordination must provide the DC bus voltage control, power balancing between generation and demand, and AC bus voltage and frequency adjustment. In the AC-DC-coupled hybrid microgrids, more coordination between energy storage and renewable generations in the DC bus and the AC bus is required to balance both sides' generation and demand and control AC and DC voltages. Please refer to Chapter 1 for more information about the different structures of hybrid microgrids.

The global objective of long-term energy management algorithms is to match renewable generation and energy storage total power production to the demand in an optimal way. These algorithms monitor and operate a complex electrical, thermal, and mechanical system, emphasizing desired and longer-term outcomes (day planning with load demand and renewable-based generations prediction). Recently, in energy management systems of power electronics intensive microgrids, AI-based techniques have been widely considered. The AI gives the intelligence demonstrated by machines to the energy system to replace human analysis, judgment, optimization, and decision-making by simulating the human brain and modeling human problem-solving. The AI methods have many advantages, including robust data analytical capability to process massive data and automatically extract the coupling relationship, less dependent on a specific mathematical model, strong adaptability, and capable of high dimensional, time-varying, and non-linear systems.

In general, the short-term and long-term power and energy management algorithms for the coordination of renewable-based generation and energy storage can be implemented in different control layers of smart microgrid control, called hierarchical control layers (please refer to Chapter 6 for detailed discussions on hierarchical control), in which efficient and reliable communications and cyber systems are crucial. In general, in smart microgrids, smart IFCs of renewable generation and energy storage communicate by wireline (such as power line communication and low-bandwidth communication) or wireless technologies (such as a ZigBee, WiFi, and cellular communication networks).

The proper size of renewable-based generation and energy storage also affect their coordination. In general, the size of power sources in microgrids are determined using an optimization algorithm, in which the components of microgrids are modeled from different perspectives, e.g. economic models, power models, and dynamic models. In microgrids, ESSs can be sized based on their provided service(s). For instance, utility-scale BESSs can provide arbitrage (charging when energy prices are low and discharging during more expensive hours), peak demand capacity, ancillary service (e.g. frequency restoration), black start, and asset upgrading deferral. Utility-scale renewable generation sizing depends on several factors, including project goals, site location and loading, and utility requirements. For example, in PV systems, project goals could be fossil-fuel consumption minimization and ensuring backup energy for critical facility functions.

The renewable-based generation and energy storage locations also profoundly affect their coordination and microgrid performance. While land requirements and availability typically constrain the renewable generation location, the energy storage location is more flexible (except PHS and compressed air energy storage). For example, BESS allocation objectives can be divided into (i) optimum battery operation and (ii) optimum microgrid operation [8]. In the first category, BESS installation, maintenance, and operational objectives are considered. However, these methods are not commonly used in power systems since they do not consider the power system's operational constraints. In the second category, different objectives such as transient stability improvement, power loss minimization, voltage stability/profile improvement, load shifting and peak shaving, and distributed generation support are considered.

2.4 Summary

The fundamentals of renewable-based power generation suitable for smart microgrids are reviewed in this chapter. In particular, detailed mathematical modeling, control, and interfacing of PV and wind power systems are provided. In addition, as complementary energy sources for renewable generation with intermittent nature, ESSs are discussed in detail. Different technologies, such as the battery, flywheel, superconductor, and electrolyzer-fuel cell systems, are reviewed. Finally, the integration of renewable energy and energy storage into the grid and their coordination for smart microgrids operation are presented in this chapter.

References

1 Renewable Energy Statistics 2018 (2018). The International Renewable Energy Agency (IRENA).

2 Global Market Outlook For Solar Power 2019-2023 (2019). Solar Power Europe, Bruxelles, Belgium.

3 Global Wind Report 2019 (2019).Global Wind Energy Council (GWEC).

4 Deloitte (2015). *Energy Storage: Tracking the Technologies that Will Transform the Power Sector*. New York, United States.

5 Wang, Y., Tian, J., Sun, Z. et al. (2020). A comprehensive review of battery modeling and state estimation approaches for advanced battery management systems. *Renewable and Sustainable Energy Reviews* 131: 110015.

6 BloombergNEF (2020). *Hydrogen Economy Outlook*. Bloomberg Finance L.P.

7 Nehrir, M.H. and Wang, C. (2009). Principles of operation and modeling of electrolyzers. In: *Modeling and Control of Fuel Cells: Distributed Generation Applications*, 116–125. New York, NY, USA: Wiley IEEE Press.

8 Nejabatkhah, F., Li, Y.W., Nassif, A.B., and Kang, T. (2018). Optimal design and operation of a remote hybrid microgrid. *CPSS Transactions on Power Electronics and Applications* 3 (1): 3–13.

3

Smart Microgrid Communications

3.1 Introduction

Smart microgrids have high penetration of power electronics converters used for interfacing distributed energy resources (DERs) (DERs include distributed generation [DG] and energy storage system [ESS]), loads, and AC and DC subgrids in hybrid microgrids. Converters can ensure their safety and basic operation without exchanging information with other converters or higher-level control systems, although communication among converters can improve their performance, e.g. droop control with low bandwidth communication. However, for microgrid optimization, communications are generally required. In such systems, the physical and electrical components are interconnected by information and communication technologies, and their operations are coupled to cyber system functionality.

In Figure 3.1, a typical power electronics-intensive smart microgrid with cyber-physical networks is shown [1]. As shown in the figure, the smart microgrid's cyber-physical model can be divided into different systems: (i) a physical power system, (ii) a sensor and actuator system, (iii) a communication system, and (iv) a management and control system. Brief explanations about the four systems are provided below:

- The physical system contains the microgrid's electrical and power components, such as transformers, generators, power electronics converters, circuit breakers, and loads.
- The sensor and actuator system consists of sensors, measurement devices, and devices to implement the control decisions in the management layer. The sensors and measurement devices are responsible for measuring information about the system's state, including voltage, frequency, current, and circuit breaker status. The actuators and control devices include generator controllers, DG controllers, and relays of circuit breakers.

Smart Hybrid AC/DC Microgrids: Power Management, Energy Management, and Power Quality Control, First Edition. Yunwei Ryan Li, Farzam Nejabatkhah, and Hao Tian.

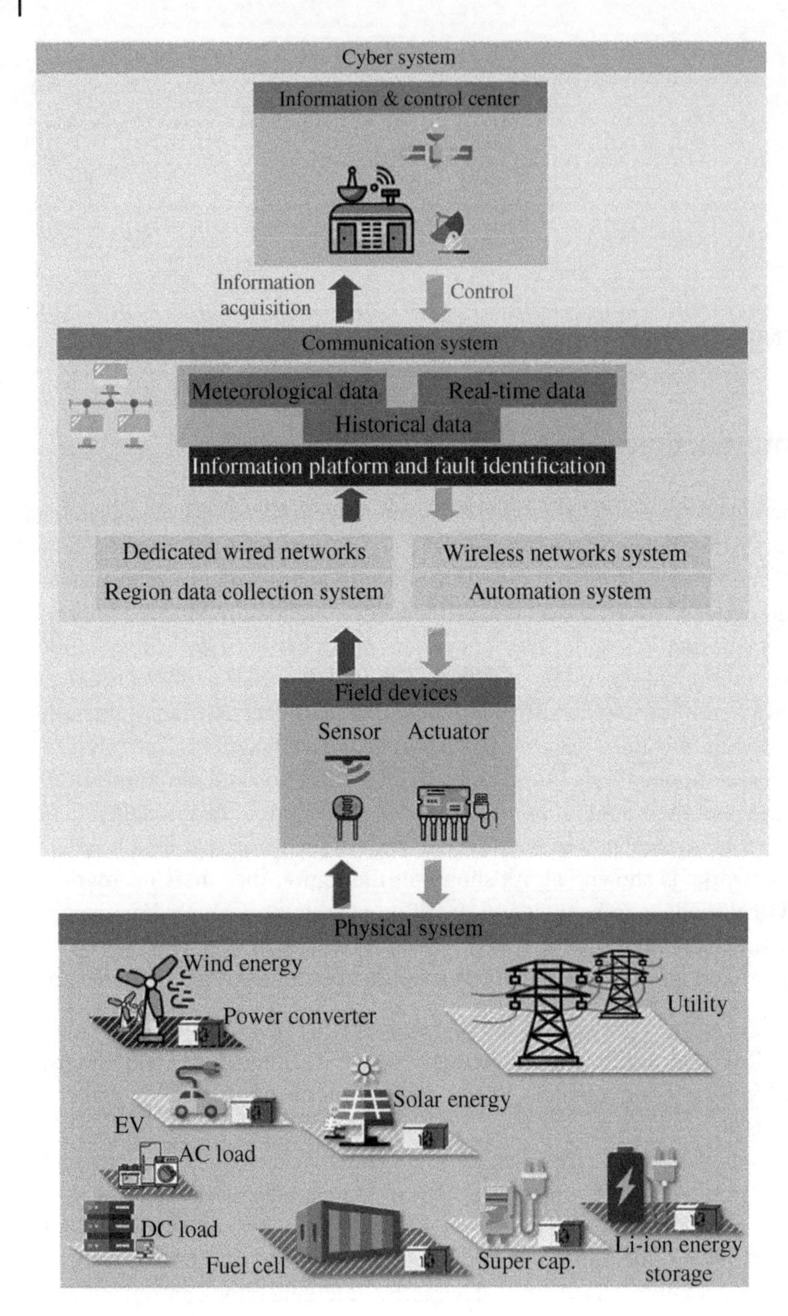

Figure 3.1 A typical cyber-physical smart microgrid.

- The communication system consists of devices including routers, switches, and the communication medium, and is responsible for information exchange among relevant systems. In smart microgrids, the communication system can be wired or wireless, depending on system requirements.
- The management system is a central control system that is responsible for the microgrid operation under different conditions. This system receives measurement layer data through the communication system and produces control signals for optimal operation of the smart microgrids. Then, the control signals are sent to actuators through the communication system again.

In the following sections, the basic concepts in communication techniques and structure of the cyber system of the microgrid will be discussed.

3.2 Communication Technique for Smart Microgrids

The accurate and optimal operation of smart microgrids relies on a secure and safe communication infrastructure. The communication system is widely used to coordinate devices within the microgrid and interact with the distribution grid, such as upstream substations or the wide-area grid regulator. In this section, the basic concepts of communication are first introduced. Then, the structure, requirements, and typical techniques of communications in the smart microgrids are presented.

3.2.1 Basic Concepts of Communication Systems

Communication techniques can be described by the open systems interconnection (OSI) model, which consists of seven layers for their functions. As shown in Figure 3.2, the layers and functions are as follows.

1) Physical Layer

The physical layer is the bottom layer of the OSI model, which defines raw unstructured data bits. In other words, it defines what is considered as digital bit 0 and what is considered as 1. The specifications can include voltages, pin layout, cabling, radiofrequency, etc. Besides the definition of raw data bits, some information about the physical medium can also be found in the physical layer definition, including cabling, network hubs, repeaters, network adapters, or modems.

2) Data-link Layer

The data-link layer defines the format of the data on the network. In this layer, the transmitted or received data bits are packaged into communication frames, and errors at the physical layer can be corrected if the mechanism is defined. As shown

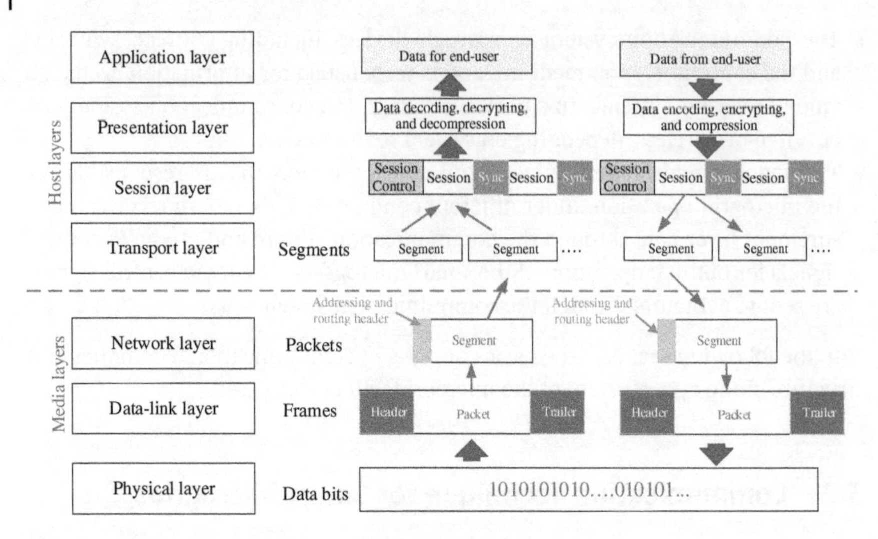

Figure 3.2 Seven-layer OSI model of communication techniques.

in Figure 3.2, headers and trailers are added to construct the frame: the header contains the source and destination addresses (e.g., MAC address), and the communication control bytes (e.g., frame start flag); the trailer generally contains error detection or error correction bits and frame stop flags.

3) Network Layer

The communication data are packets at this layer, whose headers contain the necessary addressing and routing information. So, the primary service provided by the network layer is routing – delivering the data to its destination according to addresses contained inside the package or identifying the source of the received data. For example, the internet protocol (IP) address can be one type of addressing information.

4) Transport Layer

The communication data are called segments at this layer. The transport layer manages the delivery and error-checking of data packets. Unlike the error checking in the data-link layer, error correction in the transport layer is to retry the data request and obtain the desired information. The two most common examples of the transport layer are transmission control protocol (TCP) and user datagram protocol (UDP).

5) Session Layer

The session layer manages the conversations between different communication devices. The communication session is set up, managed, and terminated at this layer. Once a communication session is built, the communication opponents can keep exchanging information until the session is ended.

6) Presentation Layer

The presentation layer is responsible for data formatting for communication. For example, it can perform encryption/decryption and data compression. This layer ensures that the application layer can identify and utilize the data; on the other hand, the lower layers can transmit the data efficiently and correctly.

7) Application Layer

This layer is the top layer of the OSI model, which provides interfaces for applications to access network services and access network services to applications directly. But it is worth noting that end user's software applications are not part of the application layer.

In practice, communication techniques do not necessarily include all seven layers. For example, the widely applied Modbus remote terminal unit (RTU) protocol only contains three layers: the physical layer, data-link layer, and application layer, while the other layers are not required. Also, some techniques may need to be combined with different standards or protocols to achieve functional systems. For example, RS-485 only defines the physical layers and must be applied with other standards, such as universal asynchronous receiver–transmitter (UART) protocol as a data-link layer.

3.2.2 Structures of Communication Networks in Smart Microgrids

Various communication techniques and protocols are applied to smart microgrids. In practice, the OSI model can help select the protocols by enabling the layer-by-layer comparisons of different communication techniques. It also facilitates the combination of technologies in different layers. As a result, the communication protocols can be selected and wisely combined for the required functionality and performance while ensuring an acceptable system cost to form a communication network. According to the coverage area and requirements on data rate, latency, quality of service, etc. the communication networks in smart grids primarily consist of three main types:

- premises network
- neighborhood area network (NAN)/field area network (FAN)
- wide area network (WAN).

These three types are shown in Figure 3.3. It should be noted that a similar structure and communication also apply to the smart microgrid.

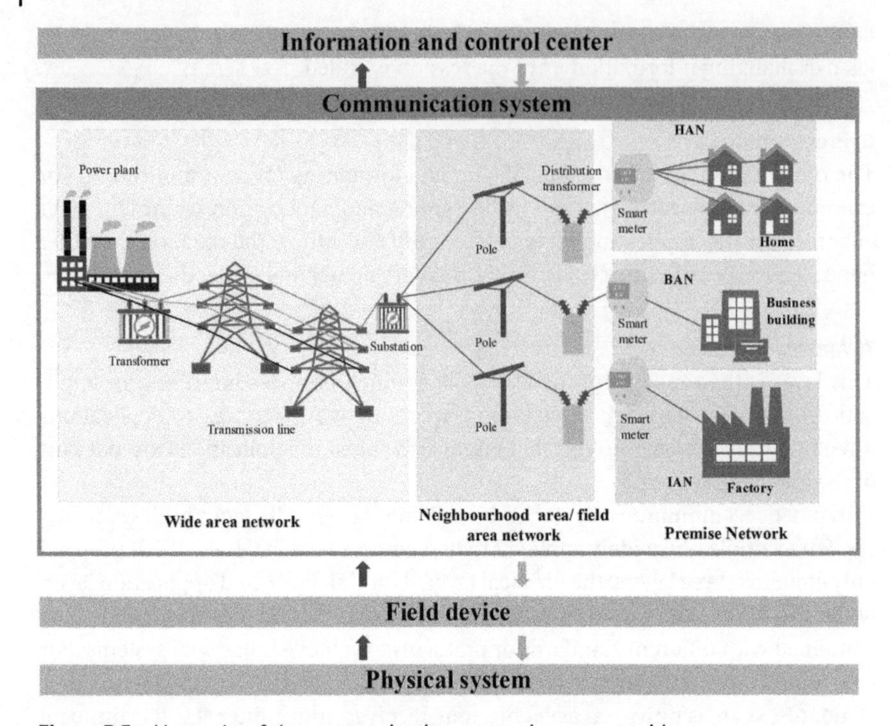

Figure 3.3 Networks of the communication system in a smart grid.

Premises Network

The premises network is deployed at the customer side and the end of the communication network. It is designed for consumer devices, such as meters, electric vehicles (EVs), home appliances, etc., and enables some key smart grid features, such as direct load control, energy efficiency improvement, smart metering, etc. Further classification of the premises network can be made: home area network (HAN), business area network (BAN), and industrial area network (IAN), depending on the needs and the opportunities presented by the residential, commercial, and industrial environments.

Neighborhood Area Network/Field Area Network

The NAN interconnects the premises network and the WAN. It is conceptualized for a network covering a neighborhood area, which can be several homes or several different factory areas. If this network is used for devices of the distribution grid, it can also be referred to as the filed area network (FAN). For example, it can connect intelligent circuit breakers, power factor correction capacitor banks, and distribution transformer terminal units (TTUs). Also, it can be part of the advanced metering infrastructure (AMI) (please refer to Chapter 6 for more discussions on AMI).

Wide Area Network

The WAN is the top level of network architecture that connects the NAN to the utility (head-end). In this network, the data exchange is not limited to a local network area but can have long-range communication, which can even provide access to the central regulator of the utility grid. In addition, smart devices in power distribution systems, such as supervisory control and data acquisition (SCADA), RTUs, phasor measurement units (PMUs), etc. can also be connected. Please refer to Chapter 6 for more discussions on advanced distribution management systems (ADMSs) and SCADA systems.

3.2.3 Requirements of Communication in Smart Microgrids

As discussed, the cyber network can be divided into several systems, while the communication system can be further divided into different networks according to the functionalities and served area. In the following, a brief discussion on the requirements of different networks is provided.

The premises network only covers a relatively small area, such as within 50–100 m. Therefore, it does not require wide-range transmission capability. Also, it generally does not require high bandwidth for microgrid applications. Instead, the ease of deployment, cost, and accessibility for various devices are the most critical concern. Some popular solutions tend to utilize existing communication systems, such as home WiFi networks, to fulfill the needs of the microgrid within a HAN even if the bandwidth is far more than enough for microgrid communication.

The NAN/FAN serves the last mile for the utility grid, where it connects substation devices to customer meters and automation devices. In the microgrid, it connects the central controller to each distributed device to monitor the system and optimize operating. This network also needs a low-cost solution but not as cost-sensitive as the premise network. Moreover, as its primary task is to serve as an AMI, deal with communications like meter reading, and perform demand response (DR) control, it does not require high bandwidth, unless the communication is related to protection, such as the IEC61850 Generic Object Oriented Substation Event (GOOSE) communications for timely operating smart relays.

Unlike the premise network and NAN/FAN, which covers a small area, the WAN generally performs long-range communication and provides data flow paths for SCADA, human–machine interface (HMI), or even video surveillance of the key components. In this case, the WAN generally focuses on high bandwidth wide-range technologies.

The microgrids contain various devices that need to exchange information. Different devices in different networks can require different specifications for the communications. For example, meter readings do not require real-time measurement, and as a result, a low rate of communication with a baud rate lower than 128 Kbps is widely used with 2–10 s latency. While for distribution line protection, the communication speed is critical as the protection devices will act before faults cause severe consequences. In this case, high speed (e.g. up to 100 Mbps) communication with latency as low as 10 ms is applied. More requirements of different applications of a smart microgrid are presented in Table 3.1, where the required bandwidth and latency are concluded.

3.2.4 Wired Communication Technologies in a Microgrid

As discussed in Section 3.2.3, various requirements can be found in different applications and different networks. This results in a wide variety of communication techniques in smart grids and microgrids. Some example techniques for microgrid communications, classified as wired communication and wireless communication according to their physical layers, are introduced in the following.

Table 3.1 Requirements of communication functions in a smart microgrid/grid.

Application	Typical function	Involved networks	Typical bandwidth	Typical latency
Meter reading	Obtain energy consumption data, access management	NAN and premise network	Up to 128 Kbps	2–10 s
Demand response	Load management, such as peak shaving and load leveling	NAN and premise network	14 K–128 Kbps	0.5 s–1 min
Operation optimization	Supervisory control for microgrids, energy management	NAN and premise network	9.6 K–56 Kbps	0.5 s–1 min
Distribution line protection	Fault protection, isolation, and recover	NAN	Up to 100 Mbps	10 ms–100 ms
Substation automation	SCADA, operation of circuit breakers	NAN and WAN	9.6 K–128 Kbps	2–4 s

Physical Mediums for Wired Communication

Twisted Pairs Twisted pairs are widely used due to their simple structure and the ability to resist electromagnetic interference (EMI). In twisted pairs, two conductors are twisted together, and, as a result, the magnetic fields generated by each conductor or other sources can be canceled. In RS485 or CAN bus communications, only two twisted pairs are used to construct differential balanced circuits. Multiple twisted pairs can be assembled and equipped with proper shields to become Ethernet cables with different categories, such as Cat 5 and Cat 6. As a low-cost option, twisted pairs generally cannot have a long length as their performance degrades with increasing length.

Digital Subscriber Lines (DSL) Digital subscriber lines (DSL) generally refer to a suite of communication technologies that enable digital data transmissions over telephone lines. The main advantage of DSL technologies is that electric utilities can interconnect residential users to control centers, avoiding the additional cost of deploying their communication infrastructure.

Optic Fiber Optic fiber offers the potential for relatively long-distance communication without the need for intermediate relays or amplification. This technology's significant advantages are (i) a long-distance coverage range with a high bandwidth (up to tens of Gbps), (ii) robustness against EMI, making it suitable for a high-voltage environment.

Power Line Communication (PLC) Power line communication (PLC) technologies offer a cost-effective option for microgrid communication by utilizing existing power cables for information exchange [2]. PLC technology has been divided into two categories, narrowband PLC (NB-PLC), which operates in transmission frequencies below 500 kHz, and broadband PLC (BB-PLC), which uses higher frequencies of up to 30 MHz with a higher data rate (up to 200 Mbps). Compared to BB-PLC, NB-PLC technology provides a lower data rate but a much higher range (150 km or more). However, the noise and interference caused by other signals such as near electric appliances or external electromagnetic sources can easily affect the signal quality. For example, when the data signals need to cross the distribution transformers, the signal attenuation and distortion can be caused by the transformers. Furthermore, the achieved data rate of PLC is inversely proportional to the wiring distance, which varies from a few hundred bps to millions of bps. Therefore, PLC is mainly used for metering applications in in-home applications.

A summary of wired communication technologies' main features is provided in Table 3.2, comparing different technologies in terms of data rate and coverage range [2–4].

Table 3.2 Comparison of different wired communication technologies.

Technology	Data rate	Coverage
Twisted pair	9600 bps–100 Mbps	100 m–1.2 km
Optic fiber	100 Mbps–2.5 Gbps	10 km–60 km
DSL	2 Mbps–85 Mbps	1.2 km–3.6 km
PLC	10 kbps–200 Mbps	1.5 km–150 km

Example Protocols of Wired Communications for Microgrids

Various communications protocols have been developed for industrial applications before microgrids are widely deployed. As a result, the microgrid can enjoy quite a high number of options for its HANs, NANs, and WANs. Here, only a few example protocols for wired communications are discussed, and a more comprehensive enumeration of the protocols for different applications can be found in Section 3.3.

Modbus Modbus is a widely used protocol and has been developed as several variants. The Modbus RTU can use RS-485, RS232, RS-422 as the physical layer with twisted pairs as the physical medium. On the other hand, Modbus TCP utilizes the TCP protocol as the network layer, the Ethernet's data-link layer, and the physical layer.

It is a kind of client–server communication, which contains a master and several slaves. Generally, the queries are always started by the master while the slaves only respond and cannot start up a conversation. However, it is possible to do simultaneous broadcasts to all system devices requesting specific action. A microgrid communication system could fit the Modbus model with the main controller or the HMI acting as a master and DERs, loads, and switches as slaves.

DNP3 DNP3 is currently the dominant master/slave protocol in electrical utility SCADA systems and is gaining popularity in other fields such as oil and gas, water, and wastewater. DNP3 messages are not restricted to transferring one data point, as is Modbus. Multiple data types (i.e. Boolean, floating-point) can be encapsulated in a single message to reduce data traffic. Timestamps and data quality information can also be included in the message. Unlike Modbus, DNP3 slave devices can send updates as values change without waiting for a poll from the master.

PROFIBUS and PROFINET Process field bus (PROFIBUS) is a standard for field communication in automation technology. There are two variations: PROFIBUS

DP (decentralized peripherals), which is the most commonly used to operate actuators and sensors via a centralized controller, and PROFIBUS PA (process automation), which is used for measuring equipment monitoring by a process control system. PROFIBUS is based on RS-485 and also works in master–slave mode. This approach saves costs by the omission of additional hardware and cabling. Also, it saves engineering time as it streamlines network installation, maintenance, and troubleshooting.

Similar to PROFIBUS, PROFINET also allows users to save in wiring and installation costs using distributed IO mechanisms. Since PROFINET networks are based on Ethernet, Ethernet devices, such as switches, can be directly applied.

Time-sensitive Networking (TSN) Time-sensitive networking (TSN) was developed to provide deterministic communication on standard Ethernet [5]. It provides a way for information traveling in a fixed and predictable amount of time. TSN comprises various standards that support improved time synchronization (TS), deterministic latency control, redundancy, and time-triggered actions. The latency guarantee, TS, and real-time control features of TSN make it a development trend of communication for microgrids.

3.2.5 Wireless Communication Technologies

Compared to wired communications, wireless communications do not rely on cables and are thus flexible and easy to deploy. Moreover, with the development of radio-frequency techniques and well-designed protocols, their reliability is already comparable with wired communications. In this subsection, some popular wireless communication technologies are presented.

ZigBee

ZigBee is a technology for HAN, which enjoys low cost, low power consumption, and easy configuration. However, several limitations exist, such as operation in license-free bands with inevitable interference, low data rates, security issues, and complexity of multi-hop networks. Its physical layer and MAC sub-layer of the data link layer are defined by IEEE 802.15.4, and the layers between the network layer and application layer are defined by Zigbee standards. The applicability of ZigBee is limited to support connectivity only between the customer premises and the data aggregators (generally within 100 m).

WiFi

WiFi operates at both layer 1 and layer 2 of the OSI model, and upper layers from the network layer to the session layer can share network protocols (such as TCP/IP). It is a mature and widely adopted wireless technology suitable for home applications in the smart grids that data rates reach Gbps. However, the main limitations for its application in power systems refer to the relatively short coverage range (for example, 70 m indoors and 250 m outdoors for the IEEE 802.11n Standard) and the use of unlicensed frequency bands.

Cellular

Cellular networks, including GPRS, 3G, 4G, and 5G networks, can enable wide-area communications for distributed grid applications because of (i) using licensed bands with better interference and security control, (ii) mature and ubiquitous coverage, (iii) high performance with the high data rate, low latency, and high system reliability, (iv) third-party operation. Therefore, cellular technology can transform the existing aging distribution grid into a modernized and fully automated power distribution system. The cellular network deployment for microgrid/smart grid applications can be costly when the public cellular network is not accessible, or a dedicated network is required due to security concerns.

Low-power Wide-area Networks (LPWAN)

Low-power wide-area networks (LPWANs) are designed to allow long-range communications at a low data rate, such as LoRa, Sigfox, eMTC, and NB-IoT, providing broad coverage at lower implementation and operation costs. It is a group of technologies created for machine-to-machine (M2M) and internet of things (IoT), and is thus very suitable for smart meters and distributed sensors in microgrids. However, it can only be used for low data-rate applications.

Satellite Communication

Satellite communication can provide extensive coverage and easy installation like other wireless communications. In this case, it can be used to build WAN without relying on the access of infrastructures like telecommunication networks, and is thus quite suitable for connecting devices installed at remote/offshore areas. Also, it can be used to building safe, dedicated networks, ultra-wide-area optimization of smart grids. However, the major disadvantage of satellite communication is the high cost of the terminal device and channel resources.

The wireless communication technologies discussed above are summarized in Table 3.3, where the main features are identified [6]. These communications feature different data rates and coverage ranges for selection according to different requirements in premise networks, NAN/FAN, and WAN. For example, Zigbee can be a good solution for industry premise networks (IANs); WiFi, whose data rate is much higher than the microgrid requirements, is widely used in homes and thus can be a good option in HAN. In addition, the cellular and LPWAN techniques can well fulfill the needs for NAN and WAN.

Table 3.3 Comparison of different wireless communication technologies.

Technology	Data rate	Coverage
ZigBee	250 Kbps	<50 m
WiFi	11 Mbps–2.4 Gbps	<100 m
WiMAX	63 Mbps	<50 km
Bluetooth	1 Mbps	<100 m
GPRS	170 Kbps	<10 km
3G	2 Mbps	<10 km
4G	100 Mbps	<50 km
5G	10 Gbps	<50 km
LPWAN	50 Kbps	<20 km
Satellite	2.4–100 Mbps	<4500 km

3.3 Standards and Protocols in Smart Microgrids

In smart grids, various devices, such as power converters, smart circuit breakers, distribution automation devices, and meters, need to exchange information for grid control purposes. These devices can be manufactured by different manufacturers and purchased/installed at different times. Therefore, it is mandatory to apply standards to coordinate the various devices, otherwise, communications will always be limited to a small group of compatible devices.

To enable the smart microgrid features, standards are built to ensure the compatibility of all the devices connected to the same networks. This generally requires a detailed definition from the physical layer to the applications layer. In addition, the end-user application on top of the communication protocols needs to be standardized. This leads to many standards and protocols, which can be categorized by their targeted type of device and their communication opponent, for example, smart meters and meter-data collecting devices. As can be seen from Figure 3.4 and Table 3.4, the communication standards can be categorized into the basic standards on communication, substation automation, control center, DG, metering, DR, electrical vehicles, cybersecurity, etc. In the following, a brief discussion of these standards will be provided.

3.3.1 Standards and Protocols for General Communication

In this section, the widely/commonly used standards and protocols for wired and wireless communication technologies are provided. These standards form the

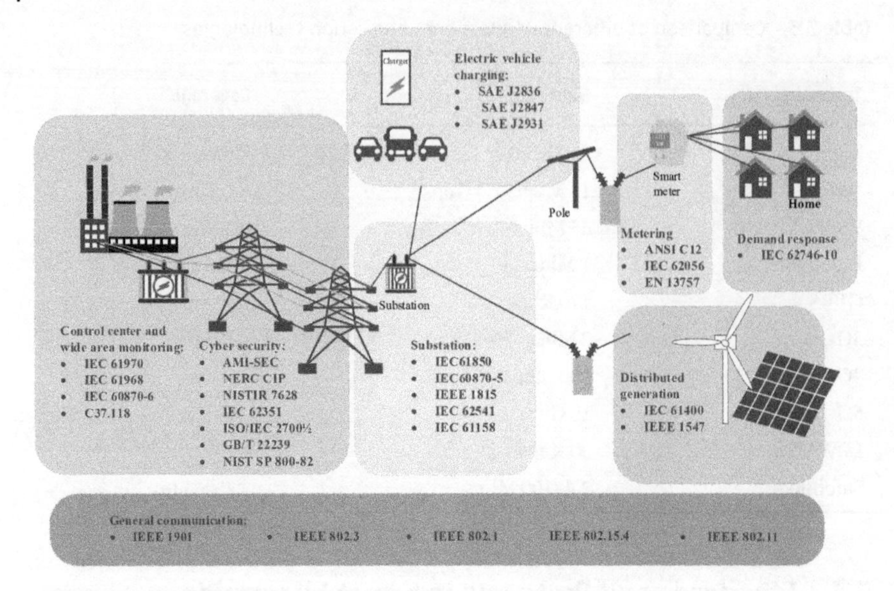

Figure 3.4 Categories of communication standards in a smart grid.

Table 3.4 Standards and protocols categories and contents.

Application	**Standards**
Communications	IEEE 1901, IEEE 802.3, IEEE 802.1, IEEE 802.15.4, IEEE 802.11
Substation automation	IEC61850, IEC60870-5, IEEE 1815, IEC 62541 IEC 61158
Control center and wide area monitoring	IEC 61970, IEC 61968, IEC 60870-6 C37.118
Distributed generation	IEC 61400 IEEE 1547
Metering	ANSI C12, IEC 62056, EN 13757
Demand response	IEC 62746-10
Electric vehicles (EV)	SAE J2836, SAE J2847, SAE J2931
Cyber security	AMI-SEC, NERC CIP, NISTIR 7628, IEC 62351, ISO/IEC 27001 and 27002, GB/T 22239, and NIST SP 800-82

foundation of communications and are commonly adopted in communications for microgrids.

IEEE 1901

IEEE 1901 is a standard for broadband PLC with transmission frequencies below 100 MHz, which generally covers a range that is shorter than 1 mile. This standard defines the physical layer and MAC in the data-link layer of the lower line communication. Currently, IEEE 901 has been widely accepted and is being endorsed by industrial organizations, accommodating a wide range of mainstream application areas of PLCs, such as smart energy applications, Electric vehicles (EV), and even audio/video/data networks in HAN.

IEEE 802.3

The IEEE 802.3 standard has a long history from the 1980s and the Ethernet technology it standardized has been one of the most widely used technologies due to the development of the internet. This standard defines the physical layer and data link layer's MAC part of wired Ethernet. It is the mainstream standard for local area networks (LAN) and generally covers a 200 m area. The communication speed can be 100 Mbps and up to 1 Gbps. Its bandwidth can well fulfill the general requirements of non-industrial users. However, the Ethernet network uses carrier-sense multiple access with collision detection (CSMA/CD) to avoid communication collision in the network, leading to unpredictable latency, from several milliseconds to several hundreds of milliseconds. This limits its usage in industry fields where deterministic communication is required.

IEEE 802.1

As discussed, IEEE 802.3 (Ethernet) cannot provide a deterministic service. To solve this problem, Ethernet-based TSN is emerging and is being standardized by the IEEE 802.1 workgroup. A series of standards, such as IEEE 802.1AS, 802.1ASrev, and 802.1Q, are defined to achieve bounded latency, low packet delay variation, and low packet loss. It focuses on the improvement of the data-link layer and adopts the widely used physical layer of Ethernet technology. The latency can be reduced to several tens of microseconds or even several microseconds.

IEEE 802.15.4

The IEEE 802.15.4 standard defines the physical layer and MAC of low-bandwidth, low-cost wireless communications. The wireless communication can use the industrial, scientific, and medical (ISM) frequency band, which does not require licenses. The typical band includes the 2.4 GHz, 915 MHz, and 868 MHz band. The data rate can be 250, 40, and 20 kbs. Other features like automatic network establishment, low latency, and low energy consumption can also be achieved with the technologies defined in this standard. The most well-known technology that adopts IEEE 802.15.4 is Zigbee, which adds higher layers to make a fully functional protocol.

IEEE 802.11

The IEEE 802.11 standard is used by the WiFi Alliance, which has been the most widely used wireless LAN technology. It uses 2.4 and 5 GHz ISM bands and also does not require a license. The data rate can be as high as 2.4 Gbps with the "WiFi 6" technology. The data rate is far beyond enough for microgrid applications like meter reading DR. Due to WiFi's easy accessibility in daily life, it is still the preferred communication in HAN of microgrids.

3.3.2 Standards and Protocols for Substation Automation

Modern substations have adopted many automatic devices which monitor and control the assets of the substation, including circuit breakers, transformers, sensors, etc. Substation automation operates these devices to ensure the safety and lifetime of assets and reduces the possibility of outages. In this section, the standards and protocols for communication services and networks and supervisory controls are provided.

IEC 61850

IEC 61850 is based on Ethernet technology and targets standardizing communication systems of power generation facilities and substations. It mainly works on the application layer and the lower layers generally employ IPs, such as TCP/IP. It also defines user applications by providing data models, such as control objectives and services. This standard series is widely adopted for power system automation architecture, called the substation automation system (SAS). The IEC 61850 defines five types of communication services: abstract communication service interface (ACSI), GOOSE, generic substation status event (GSSE), sampled measured value multicast (SMV), and TS [7]. These services can help realize functions like protection, control, measurement, and monitoring functions required by the substations, and, more importantly, with GOOSE communications IEC 61850 can realize high-speed control of the protection devices, realizing interlocking or inter tripping.

IEC 60870-5

IEC 60870 is a set of standards for dispatching power systems. IEC 60870-5 defines several "transmission protocols", such as IEC 60870-5-101/103/104. These protocols, which are also known as IEC 101/103/104, have a wide application that is not limited to the energy sector but is expanded to areas like water supply systems.

These standards focus on the application layer, and the lower layers can be based on Ethernet or serial communications. Various frame formats can be used to meet the requirements of the power system, including time-tagged messages, control commands, measurands, etc. The newer standard IEC 61850 will gradually replace it but due to the wide application of IEC 101/103/104, these protocols will co-exist for a long time.

IEEE 1815

IEEE 1815 defines the DNP3 protocol, which was discussed previously. It works as the application layer and it can adopt TCP/UDP as the session layer, IP as the network layer, and Ethernet protocol as the data-link and physical layer Alternatively, it can also employ serial communication, such as RS-485, as its physical layer. As discussed, it works under master–slave mode but the slaves can actively update their status. As an open communication protocol, it is a low-cost solution for SCADA and remote monitoring systems.

IEC 62541

IEC 62541 is a series of standards for open platform communication (OPC) unified architecture (OPC UA). It works at the application layer and utilizes IPs as lower layers. Its major feature is cross-platform – it is not bound to a specific operating system or program language. Also, security is considered an important aspect in OPC UA. Now OPC UA is mostly used in SCADA and distributed control systems (DCSs), up to manufacturing execution systems (MESs) and enterprise resource planning (ERP) systems.

IEC 61158

IEC 61158 is a comprehensive standard for fieldbus – industrial computer network protocols featuring real-time, efficient, and frequent data exchange synchronized actions between devices. The standard covers the details from the physical layer to the applications layer. It specifies the most widely used fieldbus protocols, such as Modbus, PROFIBUS, Foundation fieldbus, EtherCAT, which are widely used for the interconnection of automation and process control system components, such as HMI, programmable logic controllers (PLC), sensors, electric motors, switches, etc.

3.3.3 Standards and Protocols for Control Center and Wide Area Monitoring

The electric utility grid is a large-scale energy network constructed of various devices that account for carrying power flow, monitoring, protection, dispatching, etc. In this case, a hierarchical structure is used to realize the control and monitoring of the power system, where wide-area communications are required. The communication can happen between system operators and distribution operators, control center and substations, central control-center and regional control centers, etc. In this section, the standards and protocols associated with control centers and monitoring systems are discussed.

IEC 61970 and IEC 61968

IEC 61970 and IEC 61968 work above the communication systems and can be considered as a standard for end-user applications. To be specific, IEC 61970 defines the application program interface for EMSs while IEC 61968 defines distributed management systems. They have two major parts: (i) common information model (CIM). The CIM is part of the overall EMS-API framework, which provides an abstract model for a power application system using unified modeling language (UML) notation. (ii) Component interface specifications (CISs). The CISs specify the interfaces that a component (or application) should implement to exchange information with other components (or applications). It describes the specific events, methods, and properties that can be used by applications for this purpose. IEC 61970 and IEC 61968 can be combined with IEC 61850 as a communication system to develop a complete functional EMS or DMS.

IEC 60870-6

IEC 60870-6 defines the inter-control center communications protocol (ICCP) for WAN between utility control centers and regional control centers. It can be used for a nationwide network to realize a hierarchical control structure. The main coverage of this standard is the application layer and the most popular implementation of lower layers is TCP/IP over Ethernet. The standard defines the data model for SCADA in electrical engineering and power system automation applications.

IEEE C37.118

IEEE C37.118 defines synchronized phasor measurements and data transfer in a power system. It includes two parts: the first part is phasor measurement, such as synchro phasors, frequency, and rate of change of frequency (ROCOF) measurement under both steady-state and dynamic conditions. The second part defines the data format and the real-time communication between the PMUs and phasor data concentrators (PDCs). It specifies the application layer and the lower layers can be implemented by different communication techniques, such as TCP/IP over Ethernet or fieldbuses.

3.3.4 Standards and Protocols for Distributed Generation and Demand Response

Microgrids generally have a high penetration level of DERs. In the DER intermittence and operating schemes that maximize the active power generation, it can be challenging to maintain the stability and power balance of the microgrid. In this case, communication will play an important role in coordinating these emerging power sources to match the need of the loads. On the other hand, it is also possible to manage demand from dispatchable loads (demand respond) to match the power sources, which also rely on communications. This section discusses standards and protocols regarding the DG system and associated communication systems.

IEC 61400

IEC 61400 includes a series of standards specifying almost all aspects of wind turbines, from design to test to ensure the quality, functionality, safety, etc. The standards also require validating wind turbine quality throughout their lifetime against damage and natural disasters. IEC 61400-25 is specifically for communications of monitoring and control systems. Its focus on the communication layers above the data-link layer, and TCP/IP-based communication protocol, such as IEC 61850 MMS and DNP3, will be used. Also, it standardizes the information model and information exchange scheme, which works above the communication system.

IEEE 1547

IEEE 1547 is a standard for technical specifications of the interconnection and interoperability between utility electric power systems and DERs. It provides recommendations on the DER performance, operation, testing, safety considerations, and maintenance. Features such as grid support (reactive power compensation) and the fault ride-through capability are also involved. In IEEE 1547, interoperability is considered as a necessary function, which requires a local DER interface to exchange data with stakeholders, such as area electric power system operators and owners. To do this, SEP2 (IEEE 2030.5), DNP3, and SunSpec Modbus are recommended for the application layer of the communication system.

IEC 62746-10

IEC 62746-10 makes the Open Automated Demand Response 2.0 (OpenADR 2.0) an international standard. The OpenADR standards are developed to standardize, automate, and simplify demand response (DR) features to enable optimized coordination between utilities/aggregators. It works above the communication system as the user application and can utilize any internet communications. A typical application layer for OpenADR is HTTP protocol and does not have strict limitations on the lower layers. As a result, any communication that can be sent directly or bridged to the internet can be applied for the open ADR, such as WiFi, 3G/4G, and even Zigbee with necessary protocol conversion devices.

3.3.5 Standards and Protocols for Metering

AMI can perform bidirectional communication between the customer and the control center of the utility. Besides the timely metering data, AMI can provide functions such as outage detection, remote disconnect and connect control, monitoring of voltage, etc. In this section, standards and protocols associated with metering are provided.

ANSI C12

ANSI C12 is a comprehensive standard for meters, including the general requirements, shape, accuracy, communication, etc. As an important part of ANSI C12, the communication standards enable communication between the control center and meters from different manufacturers. The standard specifies the protocol from the physical layer to the application layer, including different requirements on a physical medium, like DSL, power line, optic fibers, and so on, and the table data. Both the point-to-point protocol and network protocol are considered.

IEC 62056

IEC 62056 is a standard focusing on meter communications. It includes the specs from the physical layer to the application layer. For bottom layers, it allows various technologies to be used for meters: Ethernet, PLCs, wireless communications, etc. It adopts device language message specification (DLMS)/companion specification for energy metering (COSEM) as the application layer and user application, which supports remote meter reading, remote control, and value-added services for metering any kind of energy, such as electricity, water, gas or heat.

EN 13757

The protocol defined in EN13757 is called Meter-Bus (M-Bus). The protocol includes the definitions of the physical layer, data-link layer, and application layer. The standard application layer enables meters from different vendors to operate in the same system. In the bus networked system, a master can be connected to multiple slave devices with only two conductors, achieving a low system cost. Both voltage signal and draining current are used to indicate raw bit 1 and 0, performing the bidirectional communication between master and slave. The wireless variant of the M-Bus operates in the 868–870 MHz Short-Range device (SRD) band.

3.3.6 Standards and Protocols for Electric Vehicle Charging

Communication plays an important role in EV charging. To charge EVs with various voltage standards and mileage ranges, communication links are built to ensure a safe and efficient charging process. In this section, the EV standards in energy transfer, communication, and interoperability requirements are provided, developed by the Society of Automotive Engineers (SAE).

SAE J2931

SAE J2931 includes a series of standards that specifies different communication protocols for EV charging, involving the PEV (plug-in electric vehicle), the electric vehicle supply equipment (EVSE), the utility or service provider, energy services interface (ESI), AMI, and HAN. The first part of the standard discusses the architecture and general requirements, including association, registration, security, and HAN requirements. The other parts define the physical and data-link layer. For example, J2931/4 defines the requirements on PLC in EV charging and J2931/5 defines internet communications. Wireless communications, such as WiFi and radio-frequency identification (RFID) are considered as solutions in J2931/6.

SAE J2847

SAE J2847 works above SAE J2931 and includes specifications and requirements on application. It includes the application layer, which is Smart Energy Profile 2.0 (SEP 2.0), and the application above the application layer – functional messages for various functions: utility/smart grid messaging (J2847/1), DC charge control (SAE J2847/2), SAE reverse energy flow (J2847/3), diagnostics (SAE J2847/4), consumer requirements (SAE J2847/5), and wireless charging (J2847/6). The vehicle-to-grid (V2G) is also considered in J2847. These above-mentioned standards also define the message cycle regarding the full charging process.

SAE J2836

SAE J2836 specifies the use cases in EV charging. It completes the messaging system by specifying the following four functions.

Enrollment: this enables the interaction between the customer and the utility, such as providing the customer accounting or the EV's ID information.

Utility programs: this enables the possibility for customers to select the way of consuming energy during the charging process. For example, customers can select if the charging process can be regulated by the utility grid to avoid overload of the distribution system.

Binding/rebinding: this specifies the method of connection between the EV and the utility: 120 V AC cord set EVSE, 240 V AC premise EVSE (240VAC), or a DC Premise EVSE.

Connection location: this specifies the location of the connections, such as the customer's home or a public area. The location alternatives for the customer to connect the PEV to the utility.

3.4 Network Cyber-security

In smart microgrids, physical components and cyber networks are tightly coupled. Therefore, information communication in the cyber system plays a critical role in smart microgrid efficient operation and reliable control. However, vulnerabilities are introduced into the system due to the tight interconnection between cyber and physical components. In detail, the smart microgrid's cyber-physical system contains complex structures, including distributed sensors and actuators, controllers, and power components and interfaces, and coordination between these components through high-precision and timely communication. Therefore, several challenges and issues, such as reliability of communication, data safety, and mass data processing, should be addressed for smart microgrids.

In general, data should meet three fundamental requirements in the cyber system: (i) availability – data should be timely and accessible, (ii) integrity – data are accurate and trustworthy, and (iii) confidentiality – data can only be viewed and used by authorized persons. In particular, the false data injection (FDI) attack, targeting data integrity, is one of the most challenging smart grid threats among different cyber-security violations. If such violations are crafted intelligently, they can penetrate the system without being detected by the conventional attack detection method. Those attacks are also called stealth attacks. The successful FDI attack could introduce major economic problems as well as steady-state and dynamic stability issues.

In smart microgrids with high penetration of power electronics converters, the cyber-security violations can be very harmful. Although optimal economical operation is not the primary concern in such microgrids, cyber-attacks could have devastating effects on microgrid stability, especially in islanded mode. In other words, due to the low inertia of such microgrids, cyber-security violations could affect the transient and steady-state stability of microgrids. Further, any cyber incidents in either AC or DC subgrid affect the other side of the hybrid AC/DC microgrids. For instance, if any cyber-attack affects the AC subgrid's frequency stability, it affects the DC voltage stability on the DC side through AC/DC subgrids interlinking power converters. Considering the increasing communication needs of smart microgrids, cyber-security will receive more and more attention in the near future.

To ensure the cyber-security of smart grids, standards are put into practice to enable mandatory security features. In this section, several significant standards and protocols associated with cyber-security are discussed. In Chapter 6, the cyber-security issues in hybrid AC/DC microgrids and their effects on control strategies will be discussed in more detail.

AMI System Security Requirements (AMI-SEC)

The AMI System Security Requirements (AMI-SEC) were established under the UCA International Users Group (UCAIug) to develop a robust security guideline for the initial AMI portion of the smart grid. The AMI-SEC supports all of the AMI system's use cases, including AMI communications network device,

AMI forecasting system, AMI head end, AMI meter, AMI meter management, and home area network. The AMI-SEC also recommends a control system and communication protection, including security function isolation, cryptographic key establishment and management, the transmission of security parameters, voice-over-IP, and many more.

NERC CIP

NERC Critical Infrastructure Protection (CIP) standards plan established the requirements for the secure operation of North America's bulk electric system. The NERC CIP plan developed a series of standards and recommendations that specify many aspects of cyber security, including critical cyber asset identification, security management controls, personnel and training, electronic security perimeters, physical security of critical cyber assets, systems security management, incident reporting, and response planning, and recovery plans for critical cyber assets. The NERC's standards for governing critical infrastructure apply to units that significantly impact the bulk power system's reliability.

NISTIR 7628

The National Institute of Standards and Technology Interagency Report (NISITR) 7628 presents an analytical framework for organizations to develop effective cyber-security strategies for their smart grid systems. The organizations in different areas of smart grids, including utilities that provide energy management services to manufacturers of EV and charging stations, can benefit from the methods and supporting information. This approach acknowledges that the electric grid is changing from a closed system to complex and highly interconnected systems, which results in multiplying and diversifying the threats to grid security. The guideline includes several parts: (i) smart grid cybersecurity strategy, architecture, and high-level requirements, (ii) privacy and the smart grid, and (iii) supportive analyses and references.

IEC 62351

IEC 62351 provides the security recommendations for different power system communication protocols of the TC 57 series, including the IEC 60870-5 series, IEC 60870-6 series, IEC 61850 series, IEC 61970 series, and IEC 61968 series. The different security objectives, such as authentication of data transfer through digital signatures, intrusion detection, eavesdropping prevention, and spoofing and playback prevention, are covered. The standard covers various aspects of the communication network and system security associated with power system operations.

Moreover, procedures and algorithms for securing manufacturing message specification (MMS) based applications are covered. Eventually, end-to-end information security, including security policies, access control, and key management, are addressed in the standard.

ISO/IEC 27001 and 27002

As the most fundamental standard of information security management, the ISO/IEC 27001 has a broad domain, including system security testing, compliance with security policies (periodical checks), and technical compliance review (contains operational systems testing to make sure that implementation of hardware and software controls are accurate). The auxiliary and practical guidance on the ISO/IEC 27001 implementation is provided in ISO/IEC 27002. ISO/IEC 27001 and 27002 can be applied to all smart grid components.

NIST SP 800-82

NIST SP 800-82 is a standard for the security of industrial control systems, which is recognized and used worldwide. The standard validates and certifies that the specified security controls are implemented correctly and operated to produce the desired outcomes. This standard also provides particular recommendations about vulnerability and penetration testing tools.

3.5 Summary

This chapter introduces information and communication networks, called cyber networks, in smart microgrids. The structure of a cyber system of the smart microgrid is first introduced, and the basic concepts of communication techniques are explained with the help of the OSI model. Then the different network types – HAN, NAN/FAN, and WAN are discussed with their different requirements explained. Typical wired and wireless communication technologies are reviewed. Also, a more comprehensive review of standards and protocols in smart grids is provided and classified according to the application areas. Finally, the importance of cyber system security and vulnerability of smart hybrid AC/DC microgrids to cyber incidents are explained.

References

1 Nejabatkhah, F., Li, Y.W., Liang, H., and Ahrabi, R.R. (2020). Cyber-security of smart microgrids: a survey. *Energies, MDPI, Open Access Journal* 14 (1): 1–27.

2 Galli, S., Scaglione, A., and Wang, Z. (2011). For the grid and through the grid: the role of power line communications in the smart grid. *Proceedings of the IEEE* 99 (6): 998–1027.

3 Bauer, M., Plappert, W., Wang, C., and Dostert, K. (2009). Packet-oriented communication protocols for Smart Grid Services over low-speed PLC. In: *2009 IEEE International Symposium on Power Line Communications and Its Applications*, Dresden, 89–94.

4 Arbab-Zavar, B., Palacios-Garcia, E.J., Vasquez, J.C., and Guerrero, J.M. (2019). Smart inverters for microgrid applications: a review. *Energies* 12: 840.

5 Cisco Systems (2017). Time-Sensitive Networking: A Technical Introduction. Cisco Systems White Paper.

6 Ancillotti, E., Bruno, R., and Conti, M. (2013). The role of communication systems in smart grids: architectures, technical solutions and research challenges. *Computer Communications* 36 (17–18): 1665–1697.

7 Mackiewicz, R. (2006). Overview of iec 61850 and benefits. In: *PSCE'06. 2006 IEEE PES*, 623–630.

Part II

Power Management Systems (PMSs) and Energy Management Systems (EMSs)

4

Smart Interfacing Power Electronics Converter Control

4.1 Primary Control of Power Electronics Converters

A typical structure of three-phase power electronic converters is shown in Figure 4.1. The converter can be an interfacing converter (IFC) from a distributed energy resource (DER) or IFCs linking AC/DC subgrids in the smart hybrid microgrids (the DER includes both distributed generations (DGs) and energy storage systems (ESSs), which can export active power to the grid). Here the converter uses an LCL filter, which is now widely used for grid-connected converters. To operate such a converter, primary control plays an important role. Primary control of power electronics converters refers to the control scheme used to regulate the basic control variables – current, voltage, power, etc. and ensure important performance like stability, robustness, settling time, etc. In other words, primary control is the fundamental of the converter control system [1]. Due to the demands on control bandwidth and dynamic performance, primary controls are generally applied in local control systems, and all the control variables are directly sampled with low latency and proper bandwidth.

Power converters can have different operating modes in a microgrid, which can be:

- Bi-directional power control mode: in this mode, the power converter follows a given AC voltage and focuses on controlling the power flow in both directions.
- DC-link voltage control mode: the power converter adjusts its active power according to requirements to regulate the DC-link voltage.
- AC-link voltage control mode: the AC voltage is directly regulated by the power converter, which directly feeds the loads or provides voltage references for other sources.

These control modes can be applied for different conditions where different primary control methods are needed. In the following, the basic techniques and

Smart Hybrid AC/DC Microgrids: Power Management, Energy Management, and Power Quality Control, First Edition. Yunwei Ryan Li, Farzam Nejabatkhah, and Hao Tian.

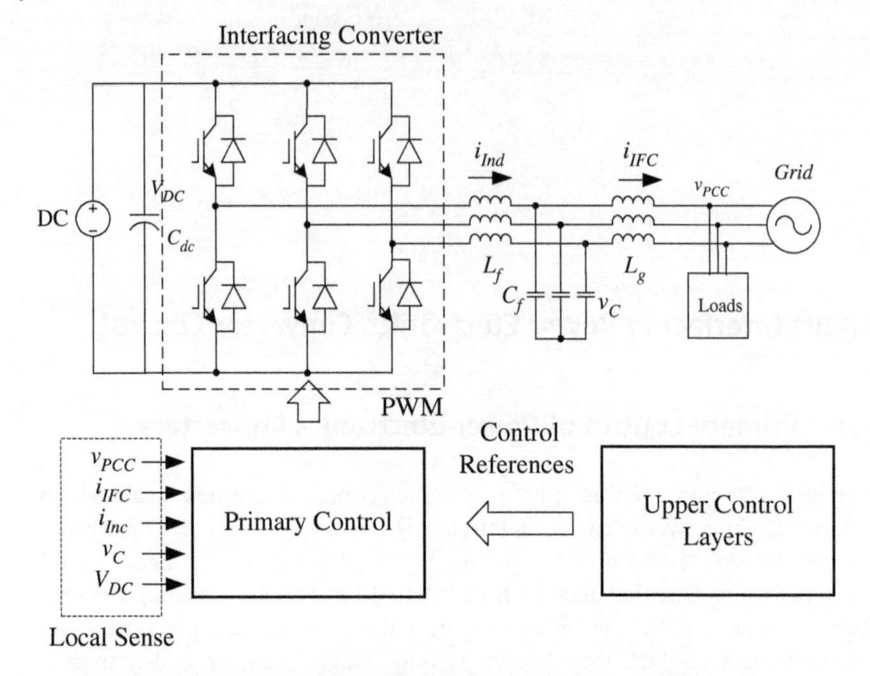

Figure 4.1 Typical structure of a three-phase converter and its control system.

different control structures [2], including the current control method (CCM) and voltage control method (VCM), will be discussed.

4.1.1 Basic Control Techniques in Power Converters

Coordinate Transformations in Converter Control

In three-phase converters, the sampled AC variables are naturally under *abc* stationary frames. Although it is straightforward to apply controllers in an *abc* frame, the variables of the three phases are not decoupled for common three-phase, three-wire systems, i.e. one of the three phasors is redundant as the sum of three-phase phasors will always be zero in a three-wire system. Also, it is not straightforward to apply the control of active and reactive components in an *abc* frame. Therefore, coordinate transformations are widely adopted in three-phase or even single-phase systems to facilitate active and reactive power control.

The Clarke transformation, or $\alpha\beta$ transformation, can be used to simplify the analysis and control of three-phase systems. The relationship between an *abc* frame and an $\alpha\beta$ frame is shown in Figure 4.2. In this orthogonal frame, the three-phase coupled variables are transformed to decoupled, fully controllable variables. The frame is still stationary and, as a result, the variables are still

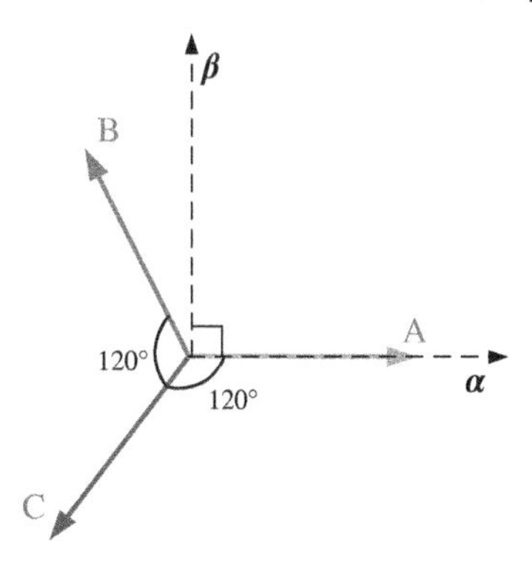

Figure 4.2 Relationships between the *abc* frame and the $\alpha\beta$ frame.

alternating with respect to time. When the three-phase system is unbalanced, a zero-axis can be added while in three-wire systems, and the component in zero-axis is always 0. The transformation and the inverse transformation are shown in (4.1) and (4.2), respectively.

$$\begin{bmatrix} \alpha \\ \beta \\ 0 \end{bmatrix} = \frac{2}{3} \begin{bmatrix} 1 & -\frac{1}{2} & -\frac{1}{2} \\ 0 & \frac{\sqrt{3}}{2} & -\frac{\sqrt{3}}{2} \\ \frac{1}{2} & \frac{1}{2} & \frac{1}{2} \end{bmatrix} \begin{bmatrix} a \\ b \\ c \end{bmatrix} \tag{4.1}$$

$$\begin{bmatrix} a \\ b \\ c \end{bmatrix} = \frac{2}{3} \begin{bmatrix} 1 & 0 & \frac{1}{2} \\ -\frac{1}{2} & \frac{\sqrt{3}}{2} & \frac{1}{2} \\ -\sqrt{\frac{1}{2}} & -\frac{\sqrt{3}}{2} & \frac{1}{2} \end{bmatrix} \begin{bmatrix} \alpha \\ \beta \\ 0 \end{bmatrix}. \tag{4.2}$$

In the $\alpha\beta$ frame, the active power and reactive power can be expressed as:

$$\begin{bmatrix} P \\ Q \end{bmatrix} = \frac{3}{2} \begin{bmatrix} v_\alpha i_\alpha + v_\beta i_\beta \\ v_\beta i_\alpha + v_\alpha i_\beta \end{bmatrix}. \tag{4.3}$$

Similar to the $\alpha\beta$ transformation, the Park transformation, or *dq* transformation also simplifies the three-phase analysis by using orthogonal coordinates. Different from stationary frames like the *abc* or $\alpha\beta$ frames, the *dq* frame rotates synchronously with the three-phase system, as shown in Figure 4.3. In this case, the rotating phasors in stationary frames will have zero relative angular speed – the positive-sequence AC signals in fundamental frequency become DC signals, which significantly simplifies the signal processing such as filtering and control.

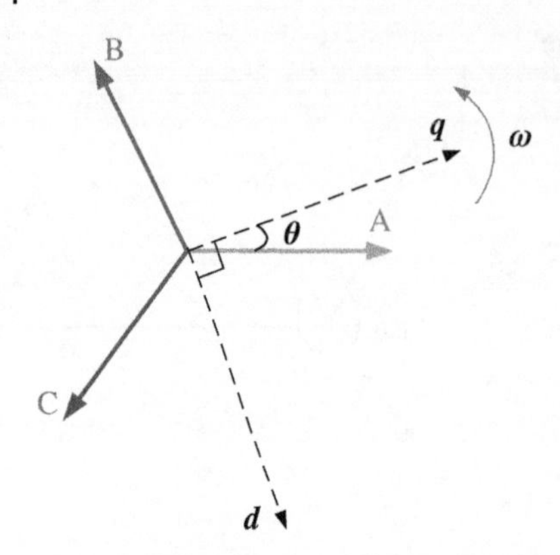

Figure 4.3 Relationships between the *abc* frame and the *dq* frame.

The Park transformation and its inverse transformation can be expressed as:

$$\begin{bmatrix} d \\ q \\ 0 \end{bmatrix} = \frac{2}{3} \begin{bmatrix} \sin\theta & \sin\left(\theta - \frac{2\pi}{3}\right) & \sin\left(\theta + \frac{2\pi}{3}\right) \\ \cos\theta & \cos\left(\theta - \frac{2\pi}{3}\right) & \cos\left(\theta + \frac{2\pi}{3}\right) \\ \frac{1}{2} & \frac{1}{2} & \frac{1}{2} \end{bmatrix} \begin{bmatrix} a \\ b \\ c \end{bmatrix} \tag{4.4}$$

$$\begin{bmatrix} a \\ b \\ c \end{bmatrix} = \frac{2}{3} \begin{bmatrix} \sin\theta & \cos\theta & \frac{1}{2} \\ \sin\left(\theta - \frac{2\pi}{3}\right) & \cos\left(\theta - \frac{2\pi}{3}\right) & \frac{1}{2} \\ \sin\left(\theta + \frac{2\pi}{3}\right) & \cos\left(\theta + \frac{2\pi}{3}\right) & \frac{1}{2} \end{bmatrix} \begin{bmatrix} d \\ q \\ 0 \end{bmatrix} \tag{4.5}$$

where $\theta = \omega t$ is the angle between the q-axis and phase A, and ω is the rotating frame frequency (typically the fundamental frequency).

The instantaneous active power and reactive power can be calculated as:

$$\begin{bmatrix} P \\ Q \end{bmatrix} = \frac{3}{2} \begin{bmatrix} v_d i_d + v_q i_q \\ v_q i_d + v_d i_q \end{bmatrix}. \tag{4.6}$$

Controllers for Reference Tracking

To reject disturbances and ensure accurate voltage/current signal regulation, closed-loop control is widely used in power converters. The selection of controllers considers the nature of the regulated signal, the required steady-state and transient performance, the required sampling rate and computation burden, etc. Some widely applied options are summarized as follows.

Proportional Integral Controller A proportional integral (PI) controller is widely used for control reference tracking in industrial applications, including in power converters. The PI controller consists of the proportional part and the integral part, and its transfer function is defined as:

$$G_{\mathrm{PI}} = K_{\mathrm{P}} + K_{\mathrm{I}}/s \tag{4.7}$$

where K_{P} is the proportional gain and K_{I} is the integral gain.

Theoretically, the PI controller has infinite gain at 0 Hz and thus can eliminate the tracking error for DC signals. This feature ensures that it can be easily applied to control DC voltage or current. Also, since the positive-sequence, fundamental-frequency AC signals become DC signals in the *dq* frame, the PI controller is also widely used to control AC voltage and current.

Proportional Resonance Controller To cope with AC signal tracking in stationary frames, the controller should have high gains at the central frequency of the signals. A proportional resonance (PR) controller is a popular option in power converters when the stationary frame is preferred. The transfer function is defined as:

$$G_{\mathrm{PI}} = K_{\mathrm{P}} + \frac{K_{\mathrm{R}} \cdot s}{s^2 + \omega_c \cdot s + \omega_o^2} \tag{4.8}$$

where K_{P} is the proportional gain and K_{R} is the resonant gain; ω_o is the angular frequency of the regulated signal. It is worth noting that the term $\omega_c \cdot s$ is not mandatory in theory but is needed in practice to damp the controller for better stability or broaden bandwidth, coping with variations of signal frequency. The resonant gain can create high gains at the desired frequency (ω_o).

In addition, the PR controller and PI controller can be combined to construct a PI+R controller to perform both fundamental frequency and harmonic control in the synchronous frame.

Deadbeat Control Deadbeat control is a discrete-time controller featuring wide control bandwidth and good dynamics. It is designed to eliminate control errors with the least time steps. For example, for an inverter with an output inductor as filter, it only requires two steps to regulate the output to follow the control reference – one step due to the control system delay and one step for the control process. The basic approach is to embed the system model into the controller.

Taking the inverter with output inductor L as the example, which is shown in Figure 4.4, the system model can be expressed as:

$$L\frac{\mathrm{d}i_{\mathrm{Ind}}}{\mathrm{d}t} + Ri_{\mathrm{Ind}} = v_{\mathrm{INV}} - v_{\mathrm{PCC}} \tag{4.9}$$

where i_{Ind} is the output current, v_{PCC} is the PCC voltage and v_{INV} is the output voltage at the inverter terminal. Discretizing (4.9) using the zero-order-hold (ZOH) method and using T_{s} as the sampling step, the following equation can be obtained:

$$i_{\mathrm{Ind}}(k+1) = \left(1 - \frac{T_{\mathrm{s}}R}{L}\right) i_{\mathrm{Ind}}(k) + \frac{T_{\mathrm{s}}}{L}[v_{\mathrm{INV}}(k) - v_{\mathrm{PCC}}(k)]. \tag{4.10}$$

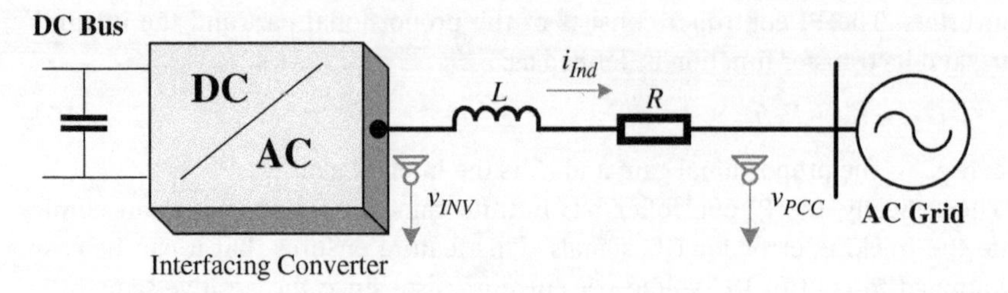

Figure 4.4 DC/AC converter with a single inductor as the filter.

If the control system delay is ignored, the voltage control reference $v_{INV}{}^{*}(k)$ that achieves $i_{ref}(k+1) = i_{Ind}(k+1)$ can be expressed as:

$$v_{INV}{}^{*}(k) = \frac{L}{T_s}\left[\left(i_{ref}\,(k+1) - \left(1 - \frac{T_s R}{L}\right) i_{Ind}(k)\right] + v_{PCC}(k). \tag{4.11}$$

When the control delay is considered, the target becomes $i_{ref}(k+2) = i_{Ind}(k+2)$, and the voltage reference needs to be derived as $v_{INV}{}^{*}(k+1)$:

$$v_{INV}{}^{*}(k+1) = \frac{L}{T_s}\left\{ i_{ref}(k+2) - \left(1 - \frac{T_s R}{L}\right)\left[\left(1 - \frac{T_s R}{L}\right) i_{Ind}(k) + \frac{T_s}{L}(v_{INV}(k) - v_{PCC}(k))\right]\right\} + v_{PCC}(k+1). \tag{4.12}$$

As can be seen, the deadbeat control is dependent on the system model and thus can be sensitive to system parameters. The implementation of deadbeat control for high-order systems (e.g. IFC with LCL filter) can be more complex and some simplifications are usually adopted in practice. For example, the filter capacitor current can be compensated by modifying the reference current while simplifying the LCL filter as an L filter in the deadbeat control model.

Model Predictive Control Similar to the deadbeat control, model predictive control (MPC) is also a type of discrete-time controller and features fast transient response. The main characteristic of the MPC controller is the use of the system model to predict control variables over a prediction horizon, which could be one or several sampling steps. Then, the applicable control action sequence is compared, while the optimal one is selected and applied to minimizes the user-defined objective or cost function. It should be noted that the prediction and selection processes discussed above are executed at each sampling step, while displacing the horizon toward the future, i.e. predicting the following one or multiple steps according to the current sampling results. Typically, the cost function is defined as:

$$g = \sum_i \lambda_i \left(x_i^* - x_i^{\mathrm{p}}\right)^2 \tag{4.13}$$

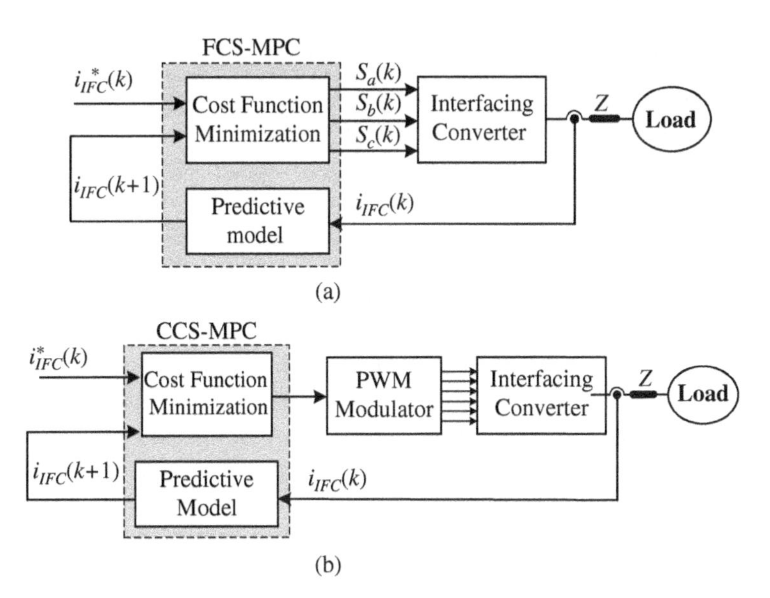

Figure 4.5 Block diagrams of typical model predictive control (MPC): (a) FCS-MPC and (b) CCS-MPC.

where x_i^* is the reference command, x_i^{p} is the predicted value for variable x_i, λ_i is a weighting factor, and index i denotes the number of variables to be controlled. Due to the use of the cost function, several control objectives can be included and achieved at the same time. Weighting factors can be assigned to each control objective to set different priorities of the corresponding control objective.

There are generally two implementation forms of MPC for power converter control, i.e. finite control set (FCS)-MPC and continuous control set (CCS)-MPC. Examples of them are shown in Figure 4.5, where only the output current i_{IFC} is controlled based on $i_{\mathrm{IFC}}(\mathrm{k}+1)$ – the current value obtained by single-step prediction. Then, the FCS-MPC will evaluate the cost function and directly select the switching state that minimizes the cost function, while CCS-MPC will generate a modulation reference to generate PWM and apply different switching states. In FCS-MPC, directly selecting switching states enables high flexibility in the control scheme, especially for some control functions that are hard to be achieved by modulation-based schemes. However, the FCS-MPC generally leads to a varying switching frequency. On the other hand, CCS-MPC guarantees a fixed switching frequency while it has reduced flexibility.

As a model-based control, the MPC can be sensitive to parameter variations and sampling errors. Also, it may have a high computation burden, especially for complex converter topologies or high order systems.

4.1.2 Current Control Method

The CCM applies direct control of IFC output currents and is very suitable for grid-following IFC operations. Grid-following operation is widely used for grid-connected converters, which synchronizes the converter AC current to a reference AC voltage and focuses on regulating the IFC output current and power flow. In islanded microgrid systems, grid-following control is also applied for converters that are not responsible for regulating the microgrid AC bus voltage. In this case, the power converter is controlled to regulate the output active and reactive powers to their reference values provided by the power management control strategy (microgrid power management will be discussed in Chapter 5).

In Figure 4.6, typical CCM control of an IFC is shown. As can be seen from the figure, both active and reactive powers are tracked in a closed-loop manner with an inner current control loop. The active power control loop produces a reference Id* for active current control, while the reactive power control loop generates the reactive current reference Iq*. With these references, the power converter output current i_{IFC} can then be controlled in the synchronous frame as shown in Figure 4.6a or in the stationary $\alpha\beta$ frame in Figure 4.6b. Then the inner current controllers can be applied to eliminate the error between control reference and

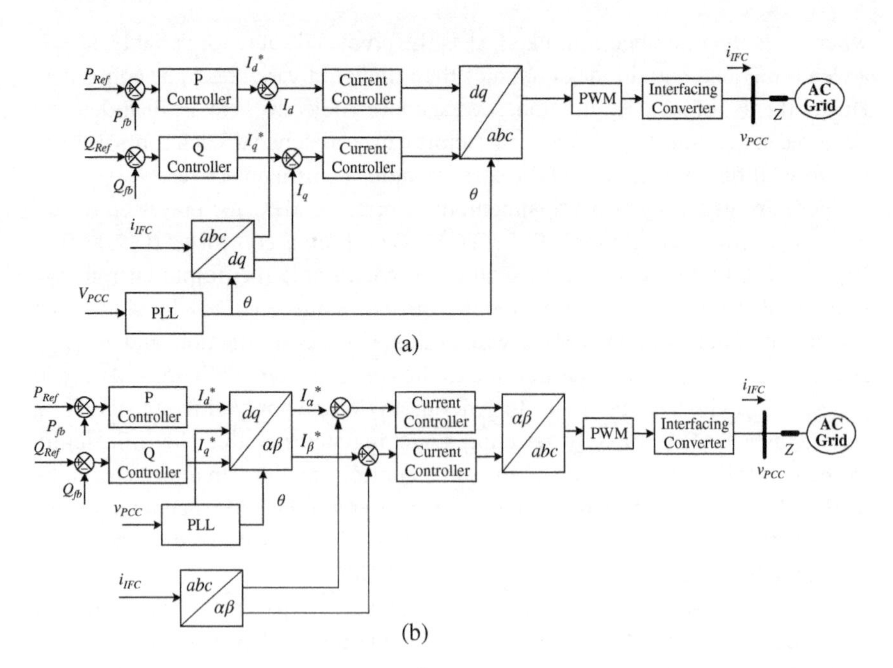

Figure 4.6 Primary control of IFCs; bi-directional power control through the current control method (CCM): (a) synchronous reference frame and (b) stationary reference frame.

actual current, forcing the output current to have the same values with references. Typical controllers that include PI, PR, deadbeat, and MPC as discussed previously can be adopted here. In this control strategy, the grid voltage angle information from phase-locked-loop (PLL) is used to synchronize the inverter output current with the grid voltage v_{PCC}.

In many grid-following applications, a real power control reference may not be given directly. For example, maximum power point tracking (MPPT) for a grid-connected PV system produces the active current reference from the MPPT scheme. Also, DC voltage control can be applied as an outer loop of CCM when the converter needs to control the DC-link voltage and ensure the power balance of a multi-stage power conversion system. A typical example is the two-stage PV inverter with a DC/DC converter followed by a grid-connected DC/AC inverter. In this system, the DC/DC converter performance MPPT and the grid side DC/AC converter maintained the DC link voltage. Under the DC link voltage control mode, the outer DC voltage control loop is applied with the inner current control loop to regulate the active current. On the other hand, the reactive current can be independently controlled as the reactive power will not impact the DC voltage. In Figure 4.7, typical examples of the DC voltage control mode, or MPPT control of the interfacing power converter are shown, where the synchronous frame control is adopted.

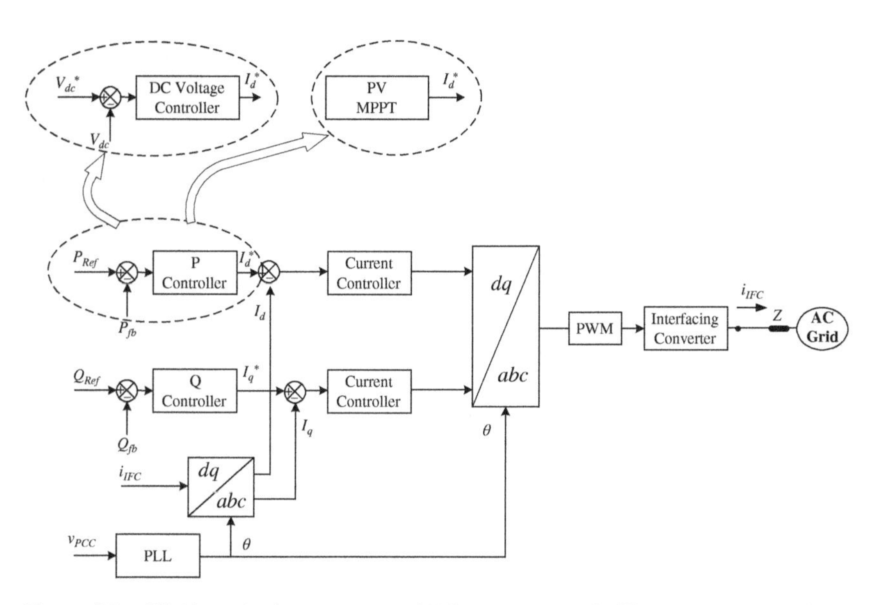

Figure 4.7 CCM-based primary control of IFCs – example of different ways to provide active power reference.

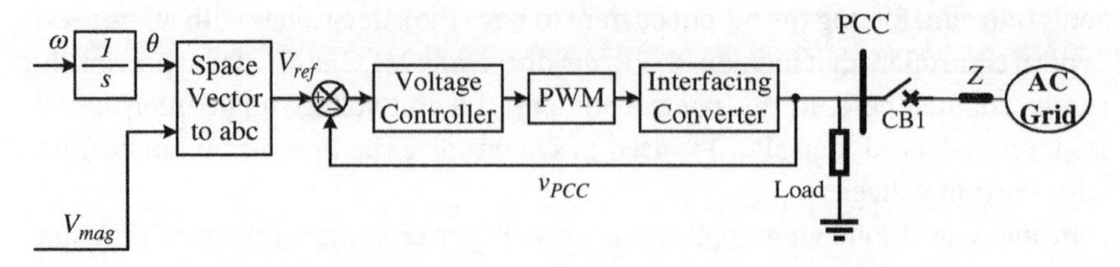

Figure 4.8 Primary control of IFCs: AC voltage control of the power converter.

4.1.3 Voltage Control Method

In general, the CCM-based power flow control strategy is popular in the grid-following operation mode where the AC bus frequency and voltage are determined by a reliable power source, such as the utility grid. However, in the stand-alone operation of a microgrid, the VCM based strategy should be applied to at least one converter, which directly provides a stable voltage for other converters to follow. This VCM is also popularly referred to as the grid-forming control or operation mode of an IFC.

A typical VCM control system is shown in Figure 4.8. The phase angle θ is produced by the integration of a reference frequency, and the voltage amplitude V_{mag} is typically the system nominal voltage. The current control loop can still be implemented as the inner loop to ensure good control loop damping and stability and reduce the sensitivity to load change under AC voltage control mode. In this case, the active and reactive power flow will be dependent on the load demand.

It is also possible to use the VCM to realize the grid-connected operation of IFC or use multiple IFCs to provide grid-forming control. The working principle can be explained as follows. Since IFCs are connected to the AC bus through the line impedance, which is assumed as mainly inductive with a high X/R ratio, and the phase angle between the converter and AC bus is typically small, according to the power flow equations, it can be concluded that the output active power of the converter is proportional to the phase angle difference between the converter output voltage and grid voltage, and the output reactive power is proportional to voltage magnitude difference. Therefore, the output active power can be controlled by the IFC output phase angle θ (or frequency ω), and the output reactive power can be controlled by its output voltage magnitude V_{mag}.

The block diagram of this control strategy is shown in Figure 4.9. This control scheme is very similar to how the traditional synchronous generators are controlled and can be applied to both grid-connected and stand-alone operation modes of microgrids. In this figure, the active and reactive power references are provided by power management strategies (see Chapter 5). Note that for low X/R feeders in the distribution system, a physical inductor or virtual inductor can be added between the converter and the grid to allow the above operation.

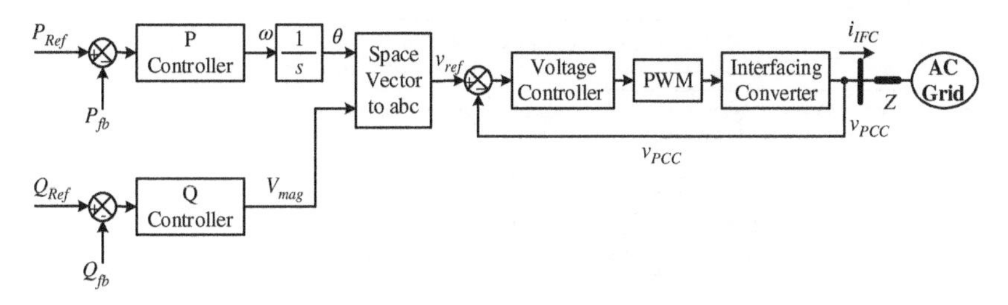

Figure 4.9 Primary control of IFCs: bi-directional power control through the voltage control method (VCM).

As discussed, in this strategy, the output voltage phase angle θ is determined by the active power controller, and the output voltage magnitude V_{mag} is controlled by the reactive power controller. The three-phase output voltages of the IFC are regulated on their reference values with a closed-loop control system. The active and reactive power controllers can be proportional controllers for realizing active power-frequency droop (P–ω) and reactive power–voltage magnitude droop (Q–V) to emulate the behavior of the synchronous generator in the steady-state (more details on droop control will be provided in Section 4.3). A more accurate virtual synchronous generator (VSG) controller can also be used to closely mimic the synchronous generator with excitation and torque dynamics during transient, which will be discussed in Section 4.4.

Compared to the CCM-based control, the main advantage of VCM-based control is that it can be used in both grid-connected and standalone operation modes, which makes the operation mode transition easy and smooth. Possible challenges when utilizing this method are mainly related to the lack of direct control of the output current to the PCC (i_{IFC}), especially during faults or grid voltage disturbances. These problems can be avoided by implementing virtual impedance control at the IFC output.

4.2 Virtual Impedance Control of Power Electronic Converters

The concept of virtual impedance control is to control the power converter to emulate the effects of having impedances that do not physically exist [3]. However, there is no energy loss associated with the implementation of virtual impedances. The virtual impedance is realized through the digitally controlled power electronics converters, and the characteristics of virtual impedance can also be adjusted dynamically through online digital control, which cannot be achieved by physical impedance.

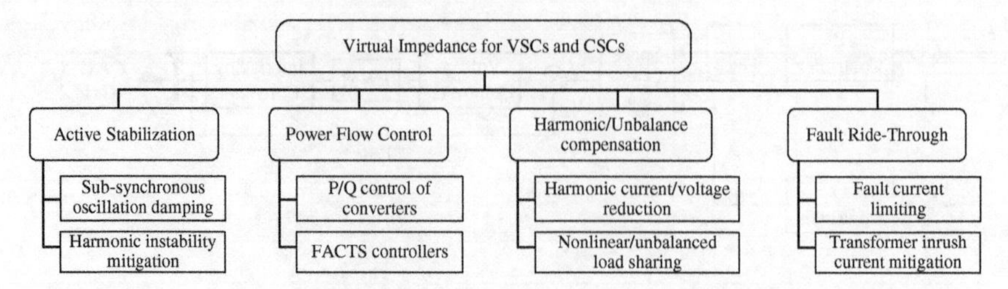

Figure 4.10 Functions of virtual impedance in a power converter [4].

The virtual impedance can be *R*, *L*, *C*, or combinations of them, or non-linear impedances, depending on the application. Due to this flexibility, the virtual impedance has wide applications. As shown in Figure 4.10, the virtual impedance can be used for the power flow control. It also enables converters to provide ancillary services, such as grid fault/disturbance ride-through, harmonic/unbalance compensation, with programmable impedances. Furthermore, the virtual impedances can improve the stability robustness of the converters against the different grid and load conditions. Another common application of the virtual impedances is the improved load sharing among the paralleled converters, such as parallel uninterruptible power supplies, and DG units.

In general, the virtual impedance can be classified as internal virtual impedance and external virtual impedance according to the position of the virtual impedance in an IFC. Their functions and implementations are discussed in the following sections.

4.2.1 Internal Virtual Impedance

The internal virtual impedance provides active damping to the converter-side LC resonance caused by an LC or LCL filter, and it is realized with high bandwidth through direct PWM regulation. The possible virtual impedance positions in an IFC can be (i) in series/parallel to the converter side inductor or (ii) in series/parallel to the filter capacitor, which is shown in Figure 4.11.

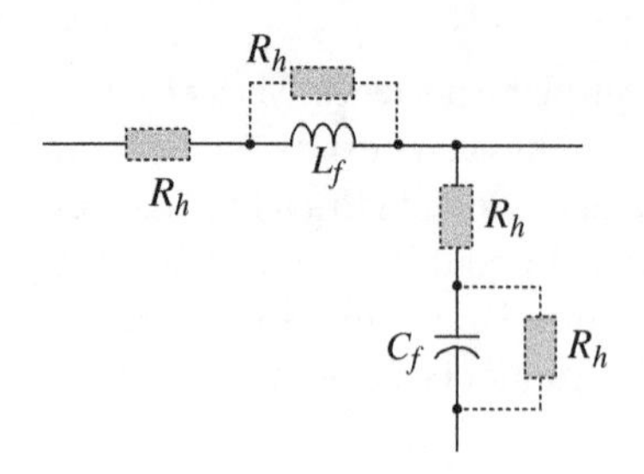

Figure 4.11 Possible internal virtual impedance positions.

The internal virtual impedance can be applied to both current source converters (CSCs) and voltage source converters (VSCs) when the filters need damping to avoid resonances. Taking a VSC as an example, it generally uses LC or LCL filters as output filters. In this case, the noises or harmonics whose frequencies are close to the filter's resonant frequency can be amplified, leading to excessive distortions or system instability.

To address this issue, two popular internal virtual impedance positions for a VSC are shown in Figure 4.12. As can be seen, the internal virtual impedance can be in series with the converter side inductor (Figure 4.12a) or parallel to the filter capacitor (Figure 4.12b). An example of implementing such internal impedance is shown in Figure 4.12c.

Taking Figure 4.12a as an example, the inductor current i_{Ind} can be measured to realize the virtual resistor R_V. If a physical resistor R_V exists in the circuit, a voltage drop will be caused when the current flows through it. To represent this behavior, the virtual voltage drop is $i_{Ind}*R_V$ and directly deducts this value from the converter terminal voltage reference.

Similarly, the paralleled resistor on the capacitor branch leads to extra current flowing to the virtual resistor. As a result, the current value v_C/R_V is then applied to the control system to inject the corresponding current. Compared to passive damping techniques using a physical resistor in the LC circuit, the virtual impedance-based damping can obviously improve the system's efficiency.

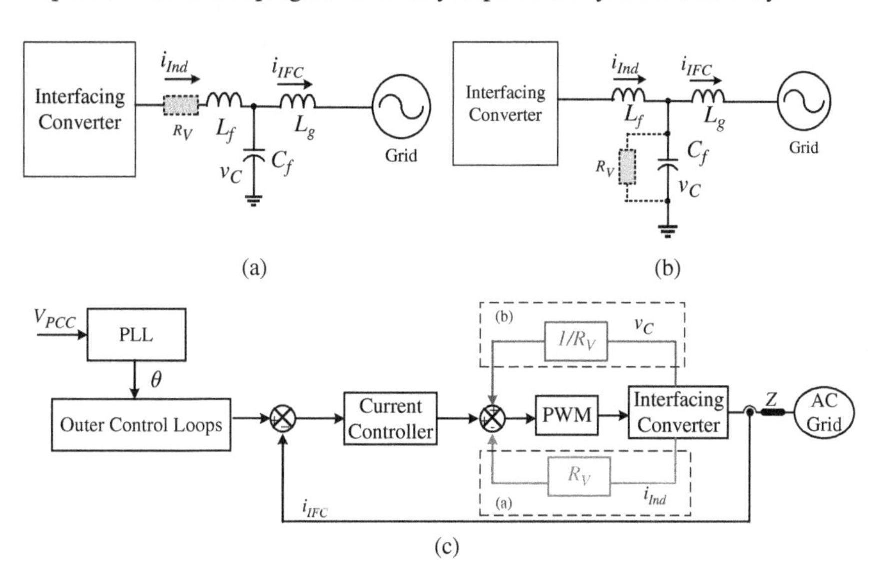

Figure 4.12 Two popular internal virtual impedance positions for a voltage source converter: (a) in series with the converter side inductor; (b) in parallel to the filter capacitor; (c) example implementations of the two types of virtual impedance.

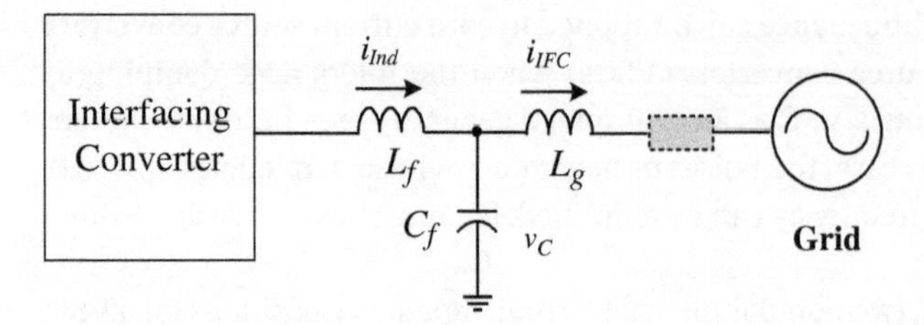

Figure 4.13 Position of the external virtual impedance of an interfacing converter.

4.2.2 External Virtual Impedance

In general, the external virtual impedance can be considered as external to the converter side filter circuit, as shown in Figure 4.13. The external virtual impedance can reshape the converter output impedance and thus provide various functions such as power flow control, voltage support, flicker mitigation, harmonics compensation, unbalanced voltage compensation, and so on.

The external virtual impedance can be realized in many ways. A few examples are shown in Figure 4.14. The virtual resistance can be easily implemented, which can be simply a real number gain. For capacitance or inductance, it is necessary to construct integral or derivative operations in the control to realize them. As shown in Figure 4.14a, the virtual inductance is implemented by the inductance value L, a derivative operation, and a low pass filter. The low pass filter avoids high spikes caused by the derivative operation when the measured value suffers from a high change rate. Note that the combination of derivative operation and a low pass filter gives an equivalent high power filter in the control loop. With this method, the virtual impedance can be applied in a range of frequencies, and the bandwidth is defined by the low pass filter.

Alternatively, if the control system is built in the orthogonal frames ($\alpha\beta$ or dq frame), the complex impedance can be implemented with the help of components

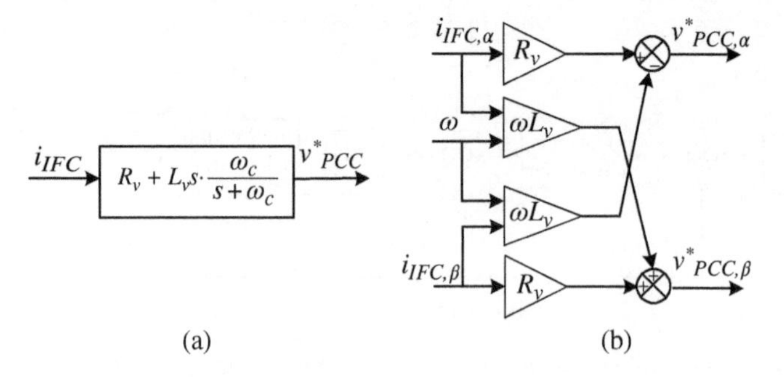

Figure 4.14 Different ways to realize complex virtual impedance: (a) LPF-based derivative controller. (b) Cross-coupling feedback of current vector.

from the orthogonal axis. For example, the virtual inductance is realized by coupling terms in the $\alpha\beta$ frame, as shown in Figure 4.14b. In this case, the virtual impedance can also be added to selected harmonic orders with the help of a band-pass filter, and the term ωL_v can help ensure an accurate virtual impedance realization at the selected frequency ω. Besides, to realize the different functions required, different forms of virtual impedance can be applied, combining the virtual resistor, virtual capacitor, and virtual inductor, and at selected frequencies or range of frequencies.

Combined with the IFC control (voltage control mode or grid-forming control), an example of external virtual impedance implementation is shown in Figure 4.15. Here, the grid side current i_{IFC} is measured, and used in the virtual impedance (Z_{V}) implementation to modify the voltage reference V_{ref}, emulating the voltage drop of a physical impedance and generate the final voltage reference v^*. According to different application requirements, the external virtual impedances Z_{V} can be applied to positive sequence, negative sequence, and zero sequence for a four-wire system. The external virtual impedance can also be defined in fundamental frequency and harmonic frequencies for each IFC. Moreover, the virtual impedance in each harmonic can be decomposed into positive, negative, and zero sequence components when both harmonics and unbalanced compensation is required. In Chapter 9, the IFC's positive, negative, and zero sequence virtual impedances at fundamental frequency are controlled for unbalanced grid voltage compensation. The control method to reshape IFC impedance at harmonic frequencies for harmonics compensation will be discussed in Chapter 10.

4.2.3 Integration of both Internal and External Virtual Impedance

As discussed, the internal impedance can provide damping effects on the IFC, enhancing its stability. On the other hand, the external impedance can reshape the equivalent impedance of the converter system, so that enhanced performance and various smart converter functions can be achieved. Since both types of virtual impedance are important for a stable and smart IFC, a method for integrating

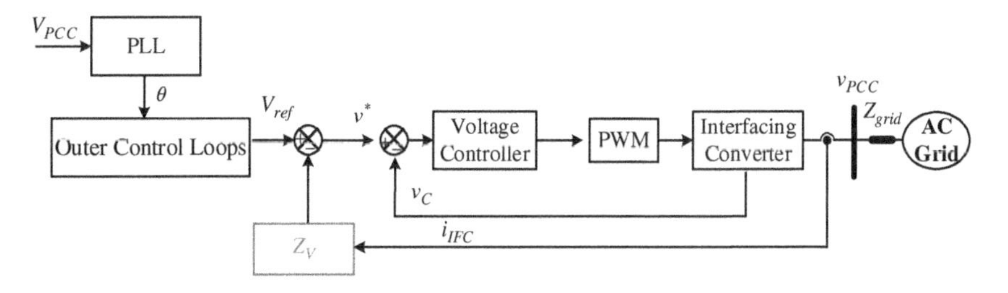

Figure 4.15 An example control scheme to implement external virtual impedance.

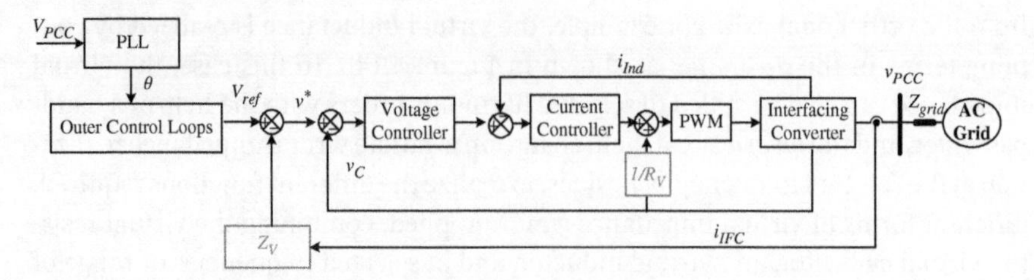

Figure 4.16 An example control system with both internal and external virtual impedance.

both internal and external virtual impedance in the same control system is presented here.

A straightforward way to implementing both external and internal virtual impedance is to combine the schemes shown in Figures 4.12c and 4.15. The combined scheme is shown in Figure 4.16, which uses the voltage-controlled IFC as an example. In this figure, the double-loop primary voltage control scheme contains an outer voltage control loop and inner current control loops. For internal virtual impedance, the capacitor voltage is measured to realize an internal virtual resistor R_V in parallel with the filter capacitor to achieve system damping. The grid current i_{IFC} is measured to reshape the IFC output impedance by applying an external virtual impedance Z_V. As the external virtual impedance control is realized by modifying the IFC voltage reference, the control bandwidth of the external virtual impedance will be limited by the voltage control loop.

A more general way of integrating internal and external impedance, i.e. general closed-loop control (GCC) [5], is shown in Figure 4.17, which uses the current-controlled IFC as an example. GCC has single closed-loop control of the reference variable and two embedded virtual impedance terms, an internal impedance and an external impedance, as expressed in the following:

$$V^*_{\mathrm{PWM}} = H_{\mathrm{G}}(s)(i_{\mathrm{ref}} - i_{\mathrm{IFC}}) - H_{\mathrm{internal}}(s)i_{\mathrm{Ind}} + H_{\mathrm{external}}(s)v_{\mathrm{PCC}} \tag{4.14}$$

where $H_{\mathrm{G}}(s)$ is the closed-loop reference tracking controller in GCC, $H_{\mathrm{internal}}(s)$ is the control term for the internal virtual impedance, and $H_{\mathrm{external}}(s)$ is the control term to produce the external virtual impedance of a VSC system. $C_{\mathrm{G,ref}}$ and

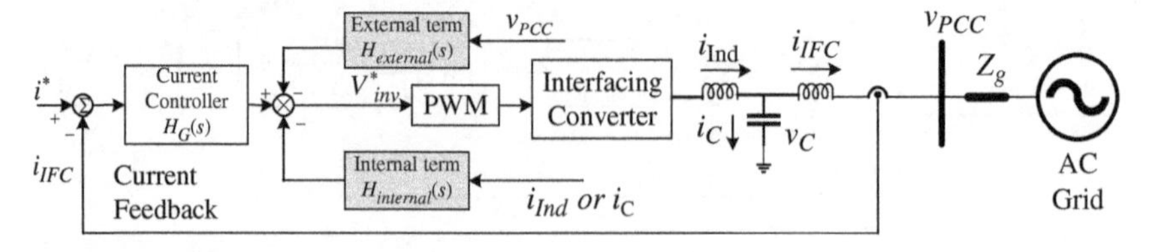

Figure 4.17 General closed-loop control with internal and external virtual impedance.

C_G are the control reference and its measured feedback (such as capacitor voltage V_C or line current i_{IFC}, see Figure 4.17), $C_{internal}$ is the selected variable (such as filter inductor current i_{Ind} or capacitor current i_C) for internal impedance control, and $C_{external}$ is the variable (such as line current i_{IFC} or grid voltage v_{PCC}) used for external impedance realization.

In the GCC scheme, the closed-loop term is used for reference tracking, which has a similar function to the single-loop control. The internal impedance is mainly used to provide sufficient damping to the filter plant. By flexibly controlling the internal impedance term, the GCC scheme can easily achieve similar performance compared to the multiloop control scheme. In addition, the selection of an internal impedance term feedback variable and the control parameter design with GCC is more straightforward and robust compared to voltage/current double-loop control. Finally, the external impedance is applied to adjust the converter system output impedance. The external virtual impedance can be at the fundamental frequency and harmonics/negative sequence for various kinds of purposes: power flow control, grid disturbance ride through, unbalance/harmonics compensation, etc. More examples of using virtual impedance are presented in the later chapters.

Besides the approaches of using closed-loop control with conventional PI or PR regulators, the implementation of virtual impedances can be flexible as various approaches can be applied. For example, deadbeat control can be used to apply virtual impedance with a high control bandwidth. Also, it can be realized by a feedforward control structure, which will be further discussed in Chapter 10.

4.3 Droop Control of Power Electronics Converters

4.3.1 Frequency and Voltage Droop Control in an AC Subgrid

A well-known method to realize the "plug and play" feature for each DER unit is to control its IFCs terminal voltage by employing the "real power versus frequency (P–ω)" and "reactive power versus voltage (Q–E)" droop control to emulate the traditional synchronous generator steady-state characteristics [6]. Simply put, this method is based on the flow of real power and reactive power (per phase) between two nodes separated by a line impedance ($Z = R + jX$) as:

$$P = \frac{E_1}{R^2 + X^2}[R(E_1 - E_2 \cos \delta) + XE_2 \sin \delta] \tag{4.15}$$

$$Q = \frac{E_1}{R^2 + X^2}[-RE_2 \sin \delta + X(E_1 - E_2 \cos \delta)] \tag{4.16}$$

where E_1 and E_2 are the magnitudes of the two voltages, and δ is the phase angle difference between the two voltages. For a mainly inductive line impedance, the line resistance (R) may be neglected. Further, considering that the phase angle difference δ is typically small, it is reasonable to assume $\sin(\delta) = \delta$ and $\cos(\delta) = 1$. Therefore, the flow of real power is proportional to the phase angle difference (δ) and the flow of reactive power is proportional to the voltage magnitude difference ($E_1 - E_2$). For this reason, the real power from each IFC unit can be controlled by varying the IFC output frequency (and hence, the phase angle), and the IFC's reactive power can be regulated by changing the IFC output voltage magnitude. This control concept could be used in both the grid-connected and islanding operation modes.

Figure 4.18 shows the P–ω droop characteristics for two DG systems interfaced by power electronics converters (note that the control strategy is equally applicable to a microgrid with more DG units). Preferably, these droop characteristics should be coordinated to make each IFC supplying real power in proportion to its power capacity, and can mathematically be expressed as:

$$\omega_i = \omega^* - SP_{Pi}\left(P_i^* - P_i\right) \tag{4.17}$$

$$SP_{Pi} = \frac{(\omega^* - \omega_{\min})}{P_i^* - P_{i\_\max}} \tag{4.18}$$

where P_i is the actual real power output of the ith IFC ($i = 1, 2, \ldots, n$), $P_{i\_\max}$ and $\omega_{\min}$ are the maximum real power of the ith IFC and the minimum allowable operating frequency, P_i^* and ω^* are the dispatched real power and operating frequency of the ith IFC in the grid-connected mode, and $SP_{pi}(<0)$ is the slope of the droop characteristics.

As shown in Figure 4.18, each power electronics interfaced DG system is initially designed to generate the dispatched real power output of P_i^* at the common base frequency of ω^* when operating in the grid-connected mode (ω^* is fixed by the stiff utility grid). Once in islanding operation, the power outputs of both IFCs must immediately be changed in accordance with their droop characteristics to supply power to all critical loads in the microgrid at a new steady-state frequency of ω.

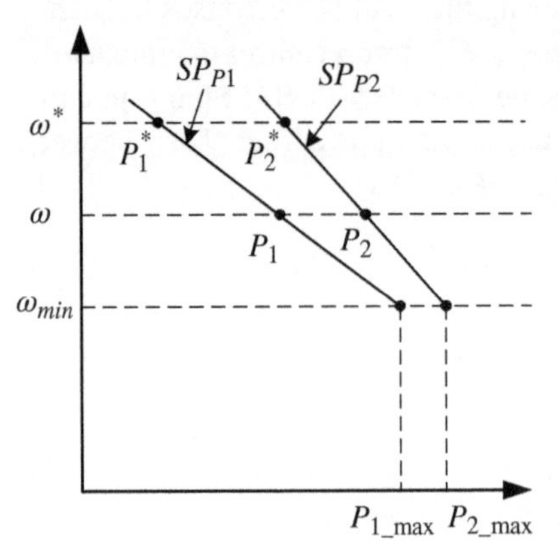

Figure 4.18 Real power sharing through frequency droop control in an AC subgrid.

This arrangement obviously allows both IFCs to share the total load demand in a predetermined manner according to their respective power ratings.

In a similar manner, the magnitude setpoint of each IFC output voltage can be tuned according to a specified Q–E droop scheme to control the flow of reactive power within the microgrid. Mathematically, the Q–E characteristics can be expressed as:

$$E_i = E^* - SP_{Qi}\left(Q_i^* - Q_i\right) \tag{4.19}$$

$$SP_{Qi} = \frac{E^* - E_{\min}}{Q_i^* - Q_{i\_\max}} \tag{4.20}$$

where Q_i is the actual reactive power output of ith IFC, $Q_{i\_\max}$ and $E_{\min}$ are the maximum reactive power output and minimum allowable voltage magnitude of the microgrid, Q_i^* and E^* are the dispatched reactive power of the ith IFC and PCC voltage magnitude when in the grid-connected mode, and SP_{Qi} is the slope of the droop characteristics. In Figure 4.19a, the ideal IFC reactive power-sharing with voltage droop control is shown. Conceptually similar to the P–ω operation, the accuracy of reactive power control and sharing is, however, subject to the voltage drop in the line impedances. As can be seen from Figure 4.19b, line impedances can cause voltage differences for each IFCs, resulting in inaccurate sharing. The details will be discussed in Section 4.3.4.

It should be again highlighted that the droop coefficients of IFCs should be regulated in inverse proportion to their power ratings to share the demand power in proportion to their power ratings. In (4.21) and (4.22), this relationship is shown:

$$SP_{P1} \cdot P_1 = SP_{P2} \cdot P_2 = \cdots = SP_{Pi} \cdot P_i \tag{4.21}$$

$$SP_{Q1} \cdot Q_1 = SP_{Q2} \cdot Q_2 = \cdots = SP_{Qi} \cdot Q_i. \tag{4.22}$$

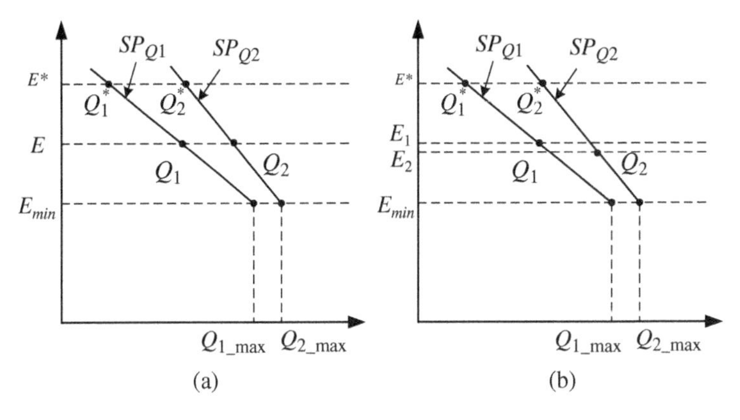

Figure 4.19 Reactive power-sharing with traditional voltage droop control: (a) ideal condition; (b) errors caused by line impedances.

4.3.2 Voltage Droop Control in DC Subgrids

Compared to AC bus droop control, the DC bus droop control is simpler since there are no reactive power, frequency, and phase angle considerations. In the droop control strategy in the DC subgrid, only the active power and voltage magnitude droop (*P*–*V* droop) is needed. The relationship between these two quantities form the droop equation for the DC subgrid as follows:

$$V_{\mathrm{DC},i} = V_{\mathrm{DC}}^{*} - SP_{\mathrm{DC},i}\left(P_{\mathrm{DC},i}^{*} - P_{\mathrm{DC},i}\right) \tag{4.23}$$

where V_{DC}^{*} is the maximum output voltage under no-load condition, and $SP_{\mathrm{DC},i}$ is the droop coefficient.

Similar to AC droop, for proportional active power-sharing to IFC power ratings, the droop coefficients should be adjusted in inverse proportion to the IFC rated power, as follows:

$$SP_{\mathrm{DC},1} \cdot P_{\mathrm{DC},1} = SP_{\mathrm{DC},2} \cdot P_{\mathrm{DC},2} = \cdots = SP_{\mathrm{DC},i} \cdot P_{\mathrm{DC},i}. \tag{4.24}$$

Figure 4.20 represents the *P*–*V* droop diagram of two IFCs connected to the DC bus. In this figure, V_{DC}^{*} is the maximum voltage of the DC bus under no-load condition, $V_{\mathrm{DC,\,min}}$ is the allowable minimum voltage of the DC bus, and $P_{\mathrm{DC},1}^{*}$ and $P_{\mathrm{DC},2}^{*}$ are the maximum real power of DGs. With this droop control, different line impedances lead to unequal DC source terminal voltages ($V_{\mathrm{DC},1}$ and $V_{\mathrm{DC},2}$ in Figure 4.20), which causes an error in the active power-sharing. This will be discussed in Section 4.3.4.

4.3.3 Unified Droop for Interlinking AC and DC Subgrids

In hybrid AC/DC microgrids, the interlinking converter that interconnects the AC subgrid and DC subgrid should perform active power balancing with both subgrids

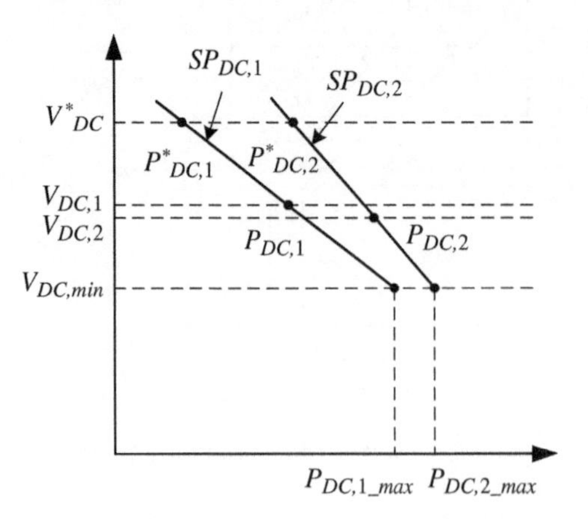

Figure 4.20 Load demand sharing in a DC subgrid using DC droop control.

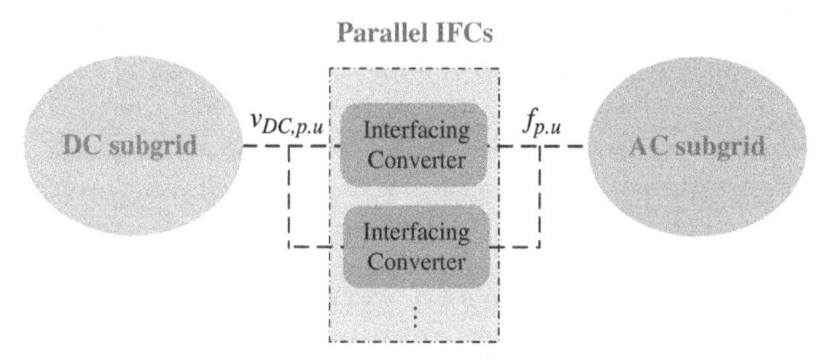

Figure 4.21 AC/DC-coupled hybrid microgrid in islanded operation mode with the unified droop method in both AC and DC subsystems and autonomous IFC control.

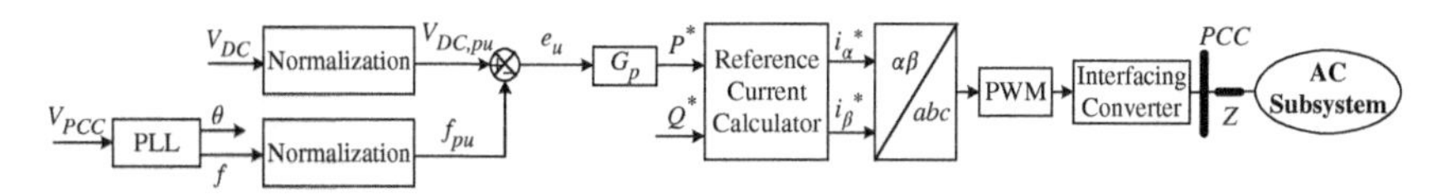

Figure 4.22 Control block diagram of one AC/DC subgrid IFC in islanded operation mode with normalized droop method in both AC and DC buses.

considered. To do this, the unified control scheme can be applied. In this scheme, per unit (p.u.) values (or "nameplate" value) of the DC subsystem voltage and AC subsystem frequency are measured, and their difference is used to adjust the power demand of the AC/DC subgrid IFC for DC or AC subgrid support. Multiple IFCs could be used in parallel with enhanced capacity and reliability. In Figure 4.21, the block diagram hybrid of the AC/DC microgrid in islanded operation mode is shown.

As shown in Figure 4.22, for proper power-sharing among IFCs to control normalized DC subsystem voltage and AC subsystem frequency, a power dispatch controller (G_P) is used for each IFC, which provides the alternative droop control technique among IFCs. These controllers are designed based on the maximum difference between DC voltage and AC frequency p.u. values (e_u) in the microgrid and each IFC's power rating. The p.u value of frequency f and DC voltage can be obtained as follows:

$$f_{pu} = \frac{f - 0.5(f_{max} + f_{min})}{0.5(f_{max} - f_{min})}$$

$$V_{DC,pu} = \frac{V_{DC} - 0.5(V_{DC,max} + V_{DC,min})}{(V_{DC,max} - V_{DC,min})}. \tag{4.25}$$

Then the normalized error e_u is defined as:

$$e_u = f_{pu} - V_{DC,pu}. \tag{4.26}$$

If the hybrid AC/DC system only contains a single IFC to carry out power exchange between the AC subgrid and the DC subgrid, the power dispatch controller G_p can be a PI controller. The power rating of the IFC is designed to be large enough to realize the power flow control and eliminate such errors. However, if the system consists of several IFCs, the power for mitigating such errors should be shared by the paralleled IFCs. In this case, a proportional gain K_p is recommended. Note that PI can still be applied if the parameters are properly designed and the IFC rating allowed to eliminate the error as mentioned earlier.

As an example, the unified droop diagram of two IFCs of the hybrid AC/DC system is shown in Figure 4.23. In this figure, the normalized error has a positive maximum value $+e_{u,max}$ and the negative maximum value $-e_{u,max}$. Similarly, the two bidirectional IFCs have maximum values at both positive and negative values, which are represented as $+P_{1,max}$, $-P_{1,max}$, $+P_{2,max}$, $-P_{2,max}$, respectively. When e_u is positive, the IFCs will control real power flowing from AC subgrid to DC subgrid (rectifying), and a negative e_u value will lead to the real power flow from DC subgrid to AC subgrid (inverting). Due to the different capacities of the two IFCs, the power flow will be shared proportionally with the help of a unified droop. As can be seen, under the same e_u value, the power flow direction is the same for both IFCs and the power is shared fulfilling (4.27):

$$SP_{eu,1} \cdot e_{u,1} = SP_{eu,2} \cdot e_{u,2} = \cdots = SP_{eu,i} \cdot e_{u,i}. \tag{4.27}$$

In this case, the unified droop can help IFCs to properly share the power and mitigate the normalized error.

Since reactive power does not participate in the power interaction between the AC subgrid and the DC subgrid, the Q–E droop is still the same as the AC voltage

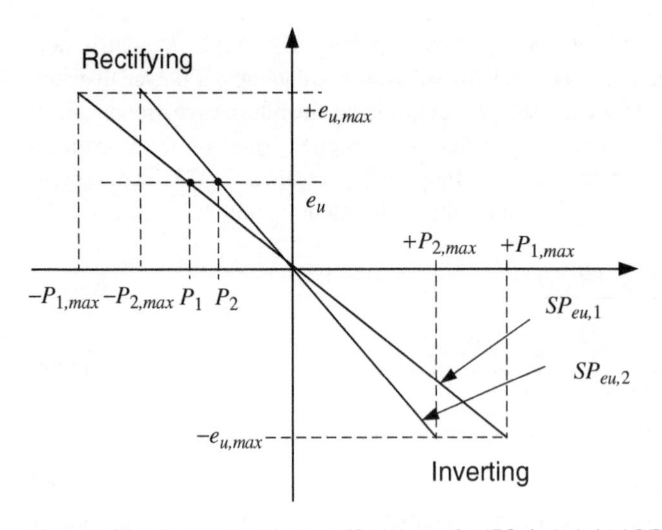

Figure 4.23 Power sharing by unified droop for IFCs in hybrid AC/DC systems.

droop systems. So, the reactive power control of IFCs can still be realized with Q–E droop control when the unified active power droop is applied.

4.3.4 Challenges of Droop Control and Solutions

Power Coupling Due to Low *R/X* Ratio and Virtual Inductance Concept

While working well in a power grid with mainly inductive line impedances, the traditional real and reactive powers control (where the line resistance is neglected) leads to concerns when implemented on a low voltage microgrid, where the feeder impedance is not inductive and the line resistance (R) should not be neglected. This is especially true for power electronics interfaced DG units without a grid-side inductor or transformer, where the output inductance is very small. In this case, a change in phase angle or voltage magnitude will influence both the real power and reactive power flows, as can be noticed from (4.15) and (4.16). As a result, controlling the power flow using the conventional P–ω and Q–E droop methods will introduce a significant coupling between the real and reactive power flows, especially during transients.

To control the decoupled real and reactive power flows in a similar manner as the conventional power system with a high X/R ratio, an effective method is controlling the interfacing inverter with an inductive external virtual impedance. Introducing a predominantly inductive impedance can effectively decouple the real and reactive power flows and requires no physical connection of any passive components at the IFC output. With a virtual inductor at each IFC output, the conventional P–ω and Q–E methods can be used which makes the power-sharing algorithm equally applicable even when the rotational machine-based DG units (where the P–ω and Q–E characteristics are determined by the mechanical governor and excitation system, respectively) are present in a microgrid.

To emulate the effect of an inductor, the line current is fed back to calculate the virtual inductor voltage drop (v_{VL}), which is then subtracted from the reference voltage (generated from the power loops) to produce the final inverter voltage reference. The virtual inductance $j\omega L_0$ can be realized in polar form with polar–rectangular transformations or through direct complex number manipulations in the $\alpha\beta$ frame, as shown in Figure 4.24.

Sharing Inaccurate Due to Line Impedance

In an AC microgrid, the system has the same voltage frequency throughout the AC bus, which ensures accurate active power-sharing under P–ω control. On the other hand, a complication with the Q–E droop control is that the terminal voltage of IFCs can be different due to the voltage drop caused by the line impedances. As a result, the Q–E droop scheme specified in (4.19) and (4.20) will lead to a reactive

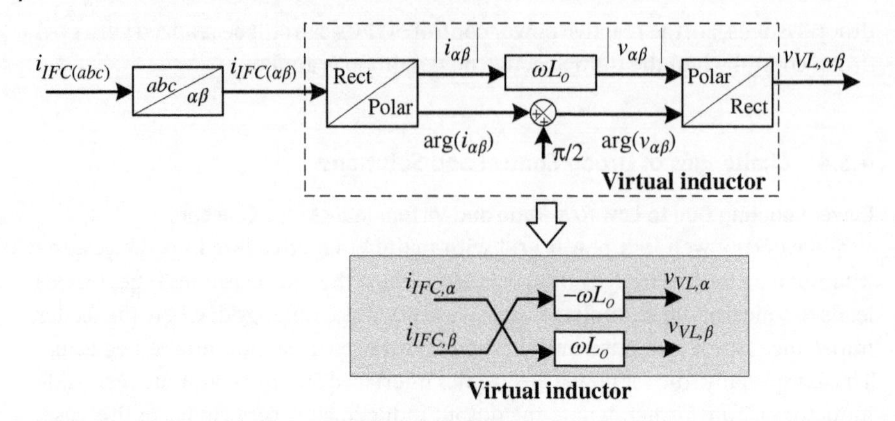

Figure 4.24 Virtual inductor realization scheme.

power control error. This phenomenon is illustrated in Figure 4.25, where the predominantly inductive line impedance is assumed which leads to an approximately linear relationship between the IFC output reactive power and the voltage magnitude difference (between IFC output voltage and PCC voltage), ΔE, as can be noticed from (4.16). This linear relationship for the ith IFC can be expressed as:

$$K_{Qi} = \frac{\Delta E}{Q_i} = \frac{X_i}{E_i} \tag{4.28}$$

where K_{Qi} is the slope of IFC output voltage magnitude difference ΔE versus reactive power. As the IFC output voltage is limited to vary only in a small range (e.g. ±10%) and the inductance between two voltages is normally a constant, it is reasonable to assume K_{Qi} as a constant slope, which can be estimated [7]. With

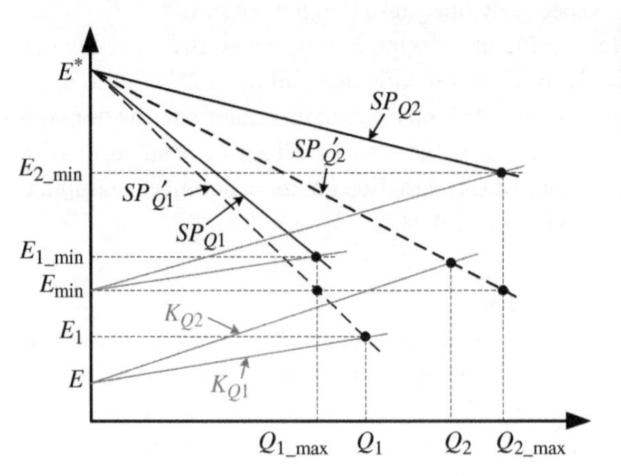

Figure 4.25 Reactive power-sharing diagram with line impedance (inductive) effects.

the consideration of the line impedance effects and the $\Delta E/Q$ slopes defined in (4.28), the voltage droop slopes can be re-defined to be SP_{Q1} and SP_{Q2} as shown in the solid lines in Figure 4.25, where the minimum allowable output voltage, power converter output voltage magnitude reference E_i^* and the voltage droop slope SP_{Qi} can be calculated as in (4.29)–(4.31) respectively:

$$E_{i\_\min} = E_{\min} + K_{Qi} \cdot Q_{i\_\max} \tag{4.29}$$

$$E_i^* = E_{i\_GC}^* - SP_{Qi}\left(Q_i^* - Q_i\right) \tag{4.30}$$

$$SP_{Qi} = \frac{E_{i\_GC}^* - E_{i\_\min}}{Q_i^* - Q_{i\_\max}}. \tag{4.31}$$

A similar challenge will also apply to the P–V droop for the DC system, where the converters at different installation points will see different voltages at their output terminals.

To solve the power-sharing accuracy problem due to impedance voltage drops, different solutions can be applied. Three examples are given as follows.

1. **Reactive power correction through low-bandwidth communication links:** Low bandwidth communication links can be used to directly correct sharing errors [8], where the system parameters are no longer needed. The sharing correction can be performed by a central controller, which will distribute correction control signals to each paralleled IFC. This approach requires the continuously updated operating information of the whole microgrid.
2. **Virtual impedance control to match p.u values:** For systems that cannot acquire complete power flow information, a way to improve sharing accuracy is to compensate the different line impedances with virtual external impedance at the fundamental frequency, matching the line voltage drops to an expected value. As shown in Figure 4.26, where the physical line impedance (L_i and R_i) is

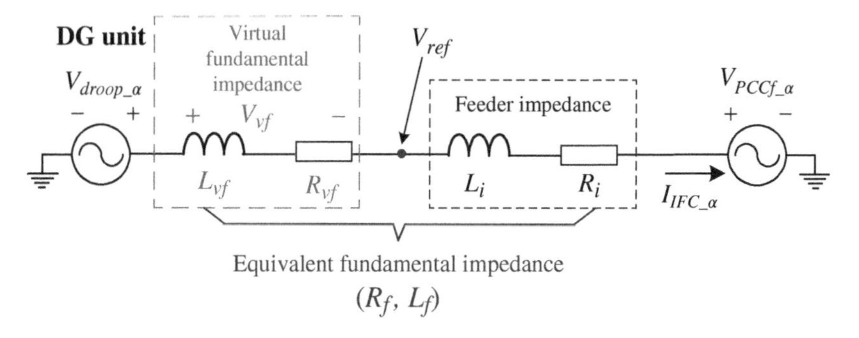

Figure 4.26 Matching line impedances with virtual impedance.

compensated with virtual impedance L_{vf} and R_{vf}. Then all the DGs paralleled in the same system can have the equivalent line impedance reverse proportional to the DG ratings. In other words, if using the DG rating as based power, all DGs should have the same p.u values of equivalent line impedance R_f and L_f. As a result, the AC *Q–E* droop or DC *P–V* droop can have better sharing accuracy.

3. **Real power disturbance and observation:** The performance of matching virtual impedance is highly dependent on the accuracy of obtained information on physical impedance. Otherwise, improper virtual impedance values will be applied and lead to unsatisfying accuracy. Alternatively, [9] proposes injecting active power disturbances to help improve reactive power sharing accuracy, whose droop control functions are modified as (4.32) and (4.33) by adding the compensation terms to conventional droop equations:

$$\omega = \omega_0 - D_p \cdot P \underbrace{-G \cdot D_Q \cdot Q}_{\text{Compensation Term}} \tag{4.32}$$

$$E = E_0 - D_Q \cdot Q \underbrace{+G \cdot (K_q/s) \cdot (P - P_{\mathrm{AVE}})}_{\text{Compensation Term}} \tag{4.33}$$

where G is a time-varying soft compensation factor. It increases slowly from 0 to 1 at the beginning of the compensation and it reduces to zero at the end of the compensation, then the IFC can observe the real power difference between P_{AVE} – the real power before disturbance is injected, and P_{LPF} – the real power after the disturbance. The differences are added to the integral term with a gain K_q, whose value shall be the same for all the IFCs in the same system.

With the modification, once the sharing accuracy enhancement is triggered ($G = 1$), any reactive power sharing errors under conventional droop control will introduce a real power disturbance, as the modified frequency droop controller (4.32) essentially enables the equal sharing of the combination power ($D_P \cdot P + D_Q \cdot Q$). For instance, the DG units providing less reactive power will experience a transient real power increase when the compensation is enabled. With the real power-sharing disturbance, the difference between the transient real power P and the saved average real power P_{ave} indirectly reflects the error of the reactive power sharing under conventional droop. Therefore, an integration of the real power difference ($P - P_{\mathrm{ave}}$) is able to eliminate the reactive power sharing error as illustrated in (4.33). Once the reactive power is shared properly, the DG unit real power flow will go back to its original value by gradually reducing the value of G back to 0.

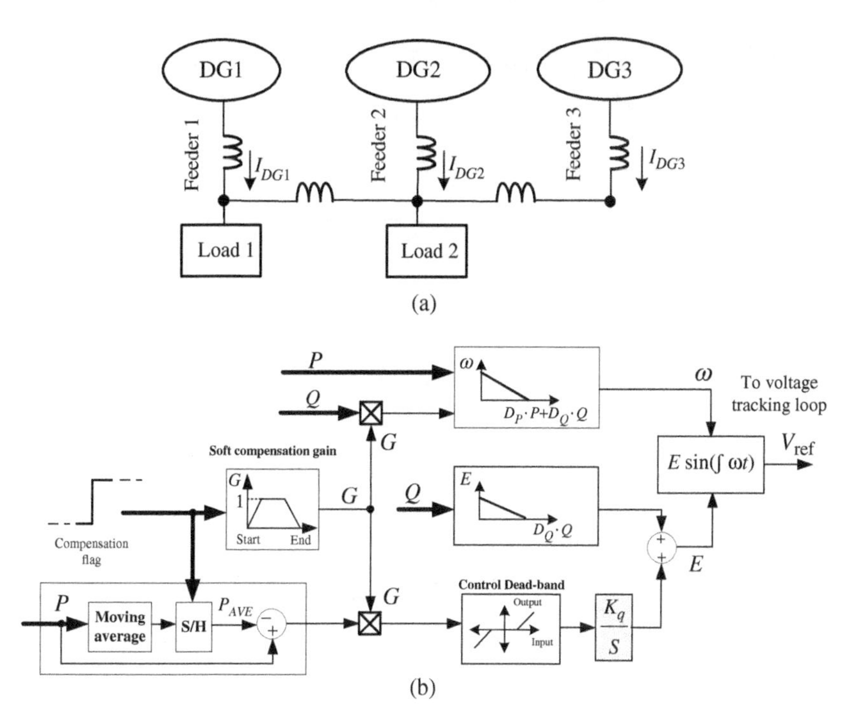

Figure 4.27 System diagram used for the case study: (a) system structure; (b) implementation of the control scheme.

> **Case Study**
>
> To demonstrate the performance of the real power disturbance method, experiments have been performed on an islanding microgrid, where two DG units at the same power rating are connected to PCC with different feeder impedances. The simplified diagram of the single-phase experimental setup is shown in Figure 4.27. The detailed circuit and control parameters of the system are provided in Table C.1 of Appendix C. Low power factor linear RL loads are considered in the system.

The corresponding power-sharing performance is presented in Figure 4.28. For the time range 0–1 s, only the conventional droop control is adopted. It is obvious that the real power sharing is accurate, but the sharing of reactive power load has some

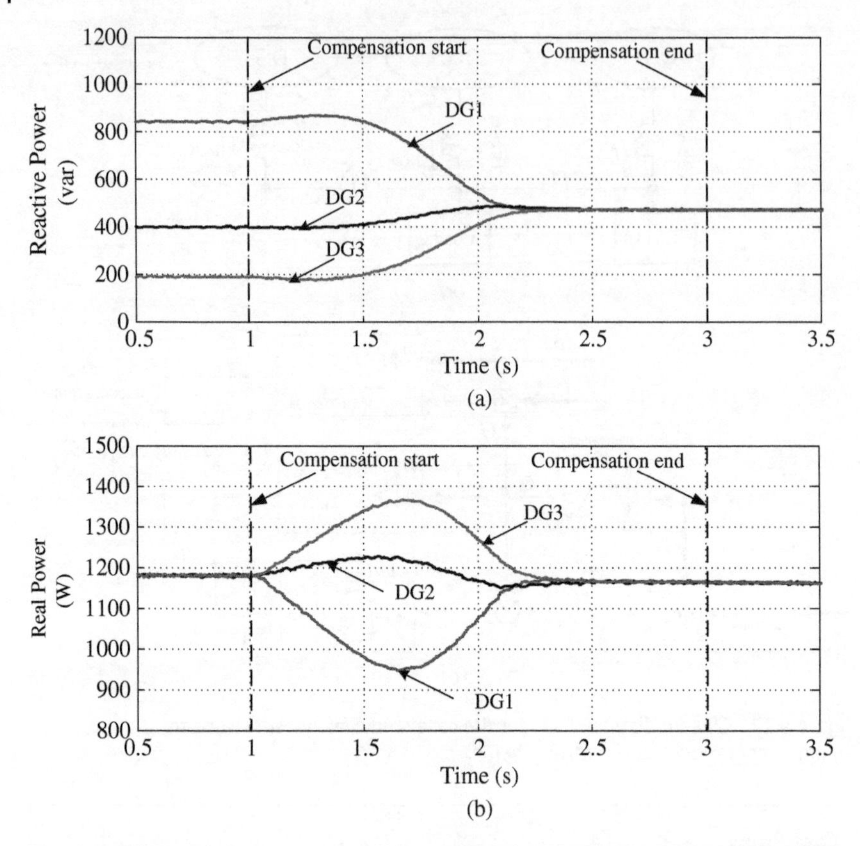

Figure 4.28 Simulation results of the case study: (a) reactive power sharing enhanced by real power disturbance, (b) real power disturbances.

errors. First, the reactive power compensation is enabled at 1.0 s, From 1.0 to 2.0 s, it can be seen that the reactive power sharing errors are effectively compensated. Note that during the compensation transient, the real power in DG units has some variations due to the injection of frequency disturbance. When the compensation comes to an end at 3 s, both real power and reactive power are evenly shared among the three DGs.

4.4 Virtual Synchronous Generator (VSG) Control of Interfacing Power Electronics Converters

As discussed earlier, renewable-energy sources are usually connected to the main grid through interfacing power electronics converters due to the incompatibility of their generated electricity with the grid. The common control method of such non-dispatchable energy sources (e.g. solar-power generators or wind) is to maximize their output powers and inject them into the power main grid as current sources. Compared with conventional power systems with synchronous machines, power electronics interfaced renewable energy sources do not have rotating inertia or damping (since they do not have a rotor, damper windings, etc.). Thus, the inertia and damping that could be provided by such energy resources are not enough, which may increase the vulnerability of power systems to power dynamics and system faults. This situation is particularly true in microgrids with a high penetration level of renewable energy resources, which makes them prone to stability issues due to the lack of inertia.

To address this issue, several control methods have been proposed for interfacing power electronics converters to enhance system stability. One important control method is a virtual synchronous machine (VSM) [10] or VSG [11], in which the dynamics and behavior of conventional synchronous machines are embedded in the power electronics converters. In other words, although such converters are physically power converters, they are mathematically operated as synchronous machines.

In this section, the basic principles of a real synchronous generator are first discussed. Then the implementation of the VSG control is illustrated, and an application example is shown.

4.4.1 Principles of VSG Control

The control system of a real synchronous generator set (genset) is shown in Figure 4.29. As can be seen, the control system includes two major controllers – engine speed governor and automatic voltage regulator (AVR) controller.

The engine speed governor monitors the rotating speed and adjusts the fuel input. As a result, the engine output power is changed to apply different torque to the rotor of the generator. The differences between the generator's mechanical torque and electromagnetic torque will lead to changes in generator frequency, which can be expressed as:

$$J\frac{\mathrm{d}\omega}{\mathrm{d}t} = T_{\mathrm{m}} - T_{\mathrm{e}} - D_{\mathrm{P}}(\omega - \omega_{\mathrm{grid}}) \tag{4.34}$$

where J is the moment of inertia of the synchronous generator, ω and ω_{grid} are rotor and grid angular frequency respectively, T_{m} and T_{e} are the mechanical torque applied to the rotor shaft and the electromagnetic torque of the stator, respectively, and D_{P} is the damping coefficient.

The speed governing control is actually controlling the output active power of the generator. To enable power sharing among multiple paralleled generators, droop slope is generally implemented in the speed governor. On the other hand,

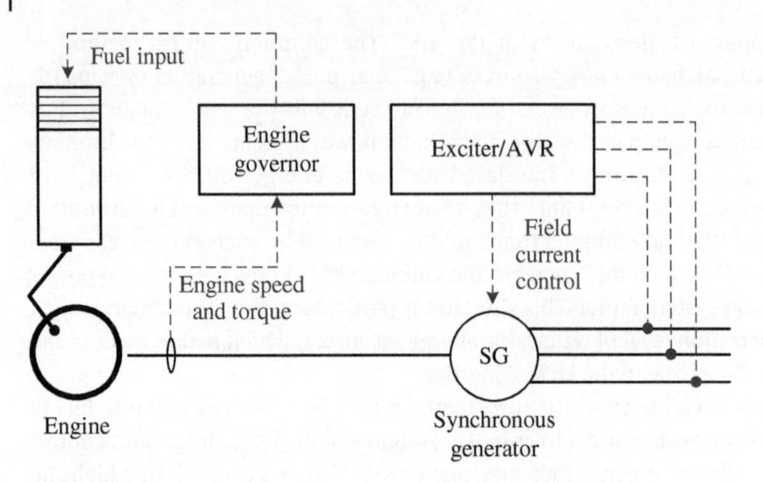

Figure 4.29 Control system of a real synchronous generator.

the AVR controller will monitor the generator voltage amplitude and adjust the excitation system. In islanded systems, the terminal voltage of the generator output will change under different output power, which requires proper excitation currents to be applied to keep the output voltage constant. In grid-connected systems, the AVR control will be used to control reactive power.

With the above-mentioned controllers, gensets can work properly with expected speed/AC frequency and terminal voltage/reactive power. In addition, the key to good robustness to external disturbance is the inertia of the whole system. The mass of the generators' rotors ensures that it will not suffer abrupt changes, even under distances in electromagnetic torque or mechanical torque, in a relatively wide range. As a result, the ability to resist grid disturbances is a favorable feature that could not be achieved in power electronic converters under traditional control. Therefore, the VSG/VSM control mimics the behaviors of the real genset in transients, implementing speed governor, AVR, and inertia in the converter systems. As a result, a VSG-controlled IFC can potentially participate in the system frequency and voltage regulations and can provide damping and inertia to the microgrid.

4.4.2 Implementation of VSG Control

The key to implement VSG control is to implement a speed governor, most importantly, inertia. It can be implemented in two ways, where the power electronics converters can act either as a current source or voltage source. Therefore, it can be classified as voltage-based VSG control and current-based VSG control.

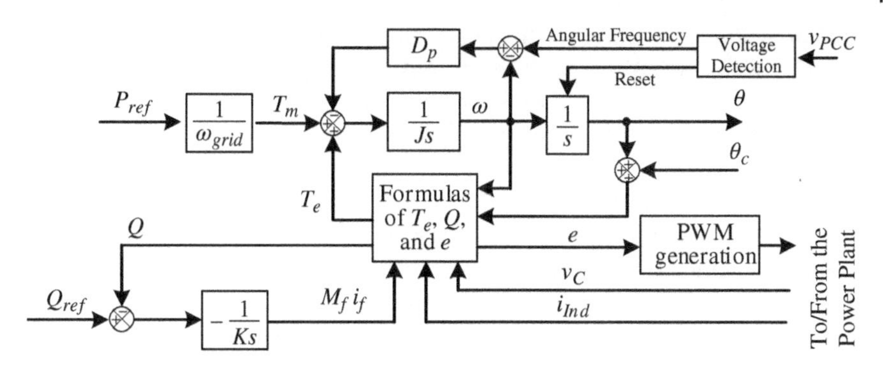

Figure 4.30 Voltage-based virtual synchronous a generator control scheme. Source: Adapted from Zhong and Weiss, 2011 [12].

Voltage-based VSG Control

The VSG control can be implemented by directly embedding the mathematical model of a synchronous generator to provide modulation references. In Figure 4.30, one example of the voltage-based VSG implementation method control is shown [12].

As shown in Figure 4.30, the mathematical model of a synchronous machine (three-phase round-rotor) is embedded in the three-phase VSG controller as the core. The mathematical model is presented in the following:

$$T_e = M_f\, i_f \left(i_a \sin\theta + i_b \sin\left(\theta - \frac{2\pi}{3}\right) + i_c \sin\left(\theta + \frac{2\pi}{3}\right)\right) \tag{4.35}$$

$$e = \omega M_f\, i_f \begin{bmatrix} \sin\theta \\ \sin\left(\theta - \frac{2\pi}{3}\right) \\ \sin\left(\theta + \frac{2\pi}{3}\right) \end{bmatrix} \tag{4.36}$$

$$\begin{aligned} P &= \omega M_f\, i_f \left(i_a \sin\theta + i_b \sin\left(\theta - \frac{2\pi}{3}\right) + i_c \sin\left(\theta + \frac{2\pi}{3}\right)\right) \\ Q &= -\omega M_f\, i_f \left(i_a \cos\theta + i_b \cos\left(\theta - \frac{2\pi}{3}\right) + i_c \cos\left(\theta + \frac{2\pi}{3}\right)\right). \end{aligned} \tag{4.37}$$

The calculated back EMF (e) by the mathematical model of the synchronous machine is used to produce PWM pulses to drive the power converter. As seen from the figure, the output current of the power stage (inductor currents) is treated as the stator current, and it is fed back to the mathematical model.

As mentioned in the frequency and voltage droop control in Section 4.3.1, real power is used to control the frequency while reactive power is used to control the voltage of the power system. In VSG control, the more accurate synchronous machine model is used to regulate the frequency and the voltage. The mechanical friction coefficient, D_P, acts as a frequency droop coefficient, which can adjust the angular frequency (ω) and produce the back EMF (e) phase angle (θ). Also, the field excitation current $M_f i_f$ is generated by reactive power control, and the voltage droop coefficient D_q is introduced to control the voltage. Thus, the VSG-controlled IFC can provide the frequency, voltage, active power, and reactive power control.

Directly implementing the synchronous generator feature and feeding the control signal to the modulation module achieves the desired features. This implementation method completely treats power electronic converters as a generator. However, the power electronic converters have a much lower overcurrent capability than generators, and the VSG method in Figure 4.30 does not provide a way of limiting the output current. This potentially leads to tripping in a large transient.

A practical solution is to use the voltage amplitude and angle generated by the VSG controller, as shown in Figure 4.31, to construct voltage reference, and embed traditional AC voltage control as inner loops (see Section 4.1.3). As can be seen, the swing function to realize virtual inertia and the reactive power control are added as the outer loop to provide the voltage references, including voltage amplitude E, and the rotating angle θ. Then the voltage reference is added to the VCM based primary control loops. In this case, the VSG control does not change the well-validated traditional control structure while still adding grid-friendly features. The well-developed control techniques, such as virtual impedance control can be easily implemented without modifying the complicated VSG model. Also, the traditional current limiting, voltage limiting features can be implemented to deal with this.

Current-based VSG Control Similar to the case of embedding AC voltage control, it is also possible to embed the CCM to implement VSG control. The current-based

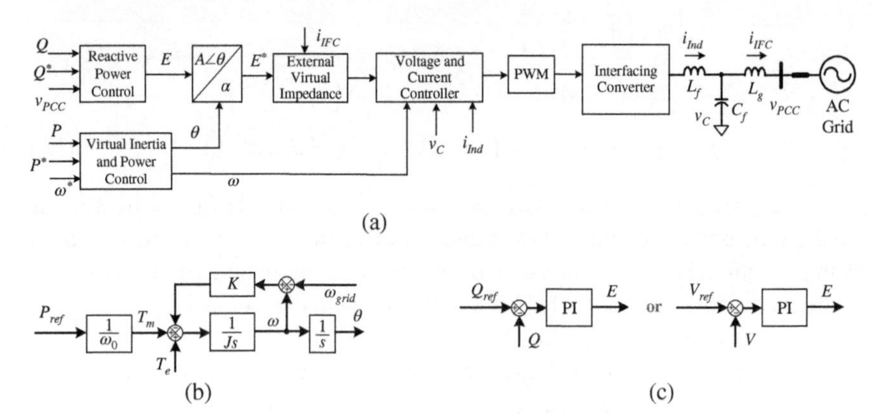

Figure 4.31 VSG control as outer loops: (a) full control scheme; (b) virtual inertial control; (c) reactive power control.

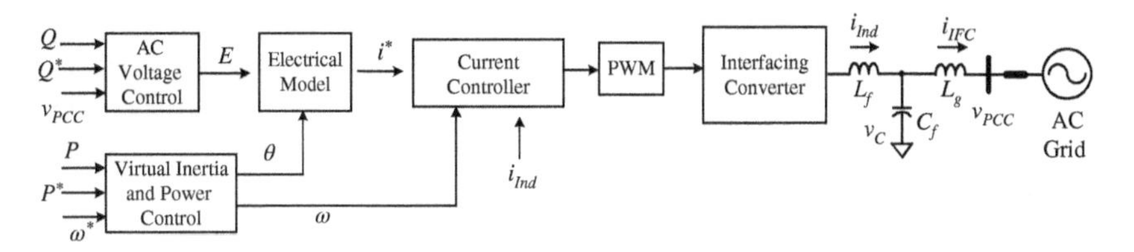

Figure 4.32 Block diagram of current-based VSG control.

VSG control scheme is shown in Figure 4.32, where the CCM-based primary IFC control is adopted. In this case, the VSG control produces proper current control references for the inner current loop. This is also implemented by the speed governor, the inertial model, and the AVR control, which are similar to the voltage-based method. Instead of voltage control loops, the control scheme directly uses an electrical model to calculate the current references [13]. This avoids cascading a voltage controller and a current controller, and also allows overcurrent limitation through reference current control.

4.4.3 Relationship Between Droop Control and VSG Control

As discussed, VSG control is grid-friendly as it can participate in grid frequency and amplitude regulation. Also, it has favorable dynamics for a system with a high penetration level of DG, where the inertia is low and prone to disturbances or faults. These features are also observed for systems with droop control applied. While droop control focuses on emulating the steady-state characteristics of a synchronous generator, it usually utilizes filters in the feedback power calculations and measurement, which adds transient dynamics to the system. Indeed, it has been discussed that VSG control has an equivalent small-signal model with P–f droop control when a low pass filter is added [13].

The swing function in (4.34) in VSG control can is shown here:

$$J \cdot \frac{\mathrm{d}\omega}{\mathrm{d}t} = T_{\mathrm{m}} - T_{\mathrm{e}} - D_{\mathrm{p}}(\omega - \omega_{\mathrm{grid}}) \tag{4.38}$$

and consider that the torques fulfill the following equations:

$$T_{\mathrm{e}} = \frac{P_i}{\omega}, T_{\mathrm{m}} = \frac{P_i^*}{\omega_n} \tag{4.39}$$

where ω_{grid} is the nominal electrical angular frequency, which is a constant value; ω is the actual electrical angular frequency of the VSG, P_i^* denotes the real power control reference, and P_i denotes the actual real power. In VSG control, the pole pair can be assumed to be 1 and thus nominal mechanical frequency $\omega_n = \omega_{\mathrm{grid}}$. Also, the actual angular frequency ω generally has minor differences with normal frequency ω_n and grid frequency ω_{grid}, so (4.38) can be rewritten as:

$$J \cdot \frac{d\omega}{dt} \approx \frac{P_i^*}{\omega_{\text{grid}}} - \frac{P_i}{\omega_{\text{grid}}} - D_p(\omega - \omega_{\text{grid}}). \tag{4.40}$$

Applying per unit values in these functions, and transforming them to the Laplace domain, (4.40) can be rewritten as:

$$J \cdot s \cdot \omega \approx P_i^* - P_i - D_p(\omega - \omega_{\text{grid}}). \tag{4.41}$$

On the other hand, the *P–f* droop with a low pass filter can be expressed as:

$$\omega = \omega_{\text{grid}} - SP_{Pi} \left[P_i^* - P_i \left(\frac{1}{T_d s + 1} \right) \right]. \tag{4.42}$$

Expanding and simplifying the droop function [14], the *P–f* droop control can be expressed as (4.43):

$$T_d \frac{1}{SP_{Pi}} \cdot s \cdot \omega = P_i^* - P_i - \frac{1}{SP_{Pi}} (\omega - \omega_{grid}). \tag{4.43}$$

Comparing (4.38) with (4.43), the following relationships can be found:

$$J = T_d \frac{1}{SP_{Pi}} \tag{4.44}$$

$$D_p = \frac{1}{SP_{Pi}}. \tag{4.45}$$

Therefore, the two control methods are similar in terms of the small-signal model, which means similar transfer functions and dynamic responses in practice. However, the concept of virtual inertia provides the physical meaning and valuable guidance for parameter design and tuning. This facilitates the design and can effectively avoid applying parameters that significantly deviates from the desirable range.

4.5 Unified Control of Power Electronics Converters

In practice, a grid-connected converter may need to work under islanded mode in a microgrid due to grid faults or maintenance requests. On the other hand, an islanded converter will also need to switch to the grid-connected mode when the microgrid is ready to be re-connected to the main grid. For IFCs in a hybrid AC/DC microgrid, the operation of the DC subgrid also needs to be considered to ensure the proper operation of both the AC subgrid and DC subgrid. When the operation mode switch is performed, seamless transitions are important, which is hard to be achieved if different control structures are applied for different operation modes. This is more challenging when islanding detection causes a delay in this transition.

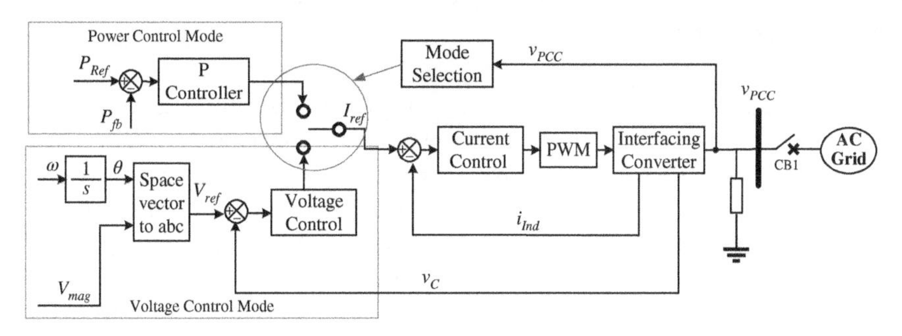

Figure 4.33 An example of a control system with a mode switch for the IFC.

In this case, a promising solution is to use the same control structures for different operating modes [15], which is referred as "unified control" in this book.

An example control system with an operation mode switch mechanism is shown in Figure 4.33. As can be seen, the control system has two different modes – power control mode for grid-connected operation and voltage control mode for islanded operation. A mode selection module must be applied to ensure the proper control mode can be applied. However, due to the detection time required to determine operation modes, the wrong control mode can be applied during the detection time when the actual microgrid operation mode has changed. This leads to an undesirable transient with the risk of system trip or even damage to the converter. Also, when changing the control structures, the converter control may generate an additional transient, which may cause a further undesirable disturbance to the system.

An example of unified control is shown in Figure 4.34, where the primary control part adopts the VCM and regulates the capacitor voltage of the LCL filter. Since the VCM can work properly under both grid-connected and islanded conditions, the primary control part does not need any change when the operating conditions change. However, to ensure proper operation under different conditions, proper voltage references must be provided. This can be achieved by droop control or VSG control, as they both can be applied on the basis of the VCM and can respond properly when circuit breaker CB1 changes its status. A detailed example and corresponding validations will be provided in Chapter 5.

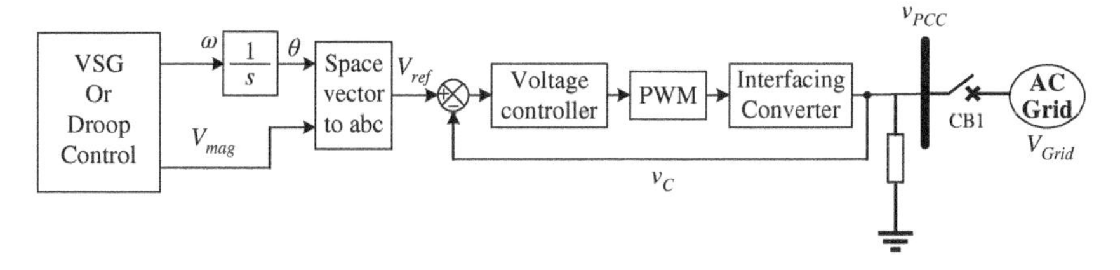

Figure 4.34 An example of unified control for an IFC.

4.6 Summary

In this chapter, control strategies of interfacing power electronics converters are discussed. After an introduction of the basic control techniques and primary controls of power converters, the virtual impedance concept for power converter control is introduced. Then, the droop control method in both AC and DC subgrids is illustrated and the challenges in practice are presented with corresponding solutions for improving the droop power-sharing performance provided. As an equivalence of droop control but with a clearer physical meaning, the VSG control strategy is studied. Finally, to cope with the transition between different control modes, a brief introduction to the concept of unified control of power electronics converters is provided, which will be further discussed in the following chapters.

References

1 Nejabatkhah, F. and Li, Y.W. (2015). Overview of power management strategies of hybrid AC/DC microgrid. *IEEE Transactions on Power Electronics* 30 (12): 7072–7089.

2 Li, Y.W. and Nejabatkhah, F. (2014). Overview of control, integration and energy management of microgrids. *Journal of Modern Power Systems and Clean Energy*.

3 He, J. and Li, Y.W. (2011). Analysis, design and implementation of virtual impedance for power electronics interfaced distributed generation. *IEEE Transactions on Industry Applications* 47: 2525–2538.

4 Wang, X., Li, Y.W., Blaabjerg, F., and Loh, P.C. (2015). Virtual-impedance-based control for voltage-source and current-source converters. *IEEE Transactions on Power Electronics* 30: 7019–7037.

5 He, J. and Li, Y.W. (2012). Generalized closed-loop control schemes with embedded virtual impedances for voltage source converters with LC or LCL filters. *IEEE Transactions on Power Electronics* 27 (4): 1850–1861.

6 Loh, P.C., Li, D., Chai, Y.K., and Blaabjerg, F. (2013). Autonomous operation of hybrid microgrid with AC and DC subgrids. *IEEE Transactions on Power Electronics* 28 (5): 2214–2223.

7 Li, Y.W. and Kao, C. (2009). An accurate power control strategy for power-electronics-interfaced distributed generation units operating in a low-voltage multibus microgrid. *IEEE Transactions on Power Electronics* 24 (12): 2977–2988.

8 He, J. and Li, Y.W. (2012). An enhanced microgrid load demand sharing strategy. *IEEE Transactions on Power Electronics* 27 (9): 3984–3995.

9 He, J., Li, Y.W., and Blaabjerg, F. (2015). An enhanced islanding microgrid reactive power, imbalance power, and harmonic power sharing scheme. *IEEE Transactions on Power Electronics* 30 (6): 3389–3401.
10 Beck, H. and Hesse, R. (2007). Virtual synchronous machine. In: *2007 9th International Conference on Electrical Power Quality and Utilisation*, Barcelona, 1–6.
11 Driesen, J. and Visscher, K. (2008). Virtual synchronous generators. In: *2008 IEEE Power and Energy Society General Meeting - Conversion and Delivery of Electrical Energy in the 21st Century*, 1–3.
12 Zhong, Q. and Weiss, G. (2011). Synchronverters: inverters that mimic synchronous generators. *IEEE Transactions on Industrial Electronics* 58 (4): 1259–1267.
13 Mo, O., D'Arco, S., and Suul, J.A. (2017). Evaluation of virtual synchronous machines with dynamic or quasi-stationary machine models. *IEEE Transactions on Industrial Electronics* 64 (7): 5952–5962.
14 D'Arco, S. and Suul, J.A. (2014). Equivalence of virtual synchronous machines and frequency-droops for converter-based microGrids. *IEEE Transactions on Smart Grid* 5 (1): 394–395.
15 Fang, F., Wei Li, Y., and Li, X. (2018). High performance unified control for interlinking converter in hybrid AC/DC microgrid. In: *2018 IEEE Energy Conversion Congress and Exposition (ECCE)*, Portland, OR, 3784–3791.

5

Power Management System (PMS) in Smart Hybrid AC/DC Microgrids

5.1 Introduction

One critical aspect of hybrid AC/DC microgrid operation is the power management system (PMS), which is essential for proper operation in grid-connected and islanding modes. In microgrids, the terms "energy management" and "power management" are different considering control tasks and time scale. The global objective of long-term energy management algorithms is to match the total power production to the demand in an optimal way. These algorithms deal with the monitoring and operation of a complex electrical, thermal, and mechanical system, emphasizing desired and longer-term outcomes. Factors like fuel costs, capital costs, maintenance costs, mission profiles, and lifetimes are considered in energy management algorithms. In general, the energy management strategies include hourly prediction of renewable energy sources, management of controllable loads, providing an appropriate level of power reserve capacity, etc. The energy management systems (EMSs) in smart hybrid AC/DC microgrids are discussed in Chapter 6.

On the other hand, the short-term PMS's objective is to control the instantaneous operational conditions toward specific desired parameters such as voltage, current, power, and frequency. The power management strategies include voltage and frequency regulations and real-time power dispatching among the different power sources of the microgrids. It should be highlighted that the power and energy management objectives are mainly realized through the control of power electronics interfacing converters (IFCs) from a distributed energy resource (DER) and IFCs linking AC/DC subgrids in the smart hybrid microgrids. The DER includes both distributed generation (DG) and energy storage system (ESS), which are capable of exporting active power to the grid.

In this chapter, the power management strategies of different hybrid AC/DC microgrid structures (AC-coupled, DC-coupled, and AC-DC-coupled) in steady-state and during the transition between grid-connected and islanding

Smart Hybrid AC/DC Microgrids: Power Management, Energy Management, and Power Quality Control, First Edition. Yunwei Ryan Li, Farzam Nejabatkhah, and Hao Tian.

modes are discussed. Also, black start in microgrids, where power generators restore the operation of the smart microgrid after a blackout in a microgrid is discussed.

5.2 Hierarchical Control of Hybrid Microgrids

Smart microgrid energy and power management strategies can be realized in different control layers, called hierarchical control layers. In general, there are three control layers in hierarchical control: primary, secondary, and tertiary layers. The primary control layer contains voltage and frequency regulations, where power generators such as DERs track their reference powers (please refer to Chapter 4 for detailed discussions on primary controls of power converters). The objectives of the second layer include system frequency restoration, unbalanced voltage compensation, harmonic compensation, etc. Considering these two layers' objectives, the PMS is realized in the primary and part of the secondary control layer. In the tertiary control, an optimization problem is usually run to achieve a global optimum operation point and determine the operating power of each power source. The tertiary control and part of the secondary control contain the EMS. In Figure 5.1, the structure of the hierarchical control in the smart microgrid is shown.

The secondary and tertiary control layers are implemented in supervisory control centers (SCCs) in smart microgrids. The SCC receives data from the IFCs and power system measurement devices and makes decisions based on defined objectives. Then, the decision signals are sent to all IFC local controllers (where the primary control of IFCs is running).

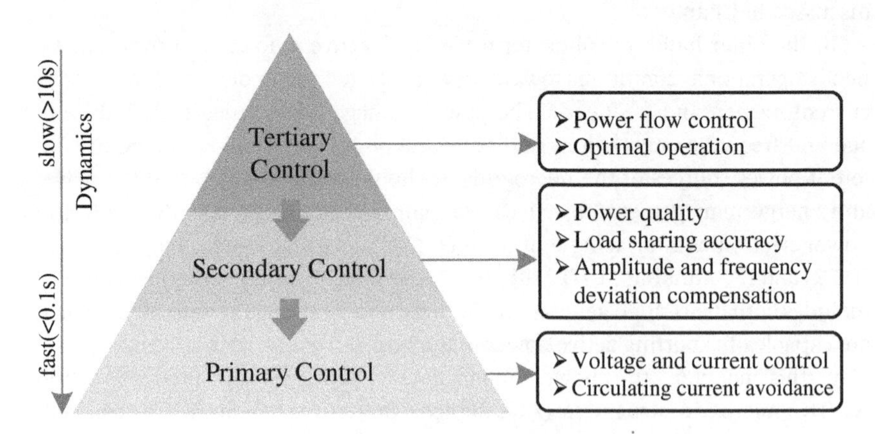

Figure 5.1 The structure of hierarchical control in smart microgrids.

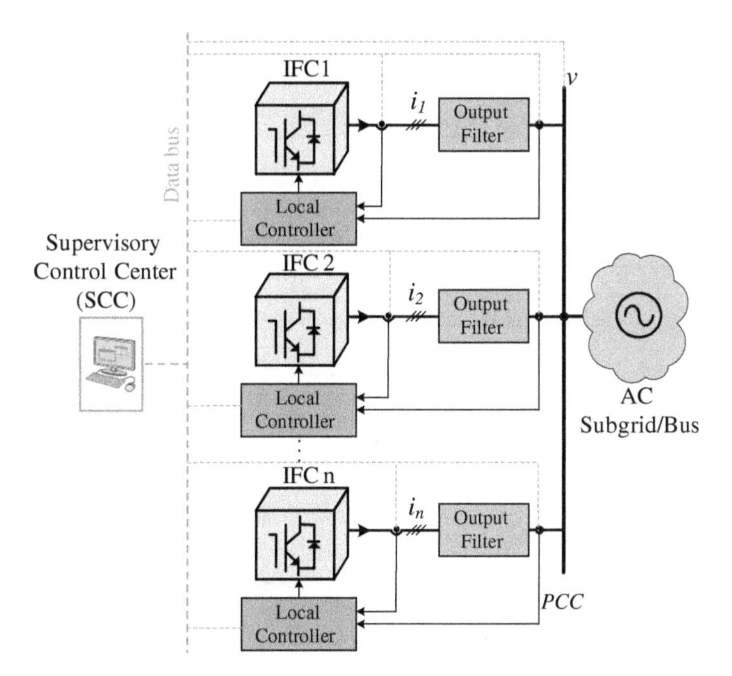

Figure 5.2 Centralized structure of supervisory control of IFCs.

In general, the structures of the supervisory control of IFCs can be categorized as centralized, distributed, and decentralized. In a centralized structure, shown in Figure 5.2, all the IFCs communicate with one SCC, and the central SCC makes decisions based on the objectives and constraints. Since all system data are gathered in one center, the global optimal operating point can be achieved. However, heavy computation burden and the possibility of communication failure (may cause overall shut down in the system) are challenges of this structure. One special control example with the centralized structure is the master–slave control, widely adopted for IFCs close to each other. In this structure, the local controller of one IFC acts as the SCC, and all IFCs communicate with each other. Overall the centralized structure is suitable for a small microgrid with the DG and ESS located physically close.

In the distributed structure, shown in Figure 5.3, a group of IFCs communicate with their SCC for local microgrid operation and optimization. Then, the SCCs of the microgrids communicate with each other and provide opportunities for system-level control and optimization. This structure is modular and is suitable for clustered interconnected microgrids, where each microgrid can have its own SCC, and microgrid clusters are coordinated with the communication among SCCs. Multi-agent control strategies can be very suitable for such a distributed structure

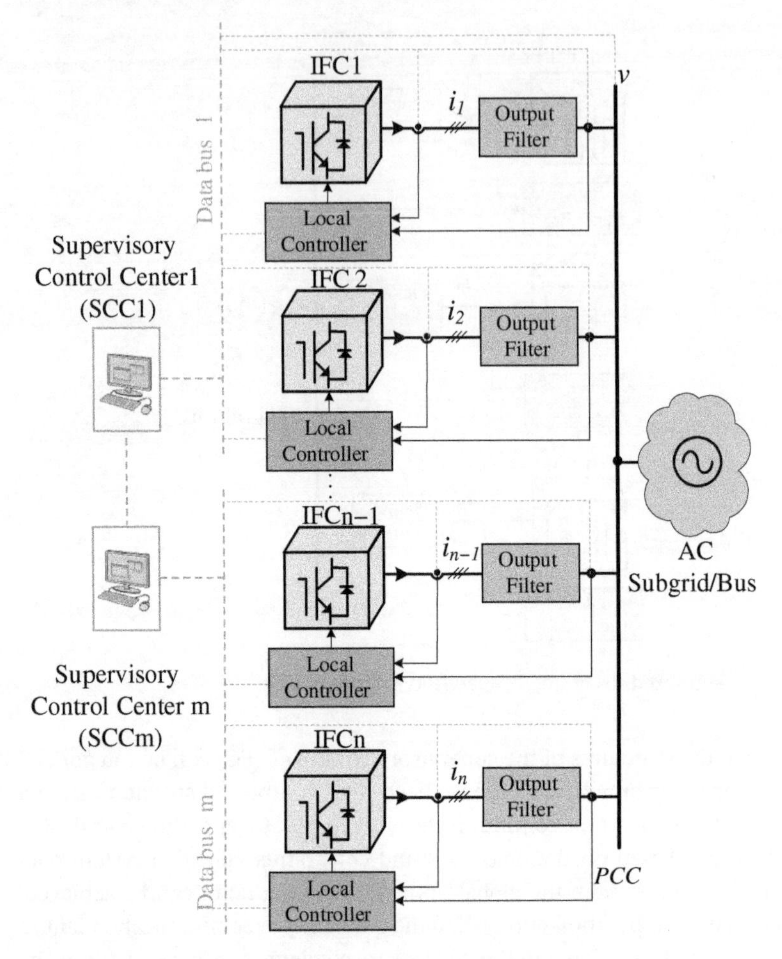

Figure 5.3 Distributed structure of supervisory control of IFCs.

for energy management control. Please refer to Chapter 6 for details on those energy management strategies. In the distributed structure, the computation burden of each SCC is reduced. The system reliability also increases with the modular system feature. For example, failure in the communication link does not affect the entire system.

In the decentralized structure, the local controllers of IFCs control their operation, and there is no communication among IFC controllers. In Figure 5.4, the decentralized structure is shown. As seen from the figure, the local measurements are used for IFC control. One well-known controller with a decentralized structure is droop control, where the IFCs use the locally available voltage and frequency

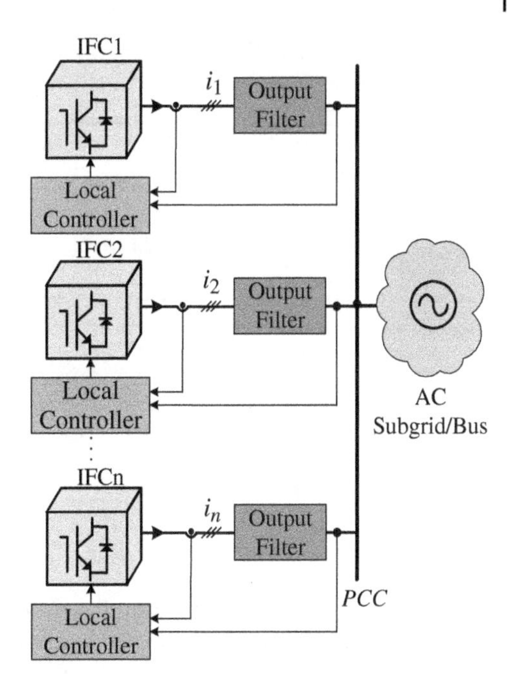

Figure 5.4 Decentralized structure of supervisory control of IFCs.

measurement to coordinate with each other for power sharing without physical communication links (please refer to Chapter 4 for details on droop control). The decentralized structure obviously does not have the reliability concern caused by the communication system failure. It is also suitable for microgrids with distributed locations of DG, ESS, and loads. However, system-level optimization is difficult in this structure due to the lack of SCC and communication system.

5.3 Power Management Systems (PMSs) in Different Structures of Hybrid Microgrids

In hybrid AC/DC microgrids, the power management strategies determine the instantaneous output active and reactive powers of IFCs and simultaneously control voltages and frequency [1]. The IFCs can be from DGs and ESSs, and they can be the AC/DC subgrids linking converters. In this section, the power management of different types of microgrids is presented in detail.

5.3.1 PMS of an AC-coupled Hybrid Microgrid

In Figure 5.5, a typical structure of an AC-coupled hybrid microgrid is shown. The power management schemes of AC-coupled hybrid microgrids are mainly

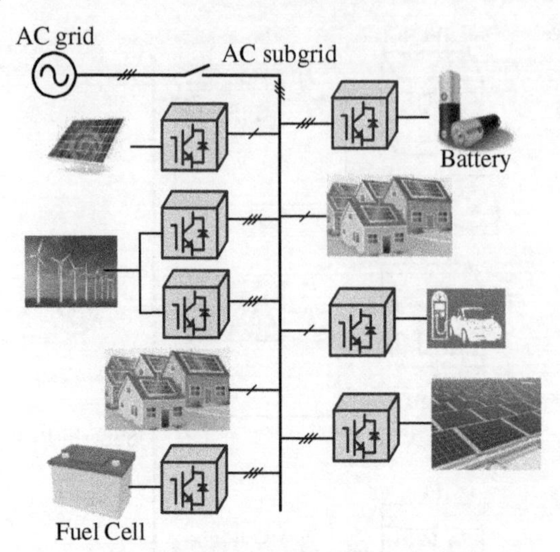

Figure 5.5 A typical structure of an AC-coupled hybrid microgrid.

focused on power balancing within the microgrid, and the AC bus voltage and frequency control, especially in islanding operation mode. These objectives are realized through the control of AC IFCs from DERs. In this structure, DERs can be treated like parallel AC voltage sources or current sources depending on their control and operation strategies. An overview of power management schemes for AC-coupled microgrids is shown in Table 5.1.

Table 5.1 Overview of the power management strategies of an AC-coupled hybrid microgrid.

PMS of an AC-coupled hybrid microgrid	Grid-connected Mode	Dispatched output power	DGs and ESSs in dispatched power control mode (power balancing)
		Undispatched output power	- DGs in MPPT control mode - ESSs typically in charging mode
	Islanding Mode	DGs and ESSs in AC bus voltage/frequency control mode and share demand power by: - Droop method - Master–slave method - Power balancing ...	

The power management strategies in the AC-coupled microgrids can be separated into grid-connected and islanding operation modes strategies. In grid-connected mode, the PMS can be classified into:

- dispatched output power mode, where the power exchange between the microgrid and the main grid is dispatched by a higher-level control/optimization scheme, and
- undispatched output power mode, where the microgrid output power is not dispatched.

In dispatched output power mode, the microgrid behaves like a controllable source or load to the main grid and can provide valuable grid support or load management functions as a whole. The ESS is a necessary part of this operation mode. In this mode, DGs and ESSs operate on power control (detailed discussion of converter primary controls can be found in Chapter 4). In this operation mode, the microgrid power balancing schemes share dispatched power among IFCs of power sources. For example, renewable-energy-based DGs can work on their maximum power point tracking (MPPT) control, and other power sources can provide power deficiency between generation and demand considering power source operation range, response time, etc.

In grid-connected undispatched power mode, all generated powers are fed to the grid; DGs work on the MPP, and ESSs are typically charged. Moreover, ESSs can be controlled to smooth microgrid output power oscillation, especially for ESSs working together with DGs with intermittent nature such as wind and PV systems.

In the grid-connected operation mode (both in dispatched and undispatched power mode), the microgrid can be used for grid support (for example, grid voltage and frequency regulations) by controlling active and reactive powers delivered to the grid. This grid support is included in the dispatched/reference microgrid power in the dispatched power control mode. The grid support function is realized individually through the DERs' IFCs in undispatched power mode. More explanation about the ancillary functions and grid supports is provided in Chapters 8–10.

In the islanding operation mode, the PMS is mainly focused on microgrid AC bus voltage and frequency control, and demand power sharing among IFCs. Various control schemes have been developed that can be used for this purpose. For example, droop control is the most common method in this group. This method is used to determine each source's output active and reactive powers to regulate the AC bus' frequency and voltage on their desired values (please refer to Chapter 4 for details on droop control). As with other alternatives, DGs can work on their MPP, and ESSs control the AC bus voltage and frequency. In other words, energy storage balances the demand and generated power to regulate the AC bus' frequency and voltage.

In AC-coupled hybrid microgrids, since DGs and ESSs are treated like parallel AC voltage sources or current sources, control strategies and power management schemes of typical parallel AC sources can be applied with some minor changes. For example, utilizing a master–slave control scheme, one DG unit, which has high output power, works on the voltage-controlled operation mode and controls AC bus voltage and frequency (master module) while the other DGs and ESSs (slave units) work on the current-controlled operation mode and track the current command provided by the master unit.

Most above-mentioned PMSs require communication mechanisms to share information among IFCs. For example, power balancing and master–slave methods are communication-based schemes. As a result, these methods depend on reliable communications. On the other hand, communication-less control methods attract more interest as they are more reliable and enable true plug-and-play of DGs and ESSs. For communication-less control, demand power is shared between DGs and ESSs without physical communication. These methods eliminate the physical location limitation of DGs and loads and improve microgrid performance. For example, the droop control method is the most popular communication-less control scheme. However, without other DGs and loads information, very accurate or optimal control of DGs and ESSs is complex in the communication-less methods. Thus, a combination of both techniques can be a better option. For instance, a combination of a low-bandwidth communication system with droop control as the backbone can provide both high performance and reliability. For detailed discussions on communication methods and standards in microgrids, please refer to Chapter 3.

5.3.2 PMS of a DC-coupled Hybrid Microgrid

A typical structure of a DC-coupled hybrid microgrid is shown in Figure 5.6. In DC-coupled hybrid microgrids, the DC link voltage control, power balancing between generation and demand, and AC link voltage and frequency control (especially in islanding mode) are the main objectives of power management. These objectives are realized by controlling the IFCs from DERs on the DC bus and DC/AC IFCs linking DC/AC subgrids. An overview of a DC-coupled hybrid microgrid's power management scheme is shown in Table 5.2. Similarly, the power management methods can be divided into grid-connected operation and islanding operation, where the grid-connected operation has a dispatched power mode and an undispatched power mode.

In the DC-coupled microgrid, the IFC that connects DC and AC subgrids can work on one of the three modes:

- bi-directional power control mode
- DC link voltage control mode
- AC link voltage control mode.

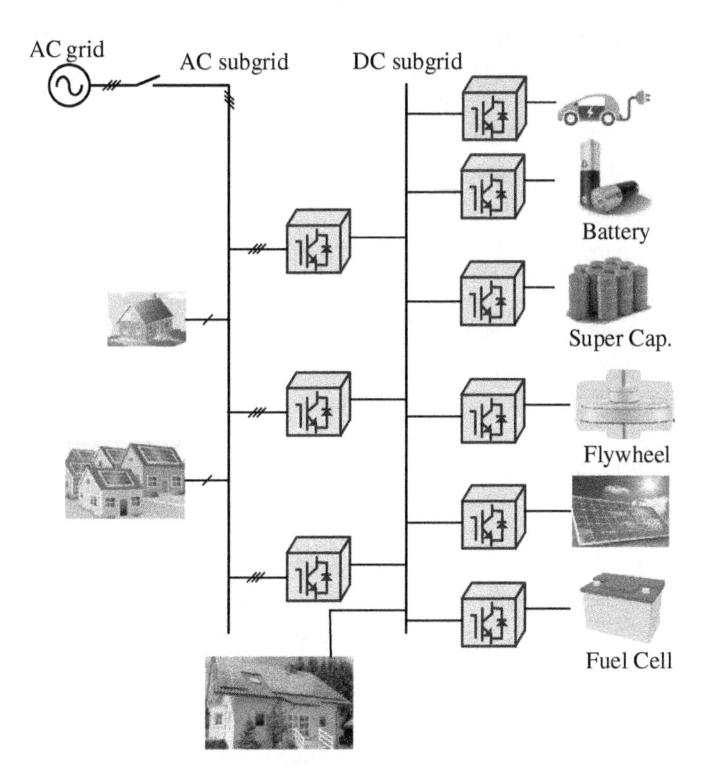

Figure 5.6 A typical structure of a DC-coupled hybrid microgrid.

Table 5.2 Overview of power management strategies of a DC-coupled hybrid microgrid.

PMS of a DC-coupled hybrid microgrid	Grid-connected Mode	Dispatched output power	- DC/AC subgrids' IFC in DC link voltage control mode - DGs and ESSs in dispatched power control mode (power balancing)
			- DGs and ESSs in DC link voltage control mode - DC/AC subgrids' IFC in power control mode
		Undispatched output power	- DC/AC subgrids' IFC in DC link voltage control mode - DGs in MPPT control mode - ESSs typically in charging mode
	Islanding Mode	DC bus voltage control	DGs and ESSs on DC link voltage control directly (e.g. droop method) or indirectly (e.g. power balancing)
		AC bus voltage control	DC/AC subgrids' IFC in AC bus voltage control mode

In power control mode, the converter output current or voltage is controlled to regulate the IFC output power to its reference value. In DC link voltage control mode, the IFC controls the DC link voltage and balances the power generation and consumption on the DC bus. The IFC's AC link voltage control mode is mainly for islanding microgrid operation, where the IFC controls the AC subgrid's voltage and frequency. For detailed schemes on IFC's different control modes, please refer to Chapter 4.

For a grid-connected DC-coupled microgrid, the DC link voltage can be controlled by two methods if it is in dispatched power mode. In the first method, the DC/AC subgrids' IFC works in DC link voltage control mode and regulates the DC link voltage at the desired value. At the same time, DGs and ESSs provide dispatched power of the hybrid microgrid through power balancing control. In the second method, ESSs on the DC bus control the DC link voltage collectively using the droop control method, and DGs can be part of the droop control or work at the MPP. In this operation mode, the DC/AC subgrids' IFC operates in power control mode and provides dispatched power to the grid.

In undispatched output power operation mode, the DC/AC subgrids' IFC operates on DC link voltage control mode. With fixed DC bus voltage, DGs work at the MPP, and ESSs are typically charged or controlled to smooth DGs' output power fluctuations.

The grid support functions can be realized in the grid-connected operation. Unlike in the AC-coupled microgrid, where the grid support can be realized by all the IFCs of DGs and ESSs, the DC/AC subgrids' IFC realizes the grid support in the DC-coupled microgrid.

In islanding operation of a DC-coupled hybrid microgrid, the DC bus voltage and AC bus voltage and frequency should be controlled simultaneously. For AC bus voltage and frequency control, the DC/AC subgrids' IFC works in AC link voltage control mode and controls the AC bus voltage and frequency. On the other hand, the DC bus voltage can be controlled directly or indirectly. In direct DC link voltage control, DGs and ESSs regulate the DC link voltage around its reference value (e.g. with droop control). In indirect DC link voltage control mode, power balancing between demand and generation regulates the DC link voltage. In islanding microgrid operation mode with parallel IFCs, it is also possible that some IFCs work on AC link voltage control mode, while the others control the DC link voltage and balance the generation and demand powers.

5.3.3 PMS of an AC-DC-coupled Hybrid Microgrid

In Figure 5.7, a typical structure of an AC-DC-coupled hybrid microgrid is shown. In AC-DC-coupled hybrid microgrids, multiple DERs are connected to AC and DC buses; thus, more coordination between the DC and AC subgrids is necessary.

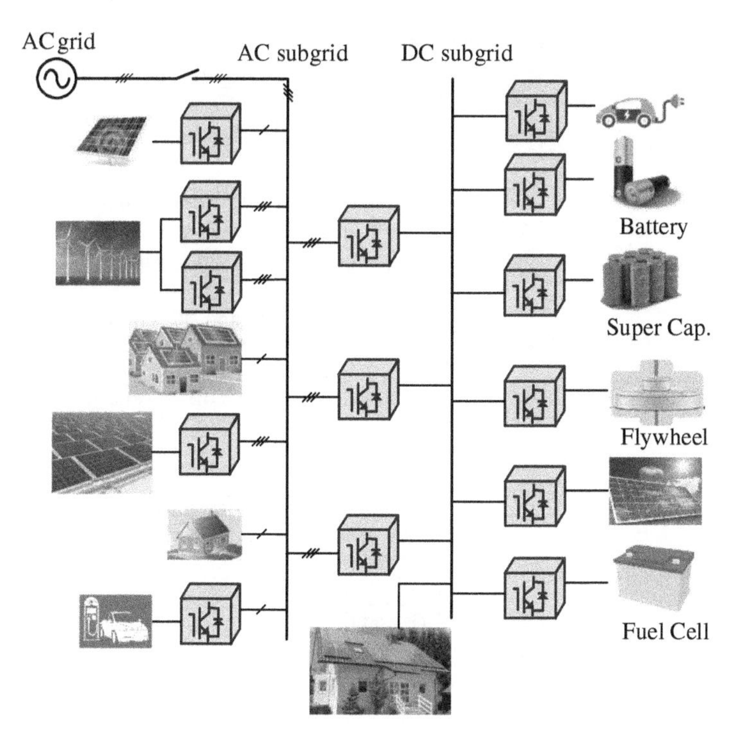

Figure 5.7 A typical structure of an AC/DC-coupled hybrid microgrid.

The control and power management schemes need to consider the power balance and voltage control in both DC and AC subgrids. In AC-DC-coupled microgrids, the control objectives are realized through the control of IFCs from DGs and ESSs connected to DC subgrid, IFCs from DGs and ESSs connected to AC subgrid, and DC/AC IFCs linking the DC and AC subgrids. An overview of power management schemes of AC-DC-coupled hybrid microgrid is shown in Table 5.3.

Similar to the DC-coupled microgrid, the IFC linking the AC and DC subgrids can be in bi-directional power control mode, DC voltage control mode or AC voltage control mode. However, the coordination between such IFCs and the AC subgrids' DGs and ESSs is necessary here in the power and AC bus control modes.

In grid-connected operation mode with dispatched microgrid output power, two methods can be used for DC link voltage control and dispatched power generation. In the first method, the AC and DC subgrids' IFC works in the DC link voltage regulation mode to set the DC bus voltage to its desired value. In this mode, coordination between DGs and ESSs on the DC bus and DGs and ESSs on the AC bus is necessary to produce the dispatched output powers. In the second method,

Table 5.3 Overview of power management strategies of the AC-DC-coupled hybrid microgrid.

<table>
<tr><td rowspan="4">PMS of AC-DC-coupled hybrid microgrid</td><td rowspan="2">Grid-connected Mode</td><td>Dispatched output power</td><td colspan="2">– AC/DC subgrids' IFC in DC bus voltage control mode
– DGs and ESSs in AC bus coordinate with DGs and ESSs in DC bus for providing dispatched power (power balancing)
– DGs and ESSs in DC bus in DC link voltage control mode
– AC/DC subgrids' IFC and DGs and ESSs in AC bus on the dispatched power control mode
– AC/DC subgrids' IFC in the power control mode</td></tr>
<tr><td>Undispatched output power</td><td colspan="2">– AC/DC subgrids' IFC in DC link voltage control mode
– DGs (in both AC and DC buses) in the MPPT control mode
– ESSs (in both AC and DC buses) typically in charging mode</td></tr>
<tr><td rowspan="2">Islanding Mode</td><td>DC bus voltage control</td><td>– DGs and ESSs in DC bus on DC link voltage control directly (e.g. droop method) or indirectly (e.g. power balancing)</td><td rowspan="2">AC/DC subgrids' IFC in AC link voltage control mode, DC link voltage control mode, or power control mode</td></tr>
<tr><td>AC bus voltage control</td><td>– DGs and ESSs in AC bus on AC link voltage control directly (e.g. droop method) or indirectly (e.g. power balancing)</td></tr>
</table>

DGs and ESSs on the DC bus regulates the DC link voltage to its reference while the AC/DC subgrids' IFC, and DGs and ESSs on the AC bus collectively provide the dispatched power. In this operation mode, the AC/DC subgrids' IFC works in power control mode.

In grid-connected operation with undispatched microgrid output power, DGs in both DC and AC buses work on MPP. In addition, ESSs can stay charged or controlled to smooth the microgrid output power injected into the grid. In this mode, the AC/DC subgrids' IFC regulates the DC link voltage to its desired value and injects all power generated by DGs and ESSs in the DC bus to the load/grid.

In AC-DC-coupled hybrid microgrids in grid-connected operation modes, similar to other structures of hybrid microgrids, DGs on the AC bus and the AC/DC subgrids' IFCs can be controlled to realize the grid support functions.

In islanding operation mode, coordination among the AC/DC subgrids' IFC, DGs and ESSs on the AC subgrid and DGs and ESSs on the DC subgrid are essential to regulate the DC bus voltage, AC bus voltage and frequency, and balance microgrid total power generation and load demand at the same time. In this operation mode, the power management strategies of AC-coupled hybrid microgrids in islanding operation mode such as droop, master–slave, etc., can be used for AC subgrid voltage and frequency regulation and demand power sharing. For the DC subgrid control, similar to the DC-coupled hybrid microgrid in islanding operation, the DC subgrid voltage can be controlled by DGs and ESSs on DC bus directly (for example, utilizing the droop control method in the DC bus) or indirectly (for example, utilizing a power balancing control strategy).

It is important to note that the AC/DC subgrids' IFC plays an important role in power management and control in islanding operation mode. Depending on the types of control strategies used in AC and DC buses, this DC/AC IFC can be used for the DC bus control mode, AC bus control mode, or output power control mode as discussed. However, coordination among these control strategies (AC bus, DC bus, and IFC control strategies) is the most critical objective in islanding operation mode. For example, in the case that the DC bus voltage is controlled by DGs and ESSs connected to the DC bus, and the AC bus voltage is controlled by DGs and ESSs connected to the AC bus, the DC/AC buses' IFC is responsible for regulating the power flow between AC and DC sides to equalize the demand and generated power. Moreover, in the case of parallel IFCs between the AC and DC subgrids (buses), the converters can work in different operation modes: some can work in DC link voltage control mode while the others work in AC link voltage control mode or power control mode.

5.4 Power Management Strategies During Transitions and Different Loading Conditions

The previous section's PMS is mainly focused on steady-state power balancing and voltage/current control of hybrid microgrids. The power management strategies during microgrid operation mode transitions and during different loading conditions are discussed in this section.

5.4.1 PMS During Transition Between Grid-Connected and Islanding Operation Modes

The transition between grid-connected and islanding operating modes should be seamless to minimize voltage and frequency disturbances and deviations and ensure proper power flows to prevent DERs overloading and circulating powers.

Table 5.4 Overview of power management during the transition between grid-connected and islanding operation modes.

Operation Mode Transition	Grid-connected to Islanding Mode	Switch of control strategy from current/power control in grid-connected mode to voltage control in islanding mode Unified control strategy in both grid-connected and islanding modes
	Islanding to Grid-connected Mode	Control strategy: – Switch of control strategy from voltage control in islanding mode to current/power control in grid-connected mode – Unified control strategy in both islanding and grid-connected modes Synchronization: – Passive synchronization – Active synchronization

Here, the control strategies for transition from the grid-connected to the islanding operation and transition from the islanding to the grid-connected operation are discussed separately. An overview of the transitions is shown in Table 5.4

Transition from Grid-Connected to Islanding Mode

There are mainly two control strategies for microgrid transition from grid-connected operation to islanding operation:

- switch of control strategies from current/power control mode in grid-connected operation to the voltage control mode in islanding operation
- unified control in both grid-connected mode and islanding mode.

In the first group, different control strategies are used in grid-connected and islanding operation modes, and the control strategy is switched between these two controllers. In particular, DGs are working in current control mode (MPPT control or dispatched power control) in grid-connected operation mode to inject power to main grids. In contrast, voltage control (droop method or conventional voltage control) is used in islanding operation mode to supply continuous power to sensitive loads and share the load demand among voltage-controlled DERs. This switching of control can be applied to DC subgrid's IFCs of DERs, AC subgrid's IFCs of DERs, as well as DC/AC subgrids' IFC. One technique is to reduce the line current of the IFC to zero before switching to islanding voltage control mode for a seamless transition. It is also possible for a faster transition without reducing the IFC line current to zero. To realize this, the current controller (in grid-connected mode) and voltage controller (in islanding mode) can be carefully coordinated to avoid transient current or voltage spikes during the transition [2].

An islanding detection algorithm (passive or active) is usually used to determine the disconnecting time instance to switch the control strategy. Then, the microgrid controller switches from grid-connected to islanding mode at this time. The passive and active islanding detection methods are discussed in detail in Chapter 8. It should be considered that seamless and smooth transitions can be difficult when islanding detection is delayed. After islanding, the voltage controller is immediately applied, and the synchronization unit works as an oscillator at a fixed frequency. As discussed earlier, droop control or other islanding operation control strategies can be applied in the presence of multiple IFCs.

In the second group, the power management control strategies are the same in both grid-connected and islanding operation modes, and it is not necessary to modify the control strategy during the transitions. Therefore, the seamless transition can be achieved while it is challenging to design and implement a robust control strategy to work on grid-connected, islanding, and transient modes. For this group of control methods, the islanding detection algorithm is unnecessary in the control strategy but is usually needed due to utility requirements or grid codes. This group may include smaller DG units that work in MPPT or current control mode in grid-connected and islanding operations. For larger DGs and ESSs, which are dominant microgrid power sources, the voltage control mode is implemented to avoid control scheme transients. However, some modifications should be applied to the voltage control strategies to be applied into in grid-connected, islanding, and transient operation modes. For example, the conventional droop method can be modified using the concept of virtual impedance to be used in both grid-connected and islanding operation modes in the presence of DGs, which is discussed in Chapter 4.

In the transition from grid-connected to islanding operation mode, the control strategy, containing the synchronization unit, should provide a stable voltage with a fixed frequency in the microgrid. Moreover, it should share the power demand among the DGs to continue supplying power to the loads within the microgrid. Transient power sharing among the DGs and ESSs are essential for both groups of control strategies, and during islanding operation, the power electronic interfaced DGs initially pick up the majority of any load change due to its quick response. In some cases, poor transient load sharing exists in the presence of DGs and conventional synchronous generators. To improve the transient power sharing, the coordination among the DGs and conventional synchronous generators is necessary, and the DGs may need to allow the voltage and frequency to swing at the expense of increased voltage and frequency dip. For example, to improve the transient load sharing, the droop slope can be modified during transient and restored to the normal in the steady-state condition. The virtual synchronous generator control of the IFC, discussed in Chapter 4, is another method that allows the power electronic interfaced DGs and ESSs to follow the conventional synchronous generators principles and help the transient load sharing.

Transition from Islanding to Grid-Connected Operation Mode

In the transition from island to grid-connected operation mode, other than the control scheme (switching from voltage control to current/power control or through the unified control scheme as discussed earlier), an important task is that the microgrid voltage should be synchronized with the grid voltage before re-connection.

There are two methods for synchronizing the microgrid to the main grid:

- passive synchronization
- active synchronization.

For passive synchronization, the microgrid voltage and main grid voltage are monitored, and the two grids are connected when they have the same phase angle. This method assumes that the microgrid and the main grid have very close voltage magnitude and slightly different frequencies (which is typically the case). Passive synchronization is very easy to implement. However, this method will lead to some transients upon re-connection of the two grids due to imperfectly matched voltage magnitude, and it does not guarantee a fast and controllable synchronization process.

On the other hand, active synchronization can achieve fast synchronization and seamless connection of the microgrid and the main grid and has attracted considerable research effort in recent years. However, since the microgrid consists of different types of DGs, electrical loads, and storage devices, the synchronization of a microgrid is quite different compared to a single traditional machine. For active synchronization, coordination of multiple DGs and ESSs is required. In some cases, the synchronization unit is embedded in the control strategies, while in others a separate synchronization unit provides the synchronization signals to prepare microgrid re-connection to the grid.

As discussed earlier, two control strategies are possible in islanding operation mode: (i) all DERs are working in the voltage control mode (for example, droop control or conventional voltage control mode applied to all DERs), (ii) some DERs are working in current control mode while the others are working in voltage control mode (for example, the master–slave control strategy or renewable generation-based DGs are working on MPPT while the controllable DGs and ESSs are working in voltage control mode). Depending on the utilized PMS, the active synchronization strategies can be classified into two groups:

- one/more DER initiate the synchronization process, and the others follow them
- all DERs take part in the synchronization process simultaneously.

The first group of synchronization strategies is mainly used in the microgrids in which some DERs are working in current control mode while the others are working in voltage control mode. The second group of synchronization strategies is mainly used in the microgrids in which all DERs are working on the voltage control mode. In both types of synchronization strategies, the communication system is essential to communicate the system information among power sources. The communication systems utilized in the hybrid microgrids are provided in Chapter 3.

5.4.2 Power Management Strategies Under Different Loading Conditions

In hybrid microgrids, loads and grid conditions can significantly impact the microgrid's performance. As a result, robust current or voltage control is needed in these conditions. In the presence of unbalanced and harmonic loads, which result in unbalanced and harmonic voltage at PCC, various control strategies have been used. In Chapters 9 and 10, details on DERs' control under unbalanced and harmonic conditions are presented.

On the other hand, constant power loads (CPLs), which are mainly interfaced by active rectifiers to the AC grid or DC/DC converter to a DC grid, cause instability problems in the microgrids because of their negative impedance characteristics. These problems can be studied using the small-signal and large-signal methods, and various solutions have been provided to ensure system stability. These include oscillation compensation techniques to increase stability margin, active-damping techniques to overcome the negative impedance instability problem, nonlinear control of AC/DC converters, just to name a few. Also, in some cases, power electronics interfaced CPLs are controlled to provide ancillary services to microgrids, such as harmonic minimization caused by power electronic nonlinear load, shunt active filter, etc.

In addition to constant power loads, frequency and voltage-dependent loads can also influence microgrid stability. It is well known that in islanding microgrid operation, the frequency and voltage deviations are dependent on the load. Therefore, microgrid unstable operation may result because of the load's voltage and frequency dependence. For example, induction motors are frequency- and voltage-dependent loads in the microgrids, where conventional P/Q control strategies may not guarantee microgrid stability. In the presence of frequency and voltage-dependent loads, accurate load models and sound control scheme design are essential to enhance stability and transient performance.

5.5 Implemented Examples of Power Management Systems in Hybrid Microgrids

5.5.1 PMS Example of an AC-coupled Hybrid Microgrid

As discussed earlier, the IFCs from DGs and ESSs connected to the AC subgrid can be used for grid support. The main objective of the grid support is to regulate the

AC voltage by controlling active and reactive powers delivered to the AC bus. In addition to the above-mentioned objectives, unbalanced compensation, harmonic control, power factor correction, etc., can be considered grid support objectives. Here, two examples of the grid support control strategies in the AC-coupled microgrids are presented: the grid voltage support and the grid unbalanced voltage compensation.

Grid Voltage Amplitude and Frequency Support

The grid-supporting IFC can be controlled as a current source (current control method) or voltage source (voltage control method). The grid-supporting IFC regulates the variations of voltage amplitude and frequency of microgrid and AC grid in the grid-connected operation. The block diagram of this grid support control strategy using current control is shown in Figure 5.8a.

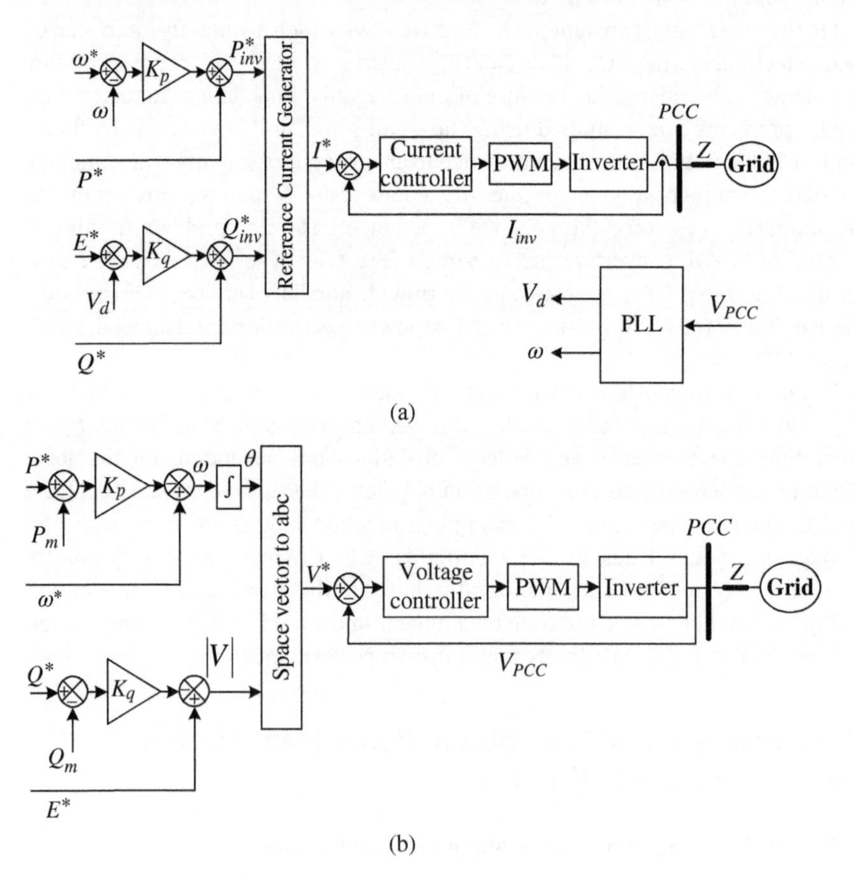

Figure 5.8 Power converter control in grid voltage amplitude and frequency support: (a) current controlled IFC (grid following), (b) voltage controlled IFC (grid forming).

In this control strategy, the frequency variation is controlled by the active power, while the reactive power controls the amplitude of the voltage. The frequency and voltage controllers are proportional controllers (K_p and K_q) for realizing the inverse P–f and Q–V droop control. Depending on the utilized reference frame (dq or $\alpha\beta$), the reference currents are generated using reference active and reactive powers and PCC voltage amplitude. For example, PI controllers with decoupling terms between the d and q axes can be used as a current controller in the dq reference frame while proportional and resonant (PR) current controllers can be used in the $\alpha\beta$ frame. Please refer to Chapter 4 for details on converter control.

In voltage controlled IFC operation (Figure 5.8b) where the power converter works as a controllable voltage source, a linking impedance will be necessary between the converter and grid, which can be a physical impedance or a virtual impedance (for details on virtual impedance control in power converters, please refer to Chapter 4). In this control scheme, the active and reactive power controllers are proportional controllers (k_P and k_q) for realizing P–f and Q–V droop control. In this strategy, the output three-phase voltages at PCC are regulated to their reference values with a closed-loop control system.

Grid Unbalanced Voltage Support

Here, an example of grid unbalanced voltage support using IFC is provided. The control block diagram is shown in Figure 5.9. In this control, the negative sequence

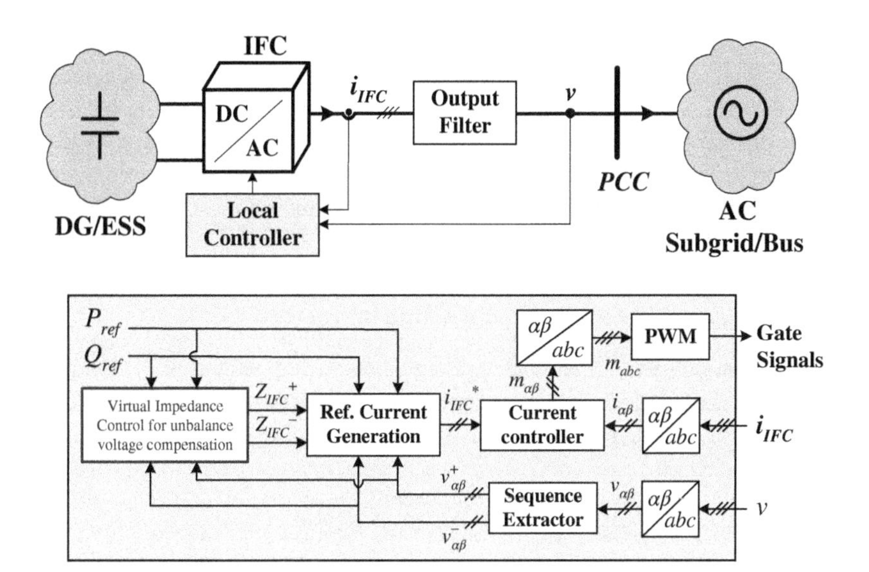

Figure 5.9 Example of a control block diagram for unbalanced voltage compensation in grid-support mode.

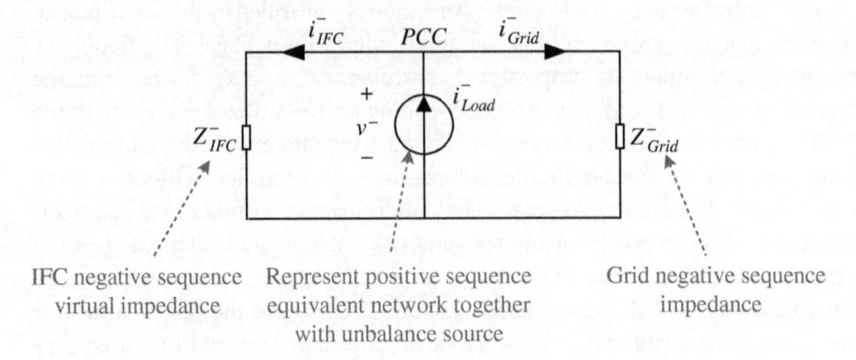

Figure 5.10 Sequential model of three-phase IFC connected to the AC subgrid.

impedance of the IFC is controlled to reduce negative sequence voltage at PCC. Figure 5.10 shows the equivalent negative sequence model of the IFC connected to the AC subgrid (no zero sequence in the system). In this figure, the positive sequence equivalent network together with the unbalanced load (which produces voltage imbalance) are represented by an equivalent current source i^-_{load}. Moreover, the IFC is represented as a virtual impedance Z^-_{IFC} in order to emulate its behavior in the negative sequence circuit. The grid negative sequence impedance (Z^-_{grid}) is shown in a parallel configuration. From the figure, the negative sequence virtual impedance of the IFC (Z^-_{IFC}) can be controlled to be much smaller than the grid impedance, directing i^-_{load} to flow to the IFC and leading to improved PCC voltage.

In Figure 5.11, an example of IFC negative sequence control in resistive microgrids for PCC unbalanced voltage support is shown. As clear from the results, under the different ratio of active–reactive reference powers ($P/Q = 4, 1, 1/4$), the virtual impedance can be controlled to absorb i^-_{load} and improve the PCC voltage. For more details on IFC unbalanced voltage control in hybrid AC/DC microgrids, please refer to Chapter 9.

5.5.2 PMS Example of a DC-coupled Hybrid Microgrid

The power balancing of a DC-coupled grid-connected wind turbine (WT)/fuel cell (FC)/supercapacitor (SC) hybrid microgrid, where all sources are connected to the common DC bus, is presented here as an example [3]. Figure 5.12 shows the block diagram of the hybrid system and the power management scheme. In this figure, subscript m represents the measured values while superscript * shows the reference values. Moreover, ω and T_{Gear} represent the rotational speed of the WT and gear torque, and m_{WT}, m_{FC}, m_{EL}, m_{SC}, m_{inv} are the modulation functions of the power converters of the WT, FC, electrolyzer (EL), SC, and grid-connected AC/DC subgrids' IFCs.

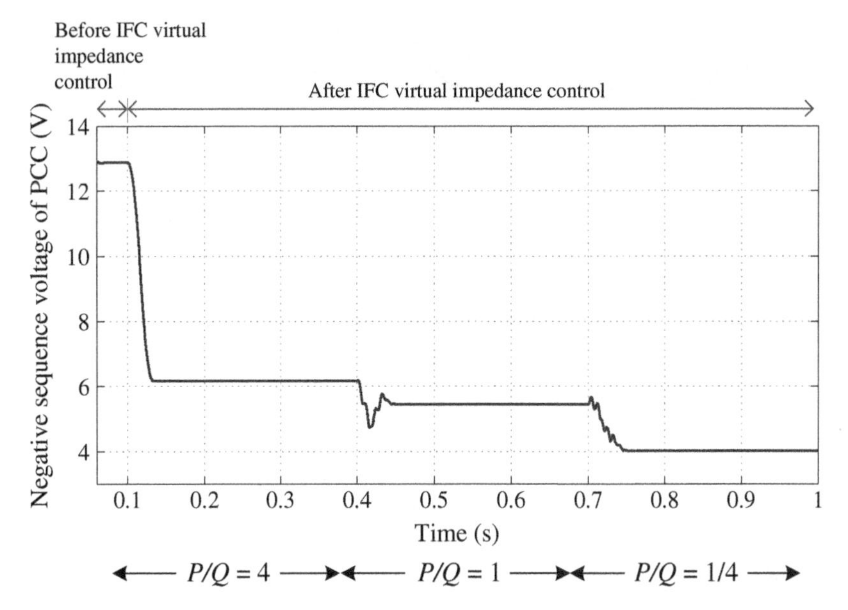

Figure 5.11 Negative sequence of PCC voltage; an example of IFC negative sequence virtual impedance control for PCC unbalance voltage compensation.

In this microgrid, the central power dispatch unit determines the output power of the microgrid ($P_{dispatched}$), the wind power system works on the MPP, and the storage devices provide power deficiency between WT output and dispatched power. In this system, the combination of FC and EL acts as a storage device responsible for long-term energy balancing and provides the low-frequency component of power deficiency between the dispatched power and wind power by producing or absorbing the necessary power with FC and EL, respectively. In addition, SC is used for short-time disturbances and buffering out the high-frequency oscillations.

In the condition that the dispatched power is lower than the wind maximum power, their low-frequency power deficiency is consumed by the EL (P_{EL}) to generate hydrogen which is stored in the tanks. It is important to consider that the FC and the EL should not work simultaneously for efficiency considerations. Therefore, a selector is used to switch the control scheme between the FC and EL based on its input power sign. Positive input power sign (low-frequency components of $P_{dispatched} > P_{WT}$) leads to FC power control mode while the EL power control mode is activated in negative input power sign (low-frequency component of $P_{dispatched} < P_{WT}$). Here, the voltage and current controllers can be simple PI controllers. In this system, the IFC between the DC and AC subgrids works in DC link voltage control mode and injects all DC-side generated power

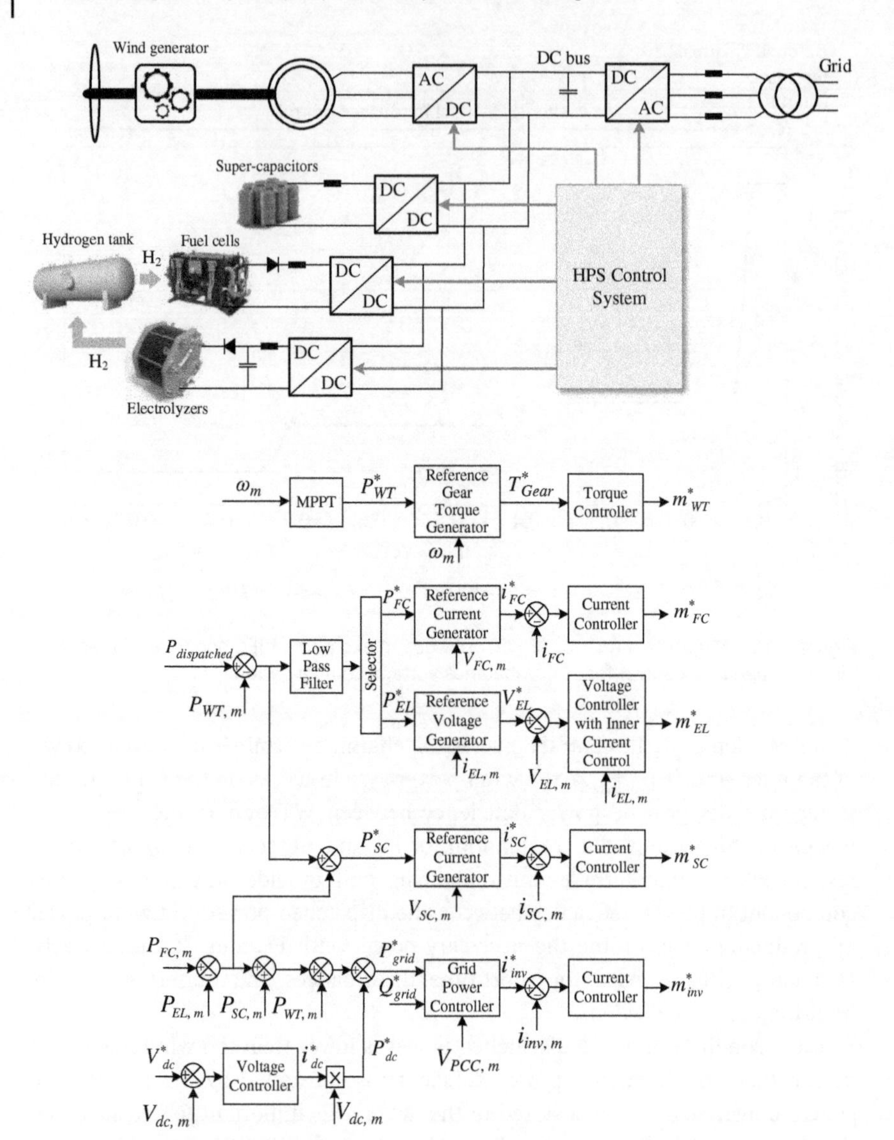

Figure 5.12 Example of power balance scheme in DC-coupled grid-connected WT/FC/SC hybrid microgrid.

$(P_{\text{FC}} + P_{\text{SC}} + P_{\text{WT}} - P_{\text{EL}})$ to the AC subgrid (P^*_{grid}). Moreover, the reactive power reference of the IFC (Q^*_{grid}) is determined considering the grid voltage conditions and control objectives.

Obviously, this power balancing control scheme needs to have access to the output power of different sources, which makes it a communications-based control strategy.

5.5.3 PMS Example of an AC-DC-coupled Hybrid Microgrid

A unified control with a unified droop for the AC and DC subgrids' IFC is provided as an example in this section. Please refer to Chapter 4 for more detail about the unified control strategy and unified droop control strategy. This controller can be used in grid-connected and islanding operation modes without mode detection and controller switching. This control strategy allows simultaneous DC subgrid voltage regulation and AC subgrid voltage and frequency adjustments.

The unified control strategy with a unified droop for the IFC in a hybrid AC/DC microgrid is shown in Figure 5.13. The control consists of three parts: power dispatch control, droop control of the IFC, and voltage control. According to their power capacity, the power dispatch control is used to dispatch power between the AC and the DC subgrids. In the AC subgrid and DC subgrid, slack terminals (controllable DERs) are included to control the AC voltage/frequency and DC voltage, respectively. For the AC and DC subgrids, the $P - \omega_s$ and $P - u$ equivalent droop characteristics are given in (5.1).

$$\begin{aligned} \omega_{pu} &= \omega_{0\_pu} - P_{s,ac\_pu}/k_{ac} \\ u_{dc\_pu} &= u_{dc0\_pu} - P_{s,dc\_pu}/k_{dc} \end{aligned} \tag{5.1}$$

where ω_{pu}, ω_{0_pu} and P_{s,ac_pu} are the actual frequency, the frequency at no load, and the active power of the slack terminal in the AC subgrid. The $u_{dc,pu}$, u_{dc0_pu}

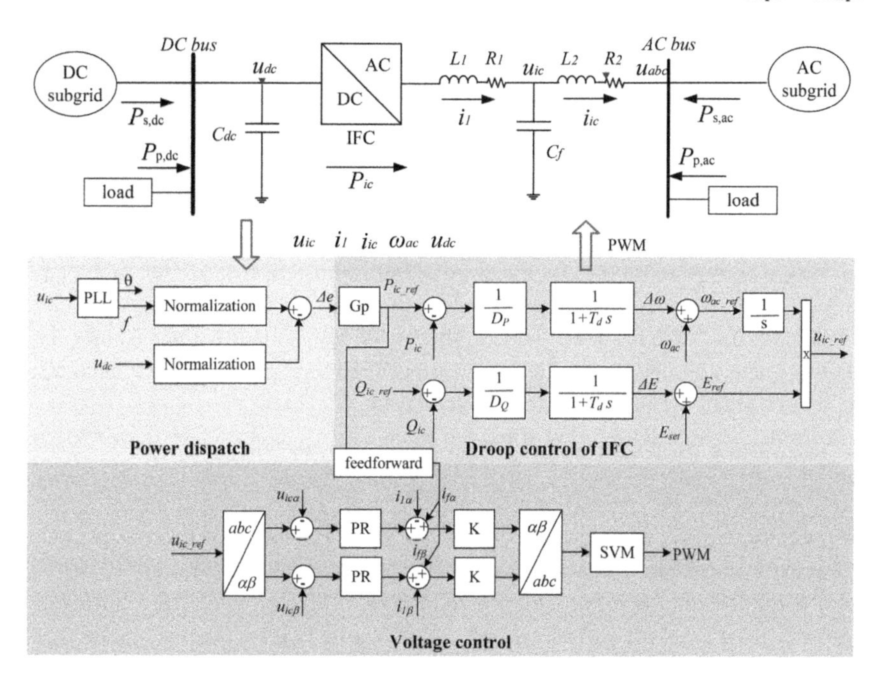

Figure 5.13 A typical unified control with a unified droop for an IFC in a hybrid AC/DC microgrid.

and $P_{s.dc\_pu}$ are the DC voltage, DC voltage at no load, and active power of the slack terminal in the DC subgrid. Also, k_{ac} and k_{dc} are the droop gains of the AC and DC subgrids, respectively. It should be mentioned that all values are per unit p.u values.

The frequency and the DC voltage can be normalized proportional to their power capacity, and the error can be achieved as:

$$\Delta e = \frac{\omega_{0\_pu} - \omega_{pu}}{\omega_{0\_pu} - \omega_{min\_pu}} - \frac{u_{dc0\_pu} - u_{dc\_pu}}{u_{dc0\_pu} - u_{dcmin\_pu}} \tag{5.2}$$

where ω_{min_pu} is the minimum frequency and u_{dcmin_pu} is the minimum DC voltage.When the error is zero, the power is shared between the AC subgrid and the DC subgrid based on their capacities. An easy way to ensure that the error is zero is to feed it to a PI controller, shown as Gp(s) in Figure 5.13. The output of this controller is the active power reference of the IFC. The PI controller can well eliminate the error when the converter's capacity is high enough; on the other hand, when multiple paralleled converters are needed to share the power, it is preferential to use the proportional gain to replace the PI controller and enable droop characteristics.

To allow bidirectional power flow in power control mode and voltage control mode seamless transition without mode detection, the conventional P–ω and Q–E droop characteristics of the IFC are:

$$\begin{aligned} \omega_{ac\_pu} &= \omega^*_{ac\_pu} - P_{ic\_pu}/D_P \\ E_{ac\_pu} &= E^*_{ac\_pu} - Q_{ic\_pu}/D_Q \end{aligned} \tag{5.3}$$

where ω_{ac_pu} and E_{ac_pu} are the frequency and voltage amplitude. The $\omega^*_{ac\_pu}$ and $E^*_{ac\_pu}$ are frequency and voltage amplitude when no power is transferred. Also, the P_{ic_pu} and Q_{ic_pu} are active and reactive powers of IFC; D_P and D_Q are droop coefficients of the IFC.

There is no inertia support for the above droop control to stabilize the frequency in AC voltage control mode. A low-pass filter is added to the droop characteristics, which adds inertia to reduce the maximum frequency excursion during faults and loading transitions. With the droop control of IFC, the AC voltage reference can be obtained. Then, the PR controller can be used to track it.

It should be highlighted that since the same controller is used under different operation modes, accurate and detailed modeling of each control mode for control parameter design is crucial. For more details on modeling and parameter design, please refer to [4].

Figure 5.14 shows the case study results under different conditions. Before $t = 2$ s, the DC load is 0.1 pu and the AC load is 0.1 pu, so the IFC's transferred power is 0. Figure 5.14a shows the simulation results in power control mode. The unified control can dispatch power based on the capacity of the AC subgrid and the DC subgrid. At $t = 2$ s, a 0.1 pu DC load is added in the DC subgrid, hence,

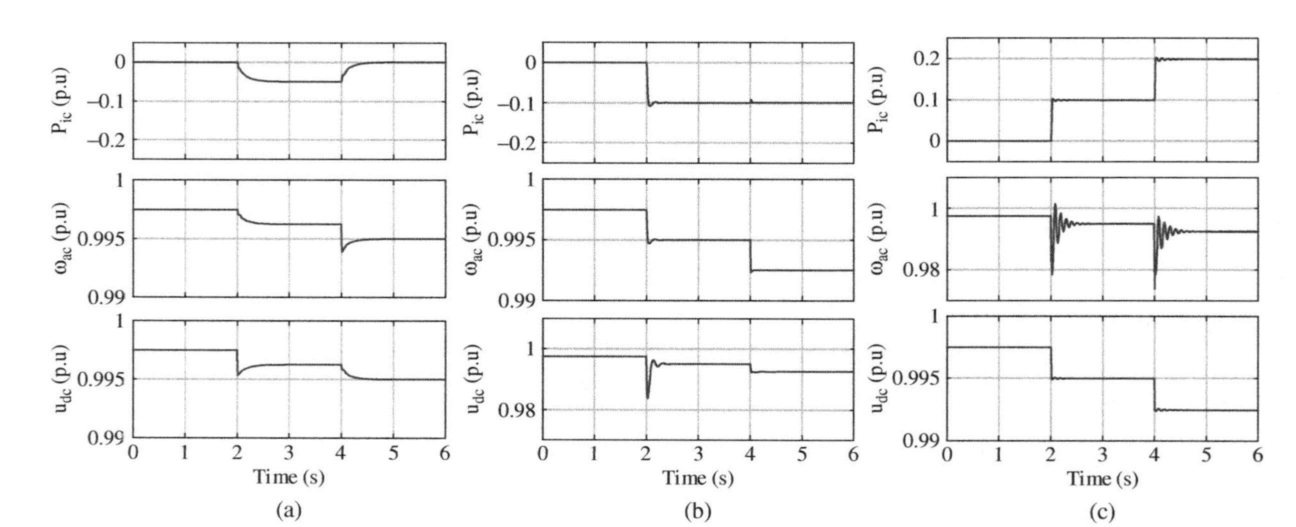

Figure 5.14 Simulation results of the case study: (a) IFC under power control mode, (b) IFC under the DC voltage support mode (DC slack terminal is unavailable), (c) IFC under AC voltage support mode (AC slack terminal is unavailable).

0.05 pu power is transferred from the AC subgrid to the DC subgrid. A 0.1 pu AC load is added in the AC subgrid at $t = 4$ s, so the transferred power is returned to 0. Figure 5.14b shows the simulation results of the DC voltage support mode. At $t = 2$ s, the DC slack terminal is unavailable. Under this condition, the DC voltage is supported by the AC subgrid via the IFC. Also, the DC load is fed by the AC subgrid, so 0.1 pu power is transferred from the AC subgrid to the DC subgrid. At $t = 4$ s, a 0.1 pu AC load is added in the AC subgrid, so the p.u values of frequency and DC voltage is reduced. Figure 5.14c shows the simulation results of the AC voltage support mode. At $t = 2$ s, the slack terminal in the AC subgrid is unavailable; therefore, the AC voltage/frequency is supported by the DC subgrid via the IFC. The power is transferred from the DC subgrid to the AC subgrid to feed the AC load, even if the AC load is increased at $t = 4$ s (0.1 pu).

5.6 Black Start in Hybrid Microgrids

Although microgrids can improve the system reliability with grid-connected and islanding operation capabilities, various uncertainties leading to power outages are inevitable, especially in the islanding microgrid. With small capacities compared to the main grid, the islanding microgrid may also have poor stability and susceptibility to interference. Therefore, a black start is an important topic for accelerating the recovery process and reducing the power loss of the microgrids.

The black start in a microgrid means when a blackout occurs, usually due to external or internal fault, the restoration process does not rely on large power systems or other microgrids connected to this microgrid, as they may not be available. In such a microgrid, the power management strategies help DERs with black start capability drive other sources and then expand the scope of restoration to the whole system. According to the IEEE Standard 1547.4-2011, the black start capability must be provided by microgrids that have islanding mode operation capability [5].

The black start capability of the microgrid can ensure its continuous and stable operation and critical load power supply. Also, it can facilitate large power grid restoration under specific conditions. With the microgrid black start capability, the traditional system start process that may rely on remote generators, such as hydropower units, and the related self-excitation and overvoltage caused by remote power sources, can effectively be reduced. Moreover, this can shorten the entire system's recovery time and reduce the loss caused by power outages.

There are three phases in the black start process of conventional power systems: the preparation phase, the network reconfiguration phase, and the load recovery phase. In microgrids, the recovery strategy can further be divided into parallel and serial strategies. Although the parallel strategy can quickly recover the system from failure, there are some obvious shortcomings. The difficulty of connecting the subsystem to the microgrids, the large inrush current when the subsystem is connected, and the complexity of the hardware design and control process are a few challenges of the parallel strategy. The serial strategy can simplify the complexity of hardware and control design and improve the system's stability. However, the recovery time is longer than the parallel strategy. Both serial and parallel strategies are discussed thoroughly in the following sections.

5.6.1 General Requirements of Black Start in Microgrids

In general, black start in microgrids has the following requirements:

1. The secondary system of microgrids must be reliable. During the black start process, the communication system among the microgrid central controller, DER controllers, and other components should be in a good state. Once the system becomes completely black, the emergency plan must take effect immediately to ensure smooth communication.
2. The DERs should have self-start capability. In the initial stages of the microgrid's black start, DERs should provide the required frequency, voltage, and power to the sources without this capacity. Also, the microgrids should have backup ESSs, which can also stabilize the frequency and voltage of the entire system.
3. The control centers of the microgrids also need to be equipped with rapid calculation and data storage capacity for critical information. The center collects all the fault information, power generation status, and load status before the complete blackout and prepares a fast decision-making system that facilities the black start process during blackouts.

5.6.2 Microgrid Black Start Scheme

Selection of Black Start Sources

The North American Electric Reliability Corporation (NERC) defines black start resources in transmission systems as generation units that can be self-started without external support. This definition can be adapted for defining black start sources in distribution systems and microgrids. The black start sources refer to the power sources that have the self-start ability and can regulate the frequency and voltage of the system [6].

Reasonably selecting the black start sources is one of the most critical actions in the black start process of microgrids. After the system becomes completely black, black start sources first complete the power generation and ensure their operation. Then, they send power to non-black start units through the distribution line to complete the start of other distributed power sources.

In large power grids consisting of high-power rotating machine-based generators, large gas power units are usually used to start the black start process, in which the start time is around a few minutes. In such systems, thermal power plants, which also have voltage and frequency control capabilities, are the non-black start power sources. In microgrids with small-power rotating machine-based generators, such as diesel generators, and power electronics-based DERs, the diesel generators, controllable DGs, and large capacity ESSs can initiate the black start process. The start time in such systems is about a few seconds. The renewable-based DGs such as PV and wind generators typically operate as non-black start power sources in microgrids.

Based on the above discussions, the general requirements of the black start sources are as follows:

1. Capable of regulating the voltage and frequency of the system: the black start sources must be responsible for maintaining the voltage and frequency of the microgrids within a specific range to ensure the normal operation of the microgrids in the islanding mode.
2. Contain ESSs and sufficient reserve capacity: considering that the black start may not be successful in the first attempt, it may require multiple operations. Thus, the backup capacity of the black start sources should be sufficient to support multiple black start processes.
3. Sufficient power generation capacity: the rated capacity of the power source should guarantee the system losses within certain limits and absorb the transient power generated by the non-black start power supply during start-up. Thus, the rated capacity of the black start DER should be:

$$S_n > \sum_{i=1}^{m_1} S_{ei} + \sum_{j=1}^{m_2} S_{Lj} + \sum_{k=1}^{m_3} S_{sk} \tag{5.4}$$

where S_n is the rated apparent power of the black start DER, S_{ei} is the excitation loss generated on the line transformer, S_{Lj} is the loss on the power line without load, and S_{sk} is the maximum impact power generated by the non-black start DERs at the start-up stage.

Optimal Restoration Path Selection

According to the microgrid structure and black start DERs' locations, the black start restoration can have different path selections. In detail, based on the various choices of black start sources and the microgrid structure, multiple black start paths can be obtained. Therefore, first, the feasibility of each path should be strictly verified according to various operational constraints. Then, after a comprehensive analysis and evaluation of each path, the best path is selected.

The following criteria should be considered to select the optimal restoration path:

- Minimize the length of the start-up path as much as possible
- Start the source that is close to the large loads first
- Minimize the loss and try to give more significant benefit to microgrid
- Create convenience for the fast recovery of the main grid
- Give high priority to large-capacity DGs.

Serial Black Start Scheme

Serial black start recovery is also called downward recovery. In this recovery scheme, the black start source is started first. After the voltage and frequency of the microgrid are established, the remaining power supplies are started one by one and then the microgrid is networked into the main grid.

The serial black start is mainly applicable to microgrids with a few sources and loads. The advantages of serial recovery include simple control methods, high start-up safety, and easy implementation. However, the start-up speed of this method is slow, and the sources' black start capabilities in the microgrids cannot be fully utilized. Thus, this method is mainly applicable to small microgrid systems.

The specific steps of the serial black start scheme are as follows:

Disconnecting all loads: The disconnection is aimed to avoid the large frequency and voltage deviations when injecting energies into the networks.

Building a low voltage network: The microgrid central controller determines which sources should provide power to low voltage cables and distribution transformers based on the information about rated power, load percentage, and other market conditions. The safety procedure of the low voltage network grounding should be considered when constructing a low voltage network. Since the ground connection is made at the neutral point of the distribution transformers, the transformers must be powered as soon as possible. The load can only be restored after the distribution transformer is powered. When supplying power to the transformer from the low voltage side, it experiences large inrush currents, which may not be supported by the electronic components of the inverter. The transformer energization should be performed using the ramp voltage generated by the IFCs of the selected sources to overcome this problem.

Small islanding synchronization: The sources operating in islanding mode should be synchronized with the low voltage network. Synchronization conditions, such as phase sequence, and frequency and voltage difference, should be verified by the local sources to avoid large transient current and power exchanges. If there are controllable sources without black start abilities, they can also be synchronized with the low voltage network. The optimal path selections should be applied in this step and the following steps.

Connection of controllable load: If the sources operating in the low voltage network are not fully loaded, the controllable load could then be connected to the low voltage network. The amount of power to be connected should consider the available storage capacity to avoid large frequency and voltage deviations during the load connection.

Connection of non-controllable sources: Examples of non-controllable sources are PV- and wind-based DGs. At this stage, the system has sources and loads that can smooth the voltage and frequency changes due to power fluctuations in uncontrollable sources; thus, they can now be connected to the system. The non-controllable sources use constant power control, and their corporations into the microgrids can expand the microgrid's power generation capacity and load capacity. It should be noted that, in the early stages of the black start process, the system is relatively weak, and the power generation capacity is small so that any disturbance may cause voltage or frequency imbalance. At this time, the non-controllable DERs that can be frequency or voltage regulated should be connected first. These sources can effectively prevent the shocks generated during the black start process and maintain the voltage and frequency stabilities in the microgrids.

Increase of local loads in the microgrids: To feed as much load as possible, other loads can then be connected depending on production capability. According to the power constraints from equilibrium constraints and power flow constraints, the load increase should meet:

$$\sum_{i=1} S_{\text{load}i} < \sum_{j=1} S_{\text{generation}j} \tag{5.5}$$

$$U_{Gi\,\min} < U_{Gi} < U_{Gi\,\max} \ldots i = 1, 2, 3 \ldots\ldots N_G \tag{5.6}$$

$$|P_{ij}| < \overline{P_{ij}} \qquad i = 1, 2, 3, \ldots N_G, \quad i \neq j \tag{5.7}$$

where $S_{\text{load}i}$ and $S_{\text{generation}j}$ are the rated capacities of the load and DERs in the microgrids; N_G is the number of DERs; U_{Gimax} and U_{Gimin} are the upper and lower limits of the node voltage amplitude; P_{ij} is the active power of the load; and $\overline{P}_{ij}$ is the limit value of the line active power.

Microgrid synchronization: When available, the microgrid should be synchronized with the main grid. The microgrid synchronization techniques are introduced in Section 5.4.1.2. Before a general power outage, the microgrid imports or exports power to the main grid. If the microgrid imports power, it would not be possible to connect all local loads.

Parallel Black Start Scheme

The serial black start scheme only relies on a single source as a black start power supply at the initial stages, which does not fully utilize all the sources in the microgrid. Moreover, the serial recovery needs to start each source and then connect to the network, which takes a long time. This also does not favor the requirements of fast start-up and restoration speed of the black start process.

Parallel recovery refers to establishing several independent small systems by multiple black start sources and loads in the microgrid and then synchronizing them simultaneously to realize the recovery of the whole microgrid. In this process, the sources are started to form a start-up unit with the nearby load. Then, through the gradual expansion of the start-up units and the simultaneous grid connection of each start-up unit, the black start of the entire microgrid is finally be achieved.

The parallel recovery method is mainly applicable to microgrid systems with scattered and multiple sources and loads. The main advantages include fast start-up speed and full use of the black start capability of each source in the microgrids. However, the control of the parallel recovery method is more complicated than that of serial recovery, and simultaneous connection to the main grid may cause an unstable start-up and lead the system to complete blackout again. Thus, this recovery method may not be suitable for large microgrid\textbf{a} systems.

Black Start Scheme for a Hybrid AC/DC Microgrid

The serial and parallel black start processes discussed above are mainly used for the AC microgrid black start. However, those methods should be adjusted for the hybrid AC/DC microgrids that consist of DC and AC subgrids linked through IFCs. Based on the idea of the black start in AC microgrids, the black start schemes for hybrid AC/DC microgrids can be divided into:

- simultaneous restoration of both the AC and DC subgrids while they re-connect after restoration
- either AC or DC subgrid restoration in the first stage and use it as the black start power supply for the other subgrid to complete the black start process.

For simultaneous restoration of both the AC and DC subgrids, the AC and DC subgrid black start can also be done in a serial or parallel manner, as discussed previously. After both AC and DC subgrids are in operation, the IFC linking DC and AC buses can be started and participate in DC, AC, or power flow control. Overall, the DC and AC subgrids' simultaneous black start can restore the normal operating state in the shortest time.

For the second scheme of the sequential start of AC or DC subgrids, the DC subgrid can be restored first as the DC subgrid usually has low complexity, flexible reconfiguration possibility, high power quality, and low line loss. For the black start on the DC subgrid, the main DERs can control the voltage to establish stable DC bus conditions to limit the potential voltage fluctuations caused by other DGs and distributed loads. Then, after the start of the DC subgrid, it can be used as the black start main power supply of the AC subgrid to complete the whole restoration process of the hybrid microgrid. In this scheme, since the black start is carried out in serial sequence, the start-up time is longer. Therefore, it can be used in low-time requirements to achieve better stability and a safe black start process.

5.6.3 Main Issues and Related Measures of Black Starts in Microgrids

During the restoration period of the power systems, due to the relatively weak grid structure, many problems can affect the restoration process. The following types of issues may arise during the microgrid black start process:

Overvoltage

When an AC microgrid operates in the islanding mode, a distribution transformer may be installed on the DER output side. However, since the inverters have a fast dynamic response speed, excessively increasing the bus voltage amplitude may lead to overvoltage and affect the recovery of the system. The rapid increase of bus voltage may also cause magnetic flux saturation in the distribution transformers. Therefore, a suitable control method on the start-up speed of the black start sources is needed.

When considering the capacitive effect of no-load or light-load transmission lines, the voltage on the transmission lines may increase. To prevent this type of overvoltage: (i) the voltage of the black start sources should be lowered as much as possible before the power transmission, (ii) a black start DER with reactive power absorption capability can be selected to keep the system voltage stable at the initial stage of recovery.

System Voltage and Frequency Stability

To keep the voltage and frequency of the AC microgrids or AC subgrids stable (or DC voltage in a DC microgrid or subgrid stable) during the black start process, it is necessary to power various loads gradually. Generally, the restorations of the controllable loads should be prioritized first, and larger and non-critical loads are restored gradually. If the loads are recovered slowly, the system recovery time will be extended accordingly; however, if the load grows quickly, the voltage or frequency of the whole system will be disturbed. Thus, the proportion of load increment must be balanced between the recovery speed and voltage and frequency stabilization requirement.

The sources in the microgrids contain a large number of renewable energy sources with strong output power randomness. As a result, an ESS can be added to strengthen the inverter output voltage and frequency.

Transient Stability of Microgrids

During the black start process, particularly the parallel recovery, there may be problems of stable operation of multiple distributed and small independent systems. To ensure the smooth start-up in microgrids, it is necessary to carry out a stability analysis of each small independent system in the initial recovery period to avoid instability and black start failure.

Fault Detection and Protection Configuration

Compared to the traditional power system, the fault current of the microgrid system is very small, and the inrush current generated by non-black start sources or loads may be larger than the fault current of the system. This makes it difficult for the fault detection devices to determine the system state. It is the transient fluctuation or system failure that affects the operation of the protection device. Therefore, it is also necessary to strengthen the coordination of the fault detection device and the relay protection device during the black start process in the microgrids.

In summary, the main problems in the black start process of the microgrids are very different from the traditional power system. Therefore, when designing the black start of the microgrids, it is necessary to pay attention to those potential problems and adopt effective control schemes and recovery strategies to solve them, thereby ensuring the smooth completion of the black start of the microgrids and promoting the development of the smart microgrids.

5.7 Summary

Power management and energy management systems are critical aspects of hybrid AC/DC microgrids' operation. First, this chapter discusses the difference between power and energy management systems and their realization in a multi-layer hierarchical control system. Then, a thorough discussion of different power management strategies of AC-coupled, DC-coupled, and AC-DC-coupled hybrid microgrids under steady-state operation is provided. Also, this chapter addresses power management strategies during the transition between grid-connected and islanding operation modes and under different loading conditions. Finally, the black start strategies in microgrids, where power management strategies play a critical role in system operation and control, are discussed.

References

1 Nejabatkhah, F. and Li, Y.W. (2015). Overview of power management strategies of hybrid AC/DC microgrid. *IEEE Transactions on Power Electronics* 30 (12): 7072–7089.

2 Balaguer, I.J., Qin, L., Shuitao, Y. et al. (2011). Control for grid-connected and intentional islanding operations of distributed power generation. *IEEE Transactions on Industrial Electronics* 58 (1): 147–157.

3 Zhou, T. and François, B. (2011). Energy management and power control of a hybrid active wind generator for distributed power generation and grid integration. *IEEE Transactions on Industrial Electronics* 58 (1): 95–104.

4 Fang, F., Wei Li, Y., and Li, X. (2018). High Performance Unified Control for Interlinking Converter in Hybrid AC/DC Microgrid. In: *2018 IEEE Energy Conversion Congress and Exposition (ECCE)*, Portland, OR, 3784–3791.

5 IEEE Guide for Design, Operation, and Integration of Distributed Resource Island Systems with Electric Power Systems," in IEEE Std 1547.4–2011, vol., no., pp. 1–54, 20 July 2011.

6 Moreira, C.L., Resende, F.O., and Lopes, J.A.P. (2007). Using low voltage microgrids for service restoration. *IEEE Transactions on Power Systems* 22 (1): 395–403.

6

Energy Management System (EMS) in Smart Hybrid Microgrids

6.1 Energy Management in Hierarchical Control of Microgrids

6.1.1 Hierarchical Control

The microgrid consists of various distributed energy resources (DERs) and loads, operating in islanding and grid-connected modes. The DERs include both distributed generations (DGs) and energy storage systems (ESSs), which can export active power to the grid. The control and management of the microgrid have multiple objectives, including regulation of power flow, power quality, energy management of energy resources, to name a few. Hierarchical control is needed to meet the microgrid's operational requirements with different significances and time scales. The hierarchical control mainly consists of three levels: primary, secondary, and tertiary control layers. As mentioned in Chapter 5, the power management system (PMS) is realized in the primary and part of the secondary control layers, while the tertiary control and part of the secondary control contain the EMS. As discussed in Chapter 5, hierarchical control can be implemented in centralized, decentralized, and distributed structures.

Primary Control Layer

The primary control is the first layer in the hierarchical control scheme. This layer is implemented in the local controllers, e.g. local controllers of interfacing converter (IFC) based DERs, to maintain voltage/frequency stability in different operating conditions, share active and reactive powers among DERs, and avoids undesired circulating currents in the network. The primary control provides reference signals to the IFC inner control loop, such as reference current and the reference voltage. The inner control loop can be implemented in the current

Smart Hybrid AC/DC Microgrids: Power Management, Energy Management, and Power Quality Control, First Edition. Yunwei Ryan Li, Farzam Nejabatkhah, and Hao Tian.

control or voltage control method (VCM). Based on the current reference, the current control method (CCM) regulates the active and reactive powers for a grid-following converter. While in the VCM, the DERs operate as a grid-forming converter and tracks the voltage references. For details about converter primary control, please refer to Chapter 4. It should be highlighted that the primary control layer realizes most of the PMS objectives in the microgrids.

One well-known example of primary control is droop control, which can share active and reactive powers among DERs without communication links. The active power-frequency and reactive power-voltage droop controls regulate the voltage and frequency in the AC microgrid. In a DC microgrid, power-voltage droop control is used to regulate the DC voltage. Please refer to Chapter 4 for more discussions about droop control.

Secondary Control Layer

The secondary control has a slower dynamic response than the primary control layer. The secondary control layer aims to restore voltage/frequency and compensate for the deviations caused by primary control. This control layer also resynchronizes the microgrid and utility grid and improves the power quality of microgrids. For example, coordination of harmonics compensation and unbalanced voltage compensation are done in this layer. Please see Chapters 9 and 10 for details about harmonics compensation and unbalanced voltage compensation in hybrid microgrids. In addition, optimal and coordinated operation of DERs are responsibilities of the secondary control layer. As evident, the secondary control layer contains parts of both EMS and PMS objectives.

Tertiary Control Layer

The tertiary control layer is the slowest among the hierarchical control layers, controlling individual or multiple interconnected microgrids. This control layer aims to optimize the microgrids' energy management to achieve economically optimal operation and control the power flow between the microgrids and the utility grid. In detail, the tertiary layer is responsible for many energy management considerations, including but not limited to:

- the most cost-effective combination of generating units in individual microgrids to minimize power losses and maximize energized loads
- proper exchange power between multiple microgrids and utility grids
- a balance between utility-scale energy storage usage and grid electricity purchase
- coordination of energy storage charging, including EVs
- distributed optimization, such as optimizing parallel converter operation (e.g. between DC and AC subgrids) to keep their efficiency at an optimal point.

In hierarchical control, each control layer performs its objectives with different time scales, and they interact with other layers via communication channels. Therefore, information and communication technologies play a crucial role in hierarchical control, and the sound operation of microgrids depend on efficient and reliable data flows in the cyber system. The data should be timely, accessible, accurate, and trustworthy. Any delay, corruption of data, or cyber incident may affect the physical system's smooth operation and jeopardize the smart grids' efficiency, stability, and safety. More discussions about microgrid cyber-security are provided in Section 6.4.

6.1.2 Energy Management System

Microgrids Management in Different Time Scales

As mentioned, a microgrid management can be classified based on timing scale, including long-term energy management and short-term power management, as shown in Figure 6.1. The long-term energy management includes a daily/hourly energy production plan using environmental impact, cost of generation, controllable load management according to supervision requirement, power reserve capacity according to the electricity market, and load demand forecast. It is also responsible for coordination between the microgrid and the utility grid. The short-term power management includes voltage/frequency regulation and real-time power dispatching/balancing. Long-term energy management can optimize the economic benefits, while short-term power management can ensure system dynamics and security performance [1]. As mentioned in Section 6.1.1, in the hierarchical control, the power management strategies are implemented in the primary layer and part of the secondary layer, while the energy management strategies are implemented in part of the secondary layer and the tertiary layer of hierarchical control.

For a system with multiple microgrids, the EMS should manage individual microgrids operation and microgrids coordination to ensure economic and secure operation. It typically has the following objectives:

- plan daily/hourly energy production using environmental impact, cost of generation, controllable loads management, power reserve capacity, and load demand forecast

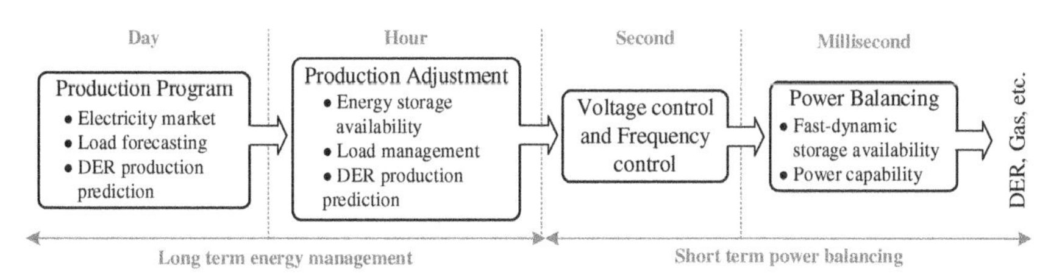

Figure 6.1 A microgrid management in different time scales.

- minimize operation cost of individual microgrids
- stabilize voltage/frequency in islanding operation mode
- manage information interactions among microgrids
- provide sustained power supply for critical loads for resilient operations during extreme events
- smooth transition between grid-connected mode and islanding mode
- manage the injection/absorption power of multiple microgrids into/from the utility grid.

The EMS of microgrids includes four main functionality modules: information interaction, control and scheduling, resilient operation, and ancillary services [2]. These modules are presented in Figure 6.2.

The information interaction module collects real-time data from remote measurement devices, processes the raw data, and distributes the processed data. This module can analyze the power flow and predict load demand and generation supply based on these data. As a result, it enables the utility to shape the demand curve by altering customer supply times to reduce the overall generation cost. This is similar to the production program in Figure 6.1.

The control and scheduling module aims to balance the power in microgrids. The voltage/frequency control is also considered to reduce voltage/frequency violations, avoid overloads/load shedding, and minimize losses. This module also controls switching functions to ensure smooth mode switching in multiple microgrids.

The resilient operation module is designed in the EMS to smooth the operation of the microgrids under disturbances, cyber incidents, and severe weather conditions. Thus, the power system can withstand severe weather disruptions, climate changes, catastrophic artificial incidents, and a combination of such incidents. Even if one or several microgrids are damaged, the remaining part can supply full or part power according to the capacities. Relay protection coordination is

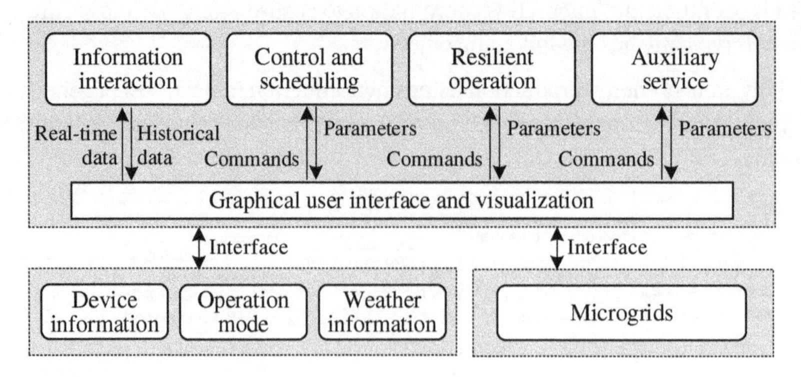

Figure 6.2 Multiple microgrid energy management.

also required in interconnected microgrids to improve reliability. The network topology frequently changes, resulting in protection coordination issues due to the magnitude and direction of fault current variations in different network topologies.

The ancillary service modules include market trading, spinning reserve support, demand response, and congestion management. This module allocates generation outputs of the committed generating units to minimize cost while meeting the system constraints such as spinning reserve. The spinning reserve is the generating capacity available to the system to meet demand if a generator goes down or another disruption to the supply. The demand response provides an opportunity for consumers to play a role in the operation of the electric grid by reducing or shifting their electricity usage during peak periods in response to time-based rates or other forms of financial incentives. When existing transmission capacity cannot accommodate all scheduled transactions simultaneously, congestion management can achieve efficient and reliable power system operation.

The energy management and power management systems are usually implemented in a centralized structure in a single microgrid. In such structures, all measurement signals and data are sent to a central supervisory control center. The central supervisory control center usually contains the hierarchical control's tertiary and secondary control layers, and it runs the energy management algorithm. Then, the generated signals, e.g. reference powers, are sent to the local controllers that contain the primary control layer and power management algorithms.

For multiple interconnected microgrids, the distributed structure can be used for control strategy implementation. In the distributed structure, each microgrid has its supervisory control center used for its EMS implementation. The supervisory control centers of microgrids are linked through the communication network (cyber system), and they work together to fulfill the multiple microgrids' energy management objectives.

Artificial Intelligence in the Energy Management System

In recent years, AI-based techniques have been widely considered for EMSs of power electronics intensive microgrids. The AI gives the intelligence demonstrated by machines to the energy system to replace human analysis, judgment, optimization, and decision-making by simulating the human brain and modeling human problem-solving. The AI methods have many advantages, including robust data analytical capability to process massive data and automatically extract the coupling relationship, less dependence on a specific mathematical model, strong adaptability, and capable of high dimensional, time-varying, and non-linear systems. The AI methods include an expert system, fuzzy logic, metaheuristic methods, and machine learning [3]. The application of AI method in power systems is shown in Figure 6.3.

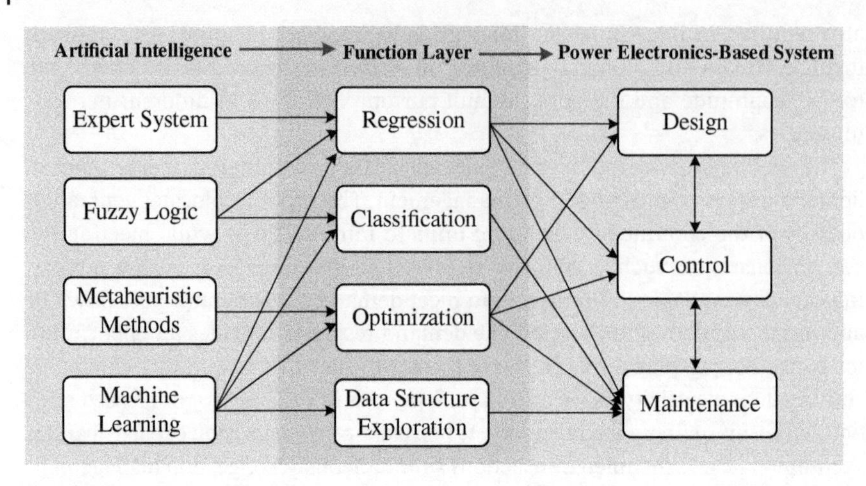

Figure 6.3 Application of AI in power systems.

The expert system is a computer system emulating a human expert to implant the expertise of a human being in a certain domain to solve the problem. An expert system has two sub-systems: knowledge base and inference engine. The knowledge base represents data, facts, and statements that support expert knowledge. The inference engine can conclude the problem that can be used to interface the knowledge base and user.

The fuzzy logic is a rule-based method. It has the capability to capture nonlinearities and uncertainties that precise mathematical models cannot describe. By employing the knowledge of specific experts, fuzzy logic can provide an effective solution or make decisions.

The metaheuristic methods are developed with inspiration from biological evolution. It contains trajectory-based methods (tabu search method, simulated annealing method) and population-based methods (genetic algorithms, particle swarm optimization, colony optimization algorithms, particle swarm optimization, immune algorithms, etc.).

Machine learning exploits data gathered from observations or experiments on a system to automatically build models predicting or explaining the behavior of the system or decision rules to interact appropriately with it. The machine learning method is designed to discover principles with experience from either collected data or interactions by trial and error. The machine learning approaches include supervised learning methods (e.g. neural network methods, probabilistic graphical method, kernel method, unsupervised learning methods), unsupervised learning methods (e.g. k-means, self-organizing maps, principal component analysis methods), and reinforcement learning methods. The supervised learning methods aim to establish relationships between the inputs and outputs. While unsupervised learning has no output, the main task is data clustering and data compression. Reinforcement learning aims to a suitable action to maximize the notion of cumulative reward.

The functions of AI include optimization, classification, regression, and data structure exploration. Optimization refers to find an optimal solution to maximize or minimize objective functions from a set of available alternatives given constraints, equalities, or inequalities that the solutions have to satisfy. The metaheuristic methods and machine learning are suitable for optimization tasks. The metaheuristic method is used for static optimization, while machine learning optimization focuses on dynamic optimization. While classification deals with assigning input information or data with a label indicating the classes, such as anomaly detection and fault diagnosis in maintenance. The fuzzy logic and machine learning methods are employed for classification tasks, but machine learning is more accurate and flexible. By identifying the relationship between input variables and target variables, the regression aims to predict the value of one or more continuous target variables given input variables. For regression tasks, the expert system, fuzzy logic, and machine learning can be applied. The data structure exploration contains data clustering that discovers groups of similar data within a dataset, density estimation that determines the distribution of data within the input space, and data compression that projects high dimensional data down to low dimensional data for feature reduction. For this task, only machine learning methods can be applied.

With the above-mentioned functions, the AI methods can be applied in the multi-objective EMS to forecast load, analyze stability, minimize cost, etc. [4]. According to application characteristics, AI technology can be divided into perception, decision, and implementation. The perception includes system prediction (forecast power generation, consumption prediction), information acquisition (model the energy system and determine pattern recognition), cyber-security (active security protection, active defense, policy configuration). The decision part includes distributed storage management, energy management, risk assessment, and fault detection. While the implementation part contains demand response, stability analysis, power quality control, generation coordination control, and storage sizing and location.

In multiple microgrids, the microgrids are coordinated to achieve stability and optimal control using AI methods. The stability of microgrids is to keep frequency and voltage in an exact range in normal operation, and after a disturbance, dispatch power among microgrids, and enhance transient stability. The optimal control is to minimize energy consumption or operation cost by optimizing power dispatch.

6.1.3 Communications in an Energy Management System

The communication network (cyber system) plays an important role in the EMS of single and interconnected microgrids. Without secure and reliable data exchange, it is hard to optimize the operation of microgrids. Both communication techniques NAN and HAN in premises network can be adopted for the EMS. If it is necessary to coordinate multiple microgrids, the WAN techniques can also be involved. Please refer to Chapter 3 for detailed information about communication techniques in smart microgrids.

As discussed in Section 6.1.2, energy management is generally performed as a long-term optimization. As a result, the requirements on communication latency are low, e.g. the 9600 bps communication can fulfill the need if a dedicated network is built for energy management. However, in some applications, the communication traffic for energy management is shared with other functions to save costs on cabling and deployment. In this case, different priorities can be assigned for different functions, and energy management can have a low priority compared to time-sensitive communications for protection, phasor synchronization, status monitoring, power management, etc.

In microgrids, cyber incidents may degrade communication security, resulting in instability of the microgrid, a communication link fault, and power imbalance. In Section 6.4, the cyber-security of microgrids will be discussed in detail.

6.2 Multi-agent Control Strategy of Microgrids

Multi-agent systems (MASs) use networked multiple autonomous agents to control smart interconnected microgrids. In such systems, the overall control is achieved using local interactions among the agents. The multi-agent control is very suitable for distributed structure implementation (please see Chapter 5 for details on distributed structure). In general, the agents in the multi-agent control strategy have the following characteristics:

- Autonomy: each agent can manage individual behaviors cooperatively or competitively without human intervention.
- Reactivity: agents can perceive the environment and respond adequately to the variable and uncertain environment. Hence, multi-agent control has strong adaptability to the uncertain environment of operation.
- Social ability and pro-activeness: agents can communicate with other agents and cooperate to achieve individuals and system goals. When one agent fails, other agents may continue the system function through interactions among agents. Hence, the multi-agent control has strong robustness to large disturbances.

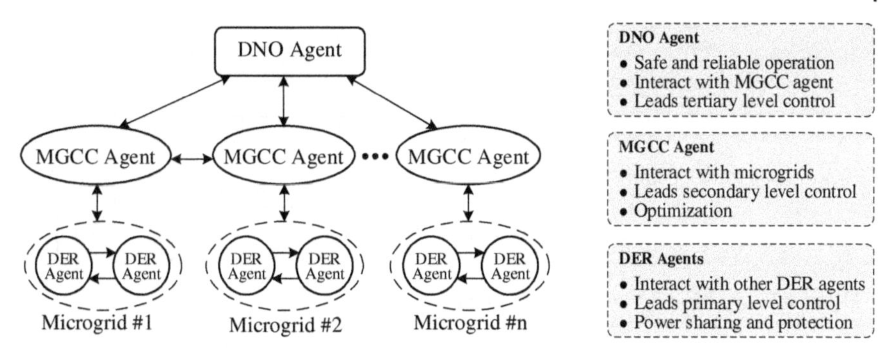

Figure 6.4 The structure of multi-agent control for multiple microgrid.

For a distributed multi-agent control structure of multiple microgrids, three types of agents are needed: a distribution network operator (DNO) agent, a microgrid central control (MGCC) agent, and a DER agent. Figure 6.4 provides the multi-agent control structure of multiple microgrids [5]. In general, the multiple microgrids' tertiary layer in hierarchical control runs in the DNO agent, while the MGCC and DER agents focus on the secondary and the primary controls.

The DNO agent is responsible for energy management considerations of multiple microgrids, as discussed in Section 6.1.2. It communicates with the MGCC agent of each microgrid and receives measured/processed data. After running the EMS algorithm, it generates the control signals for each MGCC. The DNO agent is also in charge of the safe and reliable operation of microgrids corresponding with the main grid. The MGCC agent of each microgrid is responsible for the energy management strategy of individual microgrids. The DER agent is responsible for the individual DER units, in which their primary controls and protections are handled.

Implementation Example of a Multi-agent Control System in a Hierarchical Control Structure

An example of implementing a multi-agent control system in a hierarchical control structure for overloading regulation is provided here [6]. Figure 6.5 shows the case study flow. The tertiary level control is demonstrated with an ancillary service for load regulation in the host grid feeder, and the secondary control is demonstrated with automatic generation control (AGC) inside a microgrid. In the ancillary service case, the DNO agent is responsible for the safe operation of the grid, and the microgrids intend to sell power to the host grid. In the AGC control case, the MGCC agent is responsible for frequency and voltage control, and each DER agent can achieve secondary control functions, such as restoring frequency at the nominal level.

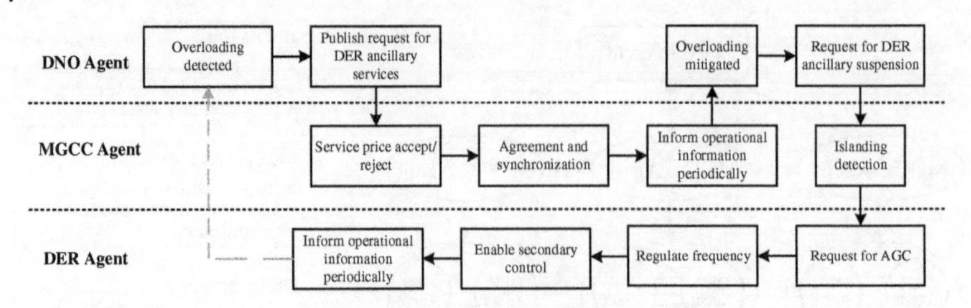

Figure 6.5 Multi-agent control case study flow chart.

Tertiary control (load regulation ancillary service): The microgrids provide load regulation and reactive power support since overload may cause overheating and voltage drops beyond permissible limits. The DNO agent continuously checks the critical current flow value through the IEC 61850 measurement function block. Once the current flow reaches its critical value, the high-alarm node of the function block becomes high. Then, the ancillary service behavior of the DNO agent is invoked, and a DER ancillary service request message is published to MGCC agents registered to the directory service (the directory service is a service mechanism where an agent can find other agents providing the services it requires). The DNO agent periodically looks up available operators from the directory facilitator.

The MGCC agents propose a service with the price. Then, the DNO agent sends the accept/reject proposal message to the MGCC agents in return. Then, the qualifying agent replies with the confirmation message, and the agent enables the synchronization behavior. After synchronization, the overloading problem can be solved with the DER services.

Secondary control (AGC control and islanding): In the case mentioned above, the microgrid was connected to the grid to solve the overloading issue, and the secondary control proposed combinations of DER powers with their cost to the DNO as candidates for overloading removal. When overloading subsides and the microgrid starts to draw power from the utility grid, the DNO agent sends an ancillary service suspension request to the utility connected microgrid. Then, the microgrid gets disconnected from the host grid. When the microgrid is disconnected, an immediate microgrid frequency dip is detected, so the islanding is detected (please refer to Chapter 8 for frequency-based islanding detection strategies). Under this condition, the MGCC agent requires DER agents to serve as frequency regulator units to restore frequency to the nominal level. First, the DER agents submit their interest with proposal messages and the generation costs at that specific time. After approval, the frequency and the terminal voltage of the inverter-based DER are compared with the corresponding reference values, ω_{ref} and E_{ref}. Then, the error signals are processed by individual controllers, and the resulting signals are sent to the primary controller of DER agents to compensate for the frequency and voltage deviations, similar to droop functions.

6.3 Advance Distribution Management Systems (ADMSs) in Smart Hybrid Microgrids

In the previous section, the EMS of microgrids was studied. In the near future, distribution power systems will likely consist of interconnected microgrids. Thus, the study of traditional distribution systems management is also essential, as it will potentially be used for future interconnected microgrid control with some improvement. In this section, an ADMS, a platform for optimized distribution system operational management, is addressed.

An ADMS could provide advanced monitoring, analysis, control, optimization, planning, and training functions for distribution systems, enabling power companies to provide more reliable, safe, and efficient power. A typical ADMS usually consists of multiple sub-systems, e.g. supervisory control and data acquisition (SCADA), geographic information systems (GIS), distribution management systems (DMSs), automated meter reading/advanced metering infrastructure (AMR/AMI), outage management systems (OMSs), distributed energy resources management systems (DERMS), EMS, customer information systems (CISs), and meter data management systems (MDMSs) [7, 8]. All the sub-systems in an ADMS are inner-connected by a shared data-bus, which facilitates the communications between the sub-systems, i.e. the sharing of measurement data, control signal, operational models, etc. The introduction of six widely used sub-systems in an ADMS is presented as follows.

6.3.1 Supervisory Control and Data Acquisition (SCADA)

SCADA system is a computer-based system for production process control and scheduling automation, which has a wide range of industrial applications. It features data gathering in real time from remote locations to control equipment and conditions. The SCADA system is the most widely used automatic control system in distribution systems. In terms of distribution system operation and control, the SCADA system makes it possible to have complete system information, improved efficiency, correct control of system operation status, speed up decision-making, quickly diagnose the system fault status, etc.

Generally, in real-world distribution systems, four components are usually included in the SCADA system: measurement/control devices, communication interfaces (interfaces for the measurement/control units), the communications network, and the SCADA host. The relationships between these four components of the SCADA system are shown in Figure 6.6.

Measurement/Control Devices

The function of the measurement devices is to real-time measure the power system parameters, e.g. voltage and current, while the actions on power networks are generated from the control devices, e.g. operating the power line circuit breaker.

Communications Interface

Two kinds of communications interfaces are usually used in power systems, i.e. remote terminal units (RTUs) and intelligent electronic devices (IEDs). The RTU is a unique computer measurement and control unit with a modular structure designed for long communication distances and harsh industrial site environments, which could connect the terminal detection instruments and actuators with the remote control centers. For modern distribution systems, the IEDs are more widely used than RTUs and are usually incorporated in the measurement/control devices themselves, enabling direct linking between the measurement/control devices and the communication network, as shown in Figure 6.6.

Communications Network The SCADA system's data and control commands generated by the SCADA host and measurement/control units are transmitted via the communications network (please refer to Chapter 3 for more information about communication networks). Multiple communications networks could exist simultaneously in one SCADA system according to its structure and communication demand.

SCADA Host As shown in Figure 6.6, the SCADA host serves as the core of the overall SCADA system. The SCADA host mainly performs two functions: one is to process the data from the measurement/control units received via the communication interfaces and communication networks for system observability; the other is to send the control commands from the distribution system operators to the corresponding communication interfaces and/or the measurement/control units.

In distribution systems, there are usually a significant number of decentralized data collection points. If the conventional SCADA system structure is used, numerous communication networks and communication interfaces are required accordingly, which is unrealistic in practice. Therefore, their SCADA system

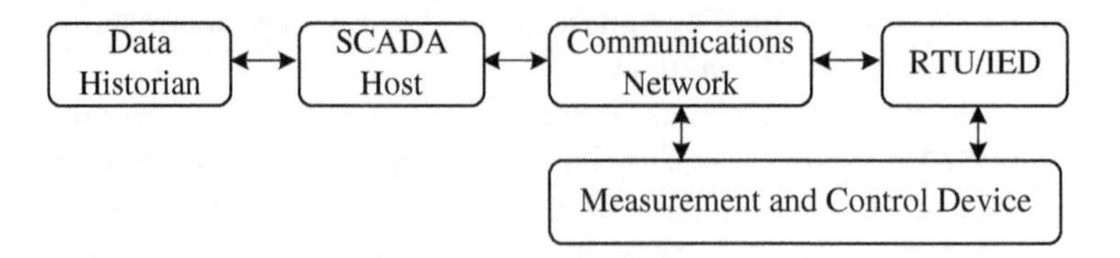

Figure 6.6 SCADA system and components.

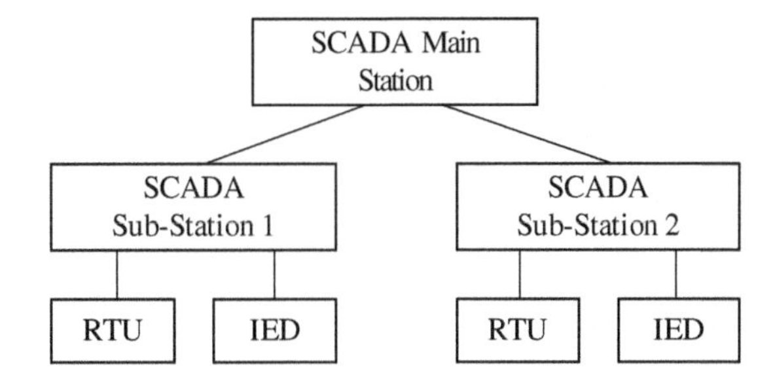

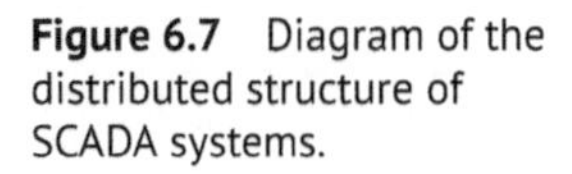
Figure 6.7 Diagram of the distributed structure of SCADA systems.

usually adopts distributed structure for power distribution systems. This structure is shown in Figure 6.7, which organizes the decentralized data collection points into several sub-systems. The data and control signals are firstly collected in each sub-system and then sent to the SCADA host in the main system via the communication network. Such a structure could greatly improve the efficiency and reliability in operating the power distribution systems.

6.3.2 Geographic Information Systems (GISs)

GISs are an emerging interdisciplinary subject that integrates computer science, geography, surveying and mapping, environmental science, economics, space science, and information and management science. With the support of computer hardware and software environment, it can collect, manage, operate, analyze, simulate and display spatial data and provide spatial and dynamic geographic information on time. For modern distribution systems, the GIS is used as a comprehensive production management information system that connects power equipment, substations, transmission and distribution networks, power users, and power loads. Thus, it could provide the power system operators with information on power equipment, power grid operating status, production management, power market, territory, natural environment information, etc. In addition, relevant data, pictures, images, maps, technical data, and management knowledge can be acquired through the GIS.

The GIS is a critical component of the ADMS in power systems, which provides connection and information for other sub-systems in the ADMS, such as the OMS, DMS, and CIS. An example of such connections is shown in Figure 6.8, where GIS provides information for the DMS and OMS systems.

6.3.3 Distribution Management System (DMS)

As mentioned before, the SCADA system could acquire the data and deliver control signals to the end-use measurement/control units via the communication networks and/or RTU/IED. A fully-established DMS could enable the power

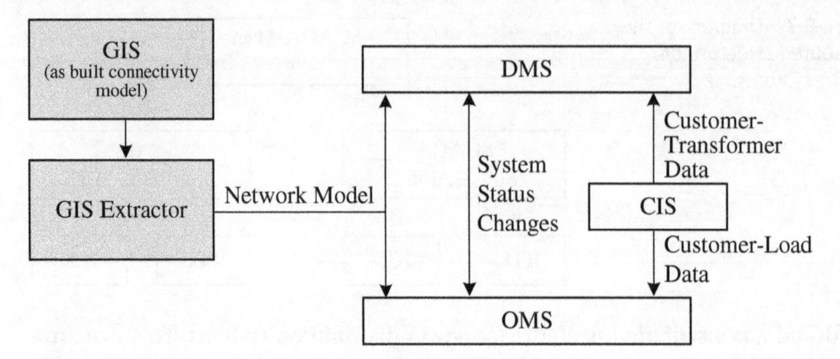

Figure 6.8 GIS and ADMS interaction architecture.

distribution system operator to control and manage the whole system actively. Based on the DMS, the distribution system operator could have real-time observations of the whole system while coordinating system operations to achieve a safe and efficient workflow. The information of the distribution systems is included in the DMS, e.g. the locations of the substations, feeders, switches, etc., while the real-time measurement data from the SCADA and AMR/AMI is also sent to the DMS. Moreover, numerous analytic applications are often associated with the DMS that operate on the network model parameters and field measurements, such as state estimation, fault location, switch management, etc.

6.3.4 Automated Meter Reading/Automatic Metering Infrastructure (AMR/AMI)

The AMR uses microelectronics and computer networks to automatically read and process meter data for each end-use customer, which has the advantages of fast reading speed, high calculation accuracy, good meter reading simultaneity, and direct networking with business computers. However, the AMR can only enable one-way communication, i.e. it only allows the measurement data to be sent from the user-side to the host-side.

Compared with AMR, AMI is an integrated system of smart meters, communications networks, and a DMS that enables two-way communication between utilities and customers. Therefore, it could provide additional functions such as transmitting pricing information, demand response signals, and outage information.

6.3.5 Outage Management Systems (OMSs)

Based on the SCADA, GIS, DMS, and CIS of the ADMS, the OMS is designed to comprehensively analyze the real-time distribution network operation information, equipment maintenance information, user information, and geographic information to determine the best power outage plan. With the help of OMS, the faulty power outages can be quickly handled, which could effectively shorten the time of power outages, improve the reliability of the power supply and the quality of service to users.

The main functions of the OMS can be summarized as follows [9].

Tracking the Power Outage

Based on the network model (e.g. user information, distribution system information, etc.), the OMS could conduct a comprehensive analysis of the system to track the locations of the power outages accurately.

Handling the Power Outages

After determining the accurate locations of the power outages, the crews could conduct timely emergency repairs and accident handling work according to the outage location displayed by the OMS. The decision for optimal dispatching of the crews is made in the computer system of the OMS.

Analyzing the Power Outages

After eliminating the power outages by the crews, the power system will resume its normal power supply. Then, the OMS is supposed to analyze the power outage reasonably and make an accident report for future queries and use.

6.3.6 Distributed Energy Resource Management System (DERMS)

With the development of smart grid technologies, DERs are increasingly integrated into modern distribution systems. DERs are electricity-producing resources that are connected to a local distribution system, which include solar panels, combined heat and power plants, ESSs, small natural gas-fuelled generators, battery storage systems, electric vehicles, etc. To efficiently manage the DERs, the DERMS has been developed as part of the ADMS (the DERMS is a part of the current EMS system in the distribution power systems).

Multiple types of DERMSs exist, which are designed to cope with the various DERs in distribution systems. In addition, the DERMS can also actively communicate with other sub-systems in the ADMS, such as the DMS, GIS, MDMS, and the EMS. For example, the DERMS can transmit the real-time DER operation information with SCADA or receive the command signals from the DMS. Figure 6.9 shows a representative architecture of a DERMS.

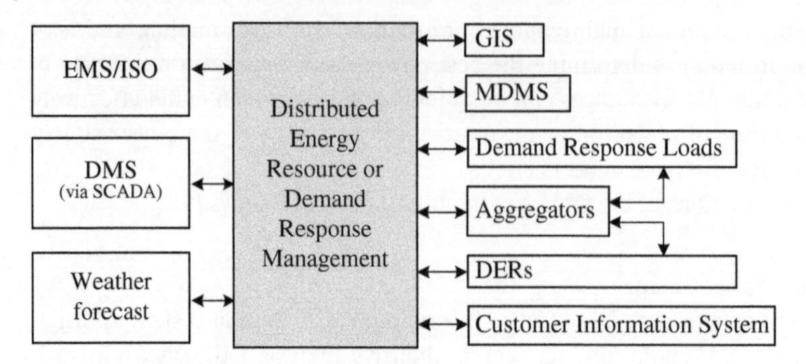

Figure 6.9 The architecture of a DERMS.

6.4 Cyber-security in Smart Hybrid Microgrids

In smart grids, the cyber system and physical process are tightly coupled. Due to cyber system vulnerabilities, any cyber incidents can have economic and physical impacts on smart grids operation. In smart microgrids with a high penetration of converter-based DERs, cyber-security violation can be very harmful. Although the optimal economical operation may not be the primary concern in such microgrids, cyber incidents could have devastating effects on microgrids stability, especially in islanding operation mode. In other words, due to the low inertia of such microgrids, cyber-security violations could affect the transient and steady-state stability of microgrids. Also, any cyber incidents in either AC or DC subgrids affect the other side of the hybrid AC/DC microgrids. For instance, if any cyber incident affects the frequency stability of the AC subgrid, it affects DC voltage stability on the DC side through AC/DC subgrid interfacing converters (IFCs). Considering the future roadmap of smart microgrids with more integration with the communication systems, cyber-security will receive more and more attention.

The cyber system in smart microgrids collects, transmits, and processes data to control physical system operation. Therefore, the data flow in the cyber system should be efficient, reliable, and timely to govern physical process operations. Cyber-security violations on smart microgrid data flow can be classified into three violations that compromise availability, integrity, and confidentiality [10, 11].

6.4.1 Different Types of Cyber-security Violations

Cyber-security Violations on Data Availability

The cyber system should guarantee that the data are timely and accessible, which is crucial for power electronics converter control in smart microgrids, especially under islanding mode and transients. Cyber-attacks whose primary purpose is to block or delay data communications are referred to as attacks on data availability. The denial of service (DoS) and distributed denial of service (DDoS) are examples of cyber-security violations on data availability. These violations can be started from one source or several sources by transferring malformed packets to the target or flooding the network/communication layer by exhausting the routers' processing capacity, network bandwidth, or servers. Moreover, data time latency cannot exceed its limit in microgrids. For example, the maximum latency of protective relay is in 4 ms, PMU-based situational awareness monitoring is sub-second, the SCADA system is in seconds, and the EMS is in minutes.

Cyber-security Violations on Data Integrity

In addition to availability, data in the cyber system should be accurate and trustworthy over their entire lifecycle and under all operating conditions. Any violation that compromises data integrity modifies the information flowing in the cyber system. These violations can be made by corrupting the measurements or command signals in the communication network and may lead to microgrid malfunctions and affect its control, including regulation of frequency and voltage, power and energy management, islanding detection, and resynchronization. A typical example of violations compromising data integrity is false data injection (FDI) cyber-attacks. The FDI attack is one of the most challenging threats for microgrids, and the impact of FDI on modern power grids can be significant. In such violations, hackers can penetrate a communication network without changing the system observability, and system operators may be unaware of any attacks. Those violations are also called stealth attacks [12]. In this section, these attacks are discussed in detail due to their importance and disruptive impacts on smart microgrids.

Cyber-security Violations on Data Confidentiality

Data confidentiality states that data should be protected from being accessed and comprehended by unauthorized parties. Cyber-security violations compromising confidentiality allow hackers to spy on the communication network to retrieve information about customers (identity and electricity usage) and microgrid operation and control strategies. Although these violations may not have a high impact on microgrids operation, the revealed information can be used by hackers to attack data availability and integrity effectively.

The impact of cyber-security violations on smart microgrid operation and the construction of cyber incidents and defensive strategies against them with a particular focus on FDI attacks are presented in this section.

6.4.2 Impacts of Cyber-security Violations on Smart Microgrids

In general, cyber-security violations can cause significant economic and technical/physical issues in smart microgrids. In the following, these impacts are reviewed.

Economic Impacts

Although much recent research has focused on cyber incident technical/physical impacts, it is also essential to study such attacks' potential financial risks. Cyber-security violations can cause significant economic problems in smart microgrids, especially in grid-connected mode with high penetration of renewable energy resources. It should be mentioned that optimal economic operation in microgrid islanding mode is not as important as the grid-connected mode (other factors such as stability are more important in an islanded mode).

Most deregulated electricity markets consist of a day-ahead market and a real-time market. In the day-ahead market, the load is forecasted, and an optimization problem is solved to minimize the cost. The optimization problem's outcome would be the predicted power generated at each bus (economic dispatch), which is used to define the locational marginal price (LMP) at each bus. The LMP is the buy/sell cost of power at different locations within electricity markets. Since FDI cyber-attacks can affect load forecasting, the day-ahead market is vulnerable to such violations.

The real-time market uses state estimation to estimate the power generated and power load at each bus, which is used to calculate the power flow through each line (for instance, optimal power flow can be used). Based on each line's calculated power, the congestion pattern is achieved (if the estimated power in each line exceeds the maximum power limit, the line is congested). In the real-time market, real-time LMP is determined based on the calculated power. Therefore, it can be seen that the state estimation is involved in congestion pattern calculations and loads and generation estimation. Thus, the FDI cyber-attacks that change the estimated state have an impact on the real-time market.

Physical/Technical Impacts

In addition to economic impacts, the FDI cyber-security violations can have physical/technical impacts on microgrids. In general, the FDI violations can impact on the transient and steady-state stability of the microgrids. In terms of steady-state stability, the FDI violations can impact voltage control of microgrids (AC or DC voltage control in AC/DC microgrids), EMSs, and demand power/current management.

In addition to the adverse effects of cyber-security violations on microgrid steady-state operation, the microgrid transient and dynamic stability can be impacted by FDI attacks. For instance, the FDI can impact on frequency control of the microgrids. Furthermore, rotor angle stability can be affected by FDI attacks in microgrids. Moreover, the violations can impact on protection system of smart microgrids. More detail of the physical/technical impacts of cyber-attacks on microgrids accompanied by construction strategies of cyber violations is discussed in the following section.

6.4.3 Construction of Cyber-security Violations in Smart Microgrids

To construct a cyber-attack, hackers usually have partial cyber–physical system information. In the case that hackers have full network information, the violation will be more effective and destructive. The hacker's knowledge of the system and the access degree determine the level of destructive impact and the possibility of detection/mitigation by the defender.

As mentioned above, the FDI attacks can target steady-state and transient operations of smart microgrids. Among several violations, the FDI attacks targeting state estimation, voltage and frequency regulations, and system protection are explained in the following due to their importance in smart microgrids.

Cyber-security Violations on State Estimation

The state estimation is used to determine the system operation status, including bus voltage magnitude and phase angles from available measurements. The primary purpose of these attacks is to introduce errors in estimating state variables in microgrids by manipulating sensor measurement data. The state estimation helps monitor and control microgrids effectively and efficiently, and it is one of the most critical tasks in microgrid operation and energy management strategies. The estimated states can also be used for contingency analysis, stability analysis, load forecasting, optimal power dispatch, bad data detection, and power market locational marginal pricing. Any FDI attacks inducing errors into estimated states can have disruptive effects on microgrid operation and performance.

There are two types of state estimation in power systems: DC state estimation and AC state estimation. Due to simple analytical models, power systems with DC state estimation have been studied more than AC state estimation in the literature. However, FDI attack construction targeting AC state estimation is gradually gaining attention.

Although research on the construction of FDI violations mostly focuses on violations targeting state estimation, FDI violations construction targeting voltage, frequency, and protection systems has also been studied.

Cyber-security Violations on Voltage Control

The smart microgrid voltage is usually controlled by IFCs from DGs and ESSs and rotational-based generators (such as diesel generators). In such systems, the voltage level and/or reactive power are measured, and the control system produces reactive reference powers for the power generation. As another option, the transformer tap changer is also controlled for microgrid voltage regulation. The FDI violations that modify sensors measured voltage and/or reactive powers data and control parameters within the control layers can impact the voltage regulation of the microgrid. Moreover, the hackers may access the microgrid multi-layer control system and modify control signals among layers (e.g. induce errors into DGs reference power signals and transformer tap changer signal).

Cyber-security Violations on Frequency Control

The violations targeting microgrids frequency are referred to as violations of transient stability. Like attacks on microgrid voltage stability, hackers can introduce errors into control signals among control layers, modify control parameters and sensor measurements, or change outputs of power sources to affect microgrid frequency stability. It should be mentioned that the microgrid frequency control is susceptible to active powers and frequency measurements, and reference signals. In microgrids, the frequency can be regulated by rotating machines. Any violations targeting rotor speed or angle measurements can affect microgrid frequency stability. ESSs are usually used for transient stability improvement in microgrids. In such systems, sensor measurements are used in the control system to actuate the storage systems to absorb and/or inject active power from the microgrid. Since energy storage systems are evolving in microgrids frequency control, the security of measurement and control signals should be guaranteed to provide stable operating conditions.

Cyber-security Violations on the Protection System

One of the main challenges of microgrids is protection system design, which should operate under grid-connected and islanding operation mode. Furthermore, the relay setting should be adjusted to the proper current level depending on the operation mode. One of the conventional approaches is adaptive protection techniques based on the IEC 61850 communication standard. In such protection systems, a secure, reliable, and fast communication network is necessary. However, communication link failures or any FDI cyber-incident may affect the protection system performance.

6.4.4 Defensive Strategies Against Cyber-attacks

The defense strategies against cyber-attacks can be classified into strategies based on protection and detection/mitigation. In the following, these two groups are discussed in detail.

Defensive Strategies Based on Protection

In the defensive strategies based on protection, meters/sensors are protected against cyber-attacks. However, since many smart sensors and meters exist in emerging smart microgrids, protecting all meters is not cost-effective. Thus, only a set of critical sensors and corresponding measurements are usually protected.

It should be mentioned that the number of meters/sensors under attack is a fundamental criterion in FDI cyber-incident detection. In some cases, the number of sensors is increased to enhance microgrid visibility; however, it increases the microgrid's vulnerability to cyber-attack. In defensive strategies based on protection, the number of protected sensors (and their locations) can be achieved considering the budget and the system's sensitivity. For example, in [13], an optimization problem is formulated to minimize the defender budget and determine the meters' number and position for protection against attacks.

Defensive Strategies Based on Detection/Mitigation

In the detection-based defense strategies, the measured data are analyzed to detect cyber-attacks and mitigate/reduce their adverse effects on microgrid operation. In general, detection strategies can be categorized into static and dynamic.

Static Detectors of Cyber-attacks The defense strategies that detect violations on steady-state stability are called static detectors. One of the well-known static detectors is detectors of attacks on state estimation. To date, several strategies have been developed to detect/mitigate FDI violations targeting state estimation, such as statistical methods, the Kalman filter, sparse optimization, state forecasting, network theory, time-series simulation, machine learning, generalized likelihood ratio, chi-square detector, and similarity matching. However, these strategies are used to recover DC state information and are suitable for FDI attacks on DC state estimation.

In AC state estimation models that are usually used in most real-world power systems, the performance of such strategies are not satisfactory. Different methods can be used to detect FDI attacks on AC state estimation, such as the Kullback–Leibler distance, information-network-based state estimation techniques, transmission line parameter variation techniques, a Bayesian detection scheme, and a discrete wavelet transform algorithm together with deep neural networks techniques.

The defense strategies against attacks targeting voltage regulations in microgrids can also be categorized as static detectors. In such cases, the detection algorithm can be embedded into the converters control system as supplementary control loops or monitors changes in sets of inferred candidate invariants (invariants are defined in terms of bounds over the output voltage and current of individual power converters) to name a few.

Dynamic Detectors of Cyber-attacks Information on system dynamics is used in dynamic detection methods to detect cyber-attacks. Various dynamic detectors mainly focus on linear systems, which cannot effectively detect attacks in real-world power systems due to nonlinearity. The detection of FDI violations targeting frequency control of microgrids is classified as the dynamic detection method. Also, FDI violations on the transient stability of the power system, such as errors on rotor speed and angle, are detected by the dynamic detection methods. Various methods such as image-processing-based techniques, adaptive control-based techniques, sliding mode observer theory, and AI-based techniques can be used in dynamic detectors.

In Table 6.1, all the discussions mentioned above are reviewed. The following implementation example of FDI violations construction and detection/mitigation in smart microgrids are provided.

6.4.5 Case Study Example: Cyber-security Violations in Power Electronics-intensive DC Microgrids

FDI cyber-security violation construction and detection in DC microgrids are presented here [12]. The studied DC microgrid is shown in Figure 6.10, in which N-number of DC power generators are connected to the DC microgrid through DC/DC converters. The power converters are controlled to adjust their output voltages to the reference values provided by their hierarchical control structure's primary and secondary control layers.

In DC microgrids, the secondary control layer uses local and neighboring measurements to tune the average voltage globally and share the currents proportionately to reduce the circulating currents. Typically, sublayers of the secondary control layer are cooperated to achieve those objectives in which the first sublayer is responsible for average voltage restoration while the current sharing is done in the second sublayer.

To regulate average voltage globally in the first sublayer, a voltage observer is used to estimate the average voltage $\bar{V}_{\mathrm{DC}_i}(k)$ for the ith converter. This value is updated by a dynamic consensus algorithm, which uses neighboring estimates $\bar{V}_{\mathrm{DC}_j}(k)\,\forall j \in N_i$ (N_i represents neighbor converters). The estimated average voltage for the ith converter is provided:

$$\begin{aligned}
&\bar{V}_{\mathrm{DC}_i}(k+1) - \bar{V}_{\mathrm{DC}_i}(k) \\
&\quad = V_{\mathrm{DC}_i}\left(k+1-\tau^i_{\mathrm{output}}\right) - V_{\mathrm{DC}_i}\left(k-\tau^i_{\mathrm{output}}\right) \\
&\qquad + \sum_{j\in N_i} a_{ij}\left(\bar{V}_{\mathrm{DC}_j}\left(k-\tau^i_{\mathrm{input}}-\tau^{ij}_{\mathrm{comm}}\right) - \bar{V}_{\mathrm{DC}_i}\left(k-\tau^i_{\mathrm{input}}\right)\right).
\end{aligned} \tag{6.1}$$

Table 6.1 Review of cyber-security violations in smart microgrids.

Cyber-security violations in smart microgrids	Impacts of cyber-security violations	Economic impacts	• Especially in grid-connected mode, optimal economic operation can be affected. • FDI violations can affect load forecasting and change the estimated state, which affects the day-ahead market and real-time market.
		Physical/ technical impacts	• FDI violations can impact on transient and steady-state stability of the microgrids. • FDI violations can impact voltage control of microgrids, energy management systems, demand power/current management, etc. • FDI violations can impact frequency control, rotor angle stability, protection system, etc.
	Constructions of cyber-security violations (main attack targets)	Violations on state estimation	• Main purpose of violations is to introduce errors into the state estimation by manipulating sensor measurement. • FDI violations inducing errors in estimated states have disruptive effects on microgrid operation. • FDI violations targeting DC state estimation has been addressed more than AC (due to simplicity).
		Violations on voltage control	• FDI violations can modify sensor measured voltage and/or reactive power data and control parameters within the control layers and impact the voltage regulation of a microgrid. • FDI violations may modify control signals among the microgrid multi-layers (for example, induce errors into DGs reference reactive power signals and transformer tap changer signal).
		Violations on frequency control	• Microgrid frequency control is very sensitive to active powers and frequency measurements. • FDI violations targeting microgrid frequency are referred to as attacks on transient stability. • Hackers can induce errors in control signals among control layers, modify control parameters and sensor measurements, or change outputs of power sources.
		Violations on protection system	• One of the main challenges of microgrids is protection system design. • Depend on the operation mode, relays setting should be adjusted to the proper current level. • FDI violations may affect the protection system performance and may lead to disaster events.
	Defensive strategies against cyber-attacks	Strategies based on protection	• Meters/sensors are protected against cyber-attacks. • Number of protected sensors and their locations are determined based on budget, the sensitivity of the system, etc.
		Strategies based on detection/ mitigation	• Measured data are analyzed to detect attacks and mitigate their adverse effects on the microgrid. • Can be classified into static detectors (detect attacks targeting steady-state stability) and dynamic detectors (information of system dynamics is used for detection).

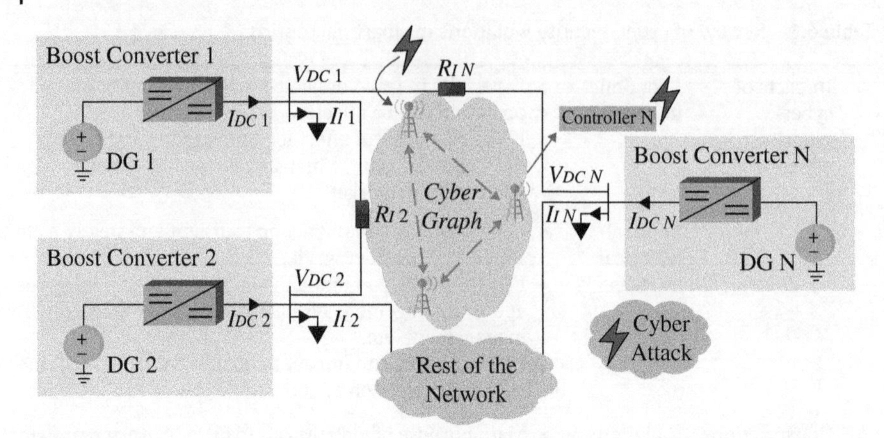

Figure 6.10 DC microgrid with the cyber–physical model.

In (6.1), τ^i_{input}, τ^i_{output}, and $V_{\text{DC}_i}(k)$ represent the input and output delays, and the measured voltage in the ith converter, and τ^{ij}_{comm} denotes the communication delay between the ith and jth converters. Further, a_{ij} is the elements of the adjacency matrix of the communication graph.

In the second sublayer, which is used to share current among converters proportionally, the ith converter normalized current regulation cooperative input is achieved by:

$$\bar{I}_{\text{DC}_i}(k) = \sum_{j \in N_i} w_i a_{ij} \left(I_{\text{DC}_j}\left(k - \tau^i_{\text{output}} - \tau^{ij}_{\text{comm}}\right) / I^{\max}_{\text{DC}_j} - I_{\text{DC}_i}\left(k - \tau^i_{\text{output}}\right) / I^{\max}_{\text{DC}_i} \right) \tag{6.2}$$

where $I_{\text{DC}_j}(k)\, \forall j \in N_i$ is the measurements of neighboring output current, and w_i, I_{DC_i}, I_{DC_j}, $I^{\max}_{\text{DC}_i}$, and $I^{\max}_{\text{DC}_j}$ denote the desired coupling gain, measured output current in the ith and jth converters, and maximum output current allowed for the ith and jth converters, respectively.

To implement the above objectives into the ith converter to regulate the output voltage, two voltage correction terms are considered as follows:

$$\Delta V^1_i(k) = K_{P1} \underbrace{\left(V^*_{\text{DC}} - \bar{V}_{\text{DC}_i}(k)\right)}_{e^i_1(k)} + K_{I1} \sum_{p=0}^{k} \left(V^*_{\text{DC}} - \bar{V}_{\text{DC}_i}(p)\right) \tag{6.3}$$

$$\Delta V^2_i(k) = K_{P2} \underbrace{\left(I^*_{\text{DC}} - \bar{I}_{\text{DC}_i}\left(k - \tau^i_{\text{input}}\right)\right)}_{e^i_2(k)} + K_{I2} \sum_{p=\tau^i_{\text{input}}}^{k} \left(I^*_{\text{DC}} - \bar{I}_{\text{DC}_i}\left(p - \tau^i_{\text{input}}\right)\right) \tag{6.4}$$

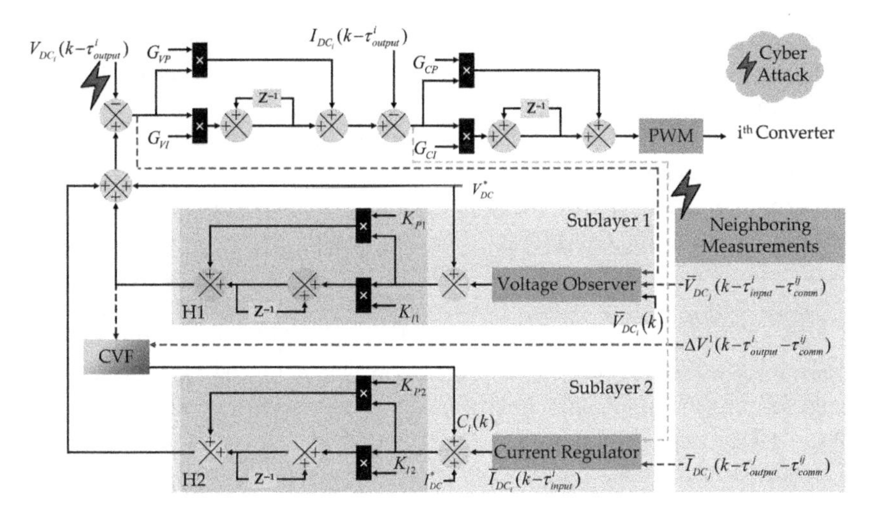

Figure 6.11 The *i*th converter controller for sensors and communication link attack detection in DC microgrids.

where K_{P1}, K_{I1}, K_{P2}, and K_{I2} are the first and second sublayer PI controller gains (see Figure 6.11). Moreover, global reference current and voltage values are represented by I^*_{DC} and V^*_{DC}, respectively.

Finally, the correction terms in (6.3) and (6.4) are added to the global reference voltage to obtain the reference value for the local voltage of *i*th converter.

$$V^*_{\mathrm{DC}_i}(k) = V^*_{\mathrm{DC}} + \Delta V^1_i(k) + \Delta V^2_i(k). \tag{6.5}$$

In such a DC microgrid, using the cooperative-based consensus algorithm, (6.1) and (6.2) will converge to

$$\lim_{k\to\infty} \bar{V}_{\mathrm{DC}_i}(k) = V^*_{\mathrm{DC}}; \quad \lim_{k\to\infty} \bar{I}_{\mathrm{DC}_i}(k) = 0\,\forall\, i \in N. \tag{6.6}$$

For cyber-security attacks in a single sensor/communication link, (6.6) will be modified as follows:

$$\lim_{k\to\infty} \bar{V}_{\mathrm{DC}_i}(k) = V^{*\prime}_{\mathrm{DC}}; \quad \lim_{k\to\infty} \bar{I}_{\mathrm{DC}_i}(k) \neq 0\,\forall\, i \in N. \tag{6.7}$$

This criterion can be used to detect cyber-attacks, including DoS and jamming. However, the stealth attacks can penetrate the system without operators' knowledge and multiple sensors/communication links. The stealth attack can bc crafted so that (6.6) is satisfied. It is proven in [12] that if a constant value P exists such that

$$\sum_{k=0}^{\infty} \left|u^a_{V_i}(k)\right| \leq P, \sum_{k=0}^{\infty} \left|u^a_{I_i}(k)\right| \leq P\,\forall\, i \in N \tag{6.8}$$

then, the state convergence (6.6) will not be affected in the presence of stealth violations. In (6.8), $u^a_{I_i}(k)$ and $u^a_{V_i}(k)$ represent the ith converter current and voltage attack vectors at the kth instant.

In the DC microgrid controlled by cooperative systems, it is challenging to detect the attacked node since the entire system is affected by the intrusion in any node. Considering Figure 6.10, each converter output current depends on voltage levels between two different points. Thus, any stealth attacks on current value (e.g. attacks on current sensors) will result in voltage variations across the DC microgrids, which leads to errors in current sharing among converters. Typically, the current sharing error could be a sufficient criterion to detect the attacks on current sensors. However, attack detection would not be easy if multiple voltage sensors/communication links were attacked stealthily. In more detail, the voltages will be manipulated, so that (6.6) still holds even under attacks.

The voltage regulation control input can be used to provide a strong stealth attack. This control input signal for the ith converter is presented as in (6.9).

$$u_i(k) = \underbrace{\sum_{j \in N_i} a_{ij}(\bar{V}_{\text{DC}_j}(k) - \bar{V}_{\text{DC}_i}(k))}_{u_{ij}(k)} + b_i e^i_1(k). \tag{6.9}$$

If a cyber-link or sensor is attacked in the ith controller, the model of the attacked control input would be as in (6.10) and (6.11), respectively.

$$u^f_{ij}(k) = u_{ij}\left(k - \tau^i_{\text{input}} - \tau^{ij}_{\text{comm}}\right) + k u^a_i(k) \tag{6.10}$$

$$u^f_i(k) = u_i\left(k - \tau^i_{\text{input}}\right) + k u^a_i(k) \tag{6.11}$$

where k shows violation presence (when $k = 1$, there is an attack in the system) and $u^a_i(k)$ represents the ith converter attack vector. From (6.10) and (6.11), local investigation of $u^f_i(k)$ can be done in each converter to detect a non-zero synchronization error with the residual output. However, since each residue comparison needs global information, this is not an appropriate criterion to detect node(s) of attacks. The controller's attempt to adjust the output to a given reference voltage is considered for violation indication to verify this case.

Using the change in PI output in sublayer 1, a cooperative vulnerability factor (CVF) is defined as in (6.12) for each converter to determine the attacked nodes accurately [12]

$$\begin{aligned} C_i(k) = c_i &\left[\sum_{j \in N_i} a_{ij}\left(\Delta V^1_j\left(k - \tau^{ij}_{\text{comm}}\right) - \Delta V^1_i(k)\right)\right] \\ &+ \left[\sum_{j \in N_i} a_{ij}\left(\Delta V^1_j\left(k - \tau^{ij}_{\text{comm}}\right) - \Delta V^1_i(k)\right)\right] \end{aligned} \tag{6.12}$$

where c_i is a positive constant value. If the calculated $C_i(k)$ for each node is a positive value, that node is the attacked node. While the non-violated nodes have the $C_i(k)$ value of zero. The above CVF is a proper criterion to detect the attacked node, especially when multiple sensor/communication links are stealthily attacked. The value of $C_i(k)$ is cross-coupled with the current sublayer to protect against attack on $C_i(k)$. In Figure 6.11, the ith converter controller to detect stealth attacks on communication links and sensors in DC microgrids is shown.

6.4.6 Future Trends of Microgrid Cyber-security

The smart hybrid AC/DC microgrids require a reliable and secure cyber system and communication network for optimal, uninterruptible, and smooth operation, and any cyber-security violations may lead to unforeseen incidents in microgrid operation. It should be emphasized that microgrids are more prone to stability issues if a cyber-attack happens due to their low inertia. Due to the tight coupling of AC and DC subgrids in hybrid AC/DC microgrids, any cyber incident in one subgrids may have destructive effects on the other side.

Here, some future trends of microgrids cyber-security are discussed.

State Estimation of AC/DC Microgrids under Cyber-security Violations

In a power system, extensive research works on the detection/mitigation of cyber-attacks on DC and AC state estimations have been done. However, in hybrid AC/DC microgrids, state estimation under cyber-security violations has not been addressed sufficiently. Thus, the hybrid AC/DC microgrids should be modeled first for estimating the state information. Then, appropriate strategies should be developed to detect the attacks and recover the state information.

Frequency Control of AC/DC Microgrids under Cyber-security Violations

In hybrid AC/DC microgrids, frequency stability is one of the main concerns due to the low inertia of power electronics-based distributed generations and energy storage. The presence of cyber-security violations will make the situation worse. It should be mentioned that any cyber violations targeting the frequency stability of the AC subgrids may jeopardize the DC voltage stability on the DC side. Therefore, a proper control strategy design to detect and mitigate cyber-attack on the frequency control of hybrid AC/DC microgrids could be an important research direction.

Voltage Regulation of AC/DC Microgrids under Cyber-security Violations

In hybrid AC/DC microgrids, any voltage variations in the AC or DC side will transfer to the other side through interfacing power electronics converters. Therefore, regulation of voltage in such hybrid microgrid is challenging, especially under cyber-attacks, and it is needed to be considered carefully.

Electric Vehicles and Cyber-security

EVs and their charging stations are increasing rapidly in recent power systems, in which they can be considered smart microgrids (i.e. EV charging stations can be considered grid-connected microgrids). Such microgrids are prone to cyber-security violations, and recently several research groups have been working on the cyber-security of EVs and EV charging stations. The cyber-security of EVs and their charging station technologies are in the early stages of development and require more study in the future.

Blockchain and Cyber-security in Modern Grids

The main purpose of blockchain technology is to achieve direct peer-to-peer electronic payments where the trusted third party does not participate. In practice, blockchain technology focuses on the financial domain. Recently, applications of blockchain technology in the power engineering sector have also been addressed, for example, in IoT and smart homes. A few types of research have been done to secure smart grid operation under cyber incidents by blockchain, and more investigation is needed in the future.

6.5 Summary

In this chapter, energy management systems in hybrid microgrids have been addressed. Different control layers of hierarchical controls have been studied, and microgrid EMS implementation in the layers has been addressed. Also, the application of AI techniques in EMS of microgrids has been discussed. In this chapter, microgrid operation with a multi-agent control strategy, with detail on each agent, has been provided. Since the future distribution system will consist of interconnected microgrids and a current distribution system management may be used with modifications, the ADMS and its different parts, currently used for distribution system control, have been discussed. Finally, cyber-security in smart hybrid microgrids, the effect of cyber-security violations on microgrids control, and defense strategies have been discussed.

References

1 Kanchev, H., Lu, D., Colas, F. et al. (2011). Energy management and perational planning of a microgrid with a PV-based active generator for smart grid applications. *IEEE Transactions on Industrial Electronics* 58 (10): 4583–4592.

2 Zhou, B. et al. Multi-microgrid energy management systems: architecture, communication, and scheduling strategies. *Journal of Modern Power Systems and Clean Energy*.

3 Zhao, S., Blaabjerg, F., and Wang, H. (2021). An overview of artificial intelligence applications for power electronics. *IEEE Transactions on Power Electronics* 36 (4): 4633–4658.
4 Sun, Q. and Yang, L. (2019). From independence to interconnection – a review of AI technology applied in energy systems. *CSEE Journal of Power and Energy Systems* 5 (1): 21–34.
5 Dimeas, A.L. and Hatziargyriou, N.D. (2005). Operation of a multi-agent system for microgrid control. *IEEE Transactions on Power Systems* 20 (3): 1447–1455.
6 Cintuglu, M.H., Youssef, T., and Mohammed, O.A. (2018). Development and application of a real-time testbed for multiagent system interoperability: a case study on hierarchical microgrid control. *IEEE Transactions on Smart Grid* 9 (3): 1759–1768.
7 Pacific Northwest National Laboratory (2016). ADMS State of the Industry and Gap Analysis.
8 VOICES of Experience: Insights into Advanced Distribution Management Systems, United States Department of Energy Office of Electricity Delivery and Energy Reliability. (2015). DOE/GO-OT-6A42-63689, Washington, D.C.
9 Vadari, M. (2013). Smart Grid, System Operations and the Management of Big Data to drive Utility Transformation. CIO Review Magazine - Technology for the Utility sector.
10 Nejabatkhah, F., Li, Y.W., Liang, H., and Ahrabi, R.R. (2021). Cyber-security of smart microgrids: a survey. *Energies* 14: 27. https://doi.org/10.3390/en14010027.
11 Li, Z., Shahidehpour, M., and Aminifar, F. (2017). Cybersecurity in distributed power systems. *Proceedings of the IEEE* 105: 1367–1388.
12 Sahoo, S., Mishra, S., Peng, J.C., and Dragicevic, T. (2019). A stealth cyber attack detection strategy for DC microgrids. *IEEE Transactions on Power Electronics* 34: 8162–8174.
13 Deng, R., Xiao, G., and Lu, R. (2017). Defending against false data injection attacks on power system state estimation. *IEEE Transactions on Industrial Informatics* 13: 198–207.

Part III

Power Quality Issues and Control in Smart Hybrid Microgrids

Part III

Power Quality Issues and Control in Smart Hybrid Microgrids

7

Overview of Power Quality in Microgrids

7.1 Introduction

A power quality problem refers to any power problems manifest in voltage, current, or frequency deviations that result in failure or misoperation of customer equipment [1]. In other words, power quality measures the quantities of voltage, current, frequency, and waveform quality. Since all grid-connected devices are designed to operate at a specific voltage, current, and frequency conditions, the deviation in these qualities may cause customer or utility equipment to malfunction or get damaged. For example, poor power quality can cause the undesirable operation of equipment, increase power losses, interfere with communication lines, etc.

The power quality problems include transients, voltage sags, swells, harmonics, short- and long-term voltage variations, and momentary power supply outages. There are various reasons for the pollution of AC supply systems, including natural ones such as lightning, flashover, equipment failure, and faults, and forced ones such as voltage distortions. Many power quality problems are related to the voltage at the point of common coupling (PCC) where the voltage problems include voltage harmonics, surge, spikes, notches, sag, swell, imbalance, fluctuations, glitches, flickers, outages. These voltage problems are related to various disturbances (such as fault) or various nonlinear loads such as furnaces and power electronics interfaced loads. Some power quality problems are related to the current drawn from the AC mains due to poor power factor, excessive reactive power, harmonic currents, unbalanced currents, and an excessive neutral current in polyphase systems due to unbalancing and harmonic currents generated by some nonlinear load.

Equipment connected to the grid can impact the system's power quality and have drastically different sensitivities to power quality problems. In a distribution

Smart Hybrid AC/DC Microgrids: Power Management, Energy Management, and Power Quality Control, First Edition. Yunwei Ryan Li, Farzam Nejabatkhah, and Hao Tian.

system and microgrids, some typical equipment that affects the power quality includes:

- adjustable-speed drives, which may produce harmonic distortion and sensitivity to transient voltages
- motors with a high starting current that can cause voltage sags
- switch-mode power supplies that generate harmonic current
- fluorescent lighting (especially with electronic ballasts) and light emitting diode (LED) lighting that produces harmonics
- power factor correction (PFC) capacitor bank that can cause harmonic resonance and switching transients
- electronic and control equipment that is sensitive to high-frequency transients.

For sensitive equipment and processes, the ride-through capability of a power quality disturbance is often one of the most important characteristics to determine. Another approach to address power quality disturbances is to monitor and improve power systems' power quality, mitigating power quality disturbances. In these cases, all the grid-connected apparatus can benefit from an improvement in power quality. In this chapter, power quality disturbances and the existing solutions will be discussed.

7.2 Classification of Power Quality Disturbances

According to various IEEE and IEC standards, the power quality disturbances can be classified into different categories according to their wave shapes. An overview of different power quality disturbances is provided in this section.

7.2.1 Transients

Transients typically last for a very short period, and are observed as high-frequency impulses or oscillations superimposed on top of the fundamental frequency voltage or current waveforms (Figure 7.1). Impulsive transients are sudden high peak events. These events can be further categorized by the very fast speed at which they occur.

An impulsive transient is a sudden, very short duration change that raises the voltage and/or current level in unidirectional polarity. Due to the high frequency involved, the shape of impulsive transients can change quickly when propagating through a network and may have significantly different characteristics when viewed from different parts of the power system. The causes of impulsive transients include lightning, poor grounding, the switching of inductive loads, and utility fault clearing. The results can range from the loss of data to

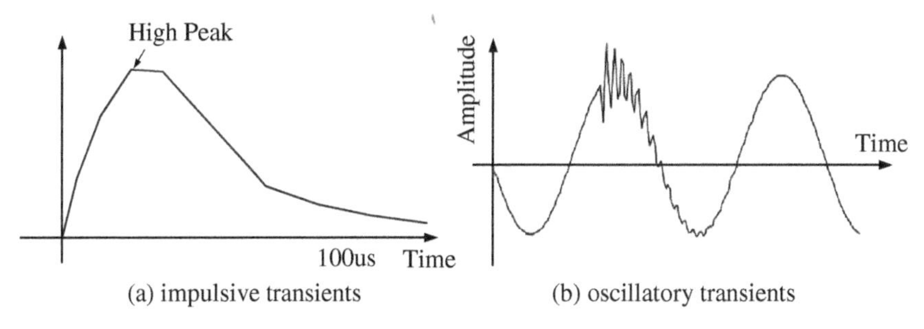

Figure 7.1 Typical waveforms of transients.

physical damage to equipment. Among these causes, lightning is probably the most damaging.

An oscillatory transient is a sudden, very short duration change in the steady-state waveforms of voltage, current, or both, including both positive and negative polarity values. An oscillatory transient usually decays to zero within a fundamental cycle. It is described by its spectral content, duration, and magnitude. Potential sources of oscillatory transients are listed as follows:

- Transients with a primary frequency component greater than 500 kHz; the local system response to an impulsive transient typically manifests as oscillatory transients at this frequency range, and the typical duration is in microseconds.
- Transients with primary frequency components between 5 and 500 kHz; these can be caused by back-to-back capacitor energization and cable switching.
- Transients with a primary frequency component between 300 Hz and 5 kHz; they are caused by many types of events, such as capacitor bank energization. Their typical duration is from 0.3 to 50 ms, which is frequently encountered in utility sub-transmission and distribution systems.
- Transient with a primary frequency component of less than 300 Hz; this type of transient can also be found in the distribution system. They are generally associated with Ferroresonance and transformer energization.

7.2.2 Short Duration Variations

Short duration voltage variations are typically caused by fault conditions, energizing large loads which require a high starting current, or intermittent loose connections in power wiring. Depending on the location and the system conditions, a fault can cause temporary voltage sags, swells, or interruptions, as shown in Figure 7.2.

1. Sag. A sag is a decrease between 0.1 and 0.9 p.u. in root-mean-square (RMS) voltage at the power frequency for a duration from 0.5 cycles to one minute [2].

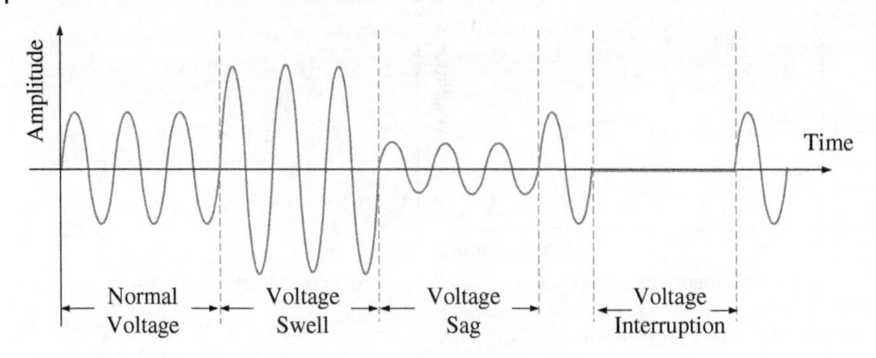

Figure 7.2 Typical waveforms of short duration variations.

Voltage sags are usually associated with system faults but can also be caused by the energization of heavy loads or the starting of large motors. A fault-caused sag depends on the fault clearing times of the circuit breakers and protection settings. Typical fault clearing times range from 3 to 30 cycles. Motor starting can also cause a noticeable voltage sag. An induction motor will draw 6–10 times its full load current during starting. If the current magnitude is large relative to the available fault current in the system, the resulting voltage sag can be significant, and the duration of such sags is a function of motor acceleration. Typical values are in the range of a few seconds.

2. Swell. A swell is defined as an increase to between 1.1 and 1.8 p.u. in RMS voltage at the power frequency for a duration from 0.5 cycles to one minute [2]. Similar to voltage sags, swells are usually associated with system fault conditions, but they are not as common as voltage sags. A common cause of swells is the temporary voltage rise on the un-faulted phases during a single line-to-ground fault. Swells can also be caused by switching off a large load or by energizing a large capacitor bank. The severity of a voltage swell during a fault condition is a function of the fault location, system impedance, and grounding.
3. Interruption. An interruption occurs when the supply voltage or load current decreases to less than 0.1 p.u. for a period not exceeding one minute. Interruption can be the result of power system faults, equipment failure, or control malfunctions. The interruptions are measured by their duration. The duration of an interruption due to a fault in the utility system is determined by the operation time of the utility protective device.

Voltage sags and interruption are probably the most significant power quality concerns. Sags and interruptions can propagate to a relatively larger area in the power system. The extent of sag propagation can be determined by using well-known short-circuit analysis methods.

7.2.3 Long Duration Variations

Long duration voltage variations, also termed overvoltage and undervoltage, encompass RMS value deviations at power frequencies for longer than one minute. Voltage abnormalities greater than one minute usually are not associated with system faults and can typically be controlled by voltage regulation equipment.

1. An over-voltage is an increase in the RMS voltage greater than 110% at the power frequency. Over-voltages are usually the result of the switching-off of a large load, energizing of a capacitor bank, or connection of a distributed generation (DG). Over-voltages occur because either the system is too weak for the desired voltage regulation or the voltage controls are inadequate, such as incorrect tap settings in transformers.
2. An undervoltage is a decrease in the RMS voltage to less than 90% at the power frequency. Under-voltages are usually caused by a load switching on or a capacitor bank switching off. Overloaded circuits and loss of major transmission support can also result in under-voltages.
3. Sustained interruption is a long-duration voltage variation (excess of one minute) when the supply voltage has been zero. Voltage interruptions longer than one minute are often permanent and require human intervention for system restoration.

7.2.4 Voltage Fluctuations

Voltage fluctuations are cycling variations in the voltage RMS values or a series of random voltage changes whose magnitude does not normally exceed the range 0.9–1.1 p.u. A common phenomenon of voltage fluctuations in the voltage is flicker.

7.2.5 Voltage Imbalance

Voltage imbalance or unbalance can be defined as the maximum deviation from the average of the three-phase voltage, divided by the average of the three-phase voltages and expressed in percentage points. Imbalance can also be defined using symmetrical components. The ratio of either the negative or zero sequence component to the positive sequence component can specify the percent imbalance.

The primary source of voltage imbalance (typically less than 2%) is the unequal distribution of single-phase loads in a three-phase circuit. Voltage imbalance can also be the result of blown fuses in one phase of a three-phase capacitor bank. Severe voltage imbalance (greater than 5%) can result from single phase failure.

7.2.6 Power Frequency Variations

Power frequency variations are the deviation in the power system fundamental frequency from its normal value. Slight variations in the frequency occur as the dynamic balance between load and generation changes. The size of the frequency shift and its duration depends on the load characteristics and the generation control system's response to loading changes. In the interconnected power system, significant frequency variations are rare. However, frequency variations are more common for microgrids, where DG supplied is isolated from the utility system. In such cases, the generator governor response to abrupt load change may not be adequate to regulate the frequency within the narrow bandwidth required by frequency-sensitive equipment.

7.2.7 Waveform Distortion

Waveform distortion is defined as a steady-state deviation from an ideal sinusoidal wave. It is characterized by the spectral content of the deviation. Primary types of waveform distortion are harmonics, notching, inter-harmonics, DC offset, and noise.

1. Harmonics are sinusoidal voltages or currents with frequencies that are integer multiples of the fundamental frequency. Distorted waveforms can be decomposed into the sum of the fundamental frequency and the harmonic components. Harmonic distortion levels are described by the complete harmonic spectrum with magnitudes and phase angles for each individual harmonic component. It is also common to use a single quantity, the total harmonic distortion (THD), to measure the effective value of harmonic distortion. Harmonic distortion originates in the nonlinear characteristics of devices and loads in the power system. Typical harmonic sources are adjustable speed drives and other power electronic-based equipment. One of the major problems related to harmonic disturbances is harmonic resonance. The resonance can magnify harmonic distortions to a level that can damage the equipment or cause equipment malfunction. PFC capacitors in the distribution system are the leading cause of harmonic resonance. Other effects of harmonics are equipment overloading, increased losses, and equipment malfunction.
2. Notching is a periodic voltage disturbance caused by electronic devices, such as light dimmers and arc welders, under normal operation. Notching can be characterized through the harmonic spectrum of the affected voltage. Although

notching is a particular case of voltage harmonics, it is generally treated as an independent disturbance.

3. Inter-harmonics are voltages or currents having frequency components that are not integer multiples of the fundamental frequency. They can appear as discrete frequencies or as a wide-band spectrum.
4. DC offset refers to the presence of a DC voltage or current in an AC power system. This phenomenon can occur as the result of a geomagnetic disturbance or the effect of half-wave rectification. DC current in AC networks can cause bias on transformer core fluxes, which results in overheating and transformer saturation. DC offset may also cause the electrolytic erosion of grounding electrodes and other connectors.
5. Noise is defined as unwanted electrical signals with broadband spectral content lower than 200 kHz, superimposed upon the power system voltage or current waveform. Noise in power systems can be caused by power electronic devices, control circuits, and arcing equipment. Noise problems are often exacerbated by improper grounding that fails to conduct noise away from the power system. Basically, noise consists of any unwanted distortion of the power signal that cannot be classified as harmonic distortion or a transient. Noise can disturb electronic devices such as microcomputers and programmable controllers.

Various power quality problems with characteristic description, generation, and effects are introduced above. For a direct review, the common power quality problems and their effects are summarized in Table 7.1.

7.3 Overview of Power Quality Standards

Given these power quality problems, many organizations such as IEC, IEEE, American National Standards Institute (ANSI), British Standards (BS), European Norms (EN), and Information Technology Industry Council (ITIC) have developed different standards. These standards specify the permissible limits of various performance indices to maintain the level of power quality to an acceptable benchmark and provide guidelines to the customers, manufacturers, and utilities to curb the various events causing the power quality problems. Table 7.2 provides a brief description of these power quality problems along with applicable industry standards. Tables 7.3–7.5 shows some important limits on voltages and currents in these standards.

The limitations on harmonics defined in IEEE 519 [3] are widely accepted as industry recommendations, which are shown in Tables 7.4 and 7.5.

Table 7.1 Power quality problems: causes and effects.

Problems	Category	Categorization	Typical effects
Transients	Impulsive	Peak, rise time, and duration	Power system resonance
	Oscillatory	Peak magnitude and frequency components	System resonance
Short duration voltage variation	Sag	Magnitude, duration	Protection malfunction, production loss
	Swell	Magnitude, duration	Protection malfunction, stress on loads
	Interruption	Duration	Production loss, malfunction of loads
Long duration voltage variation	Sustained interruption	Duration	Production loss
	Overvoltage	Magnitude, duration	Damage to loads
	Undervoltage	Magnitude, duration	Increased losses, heating generation
Voltage fluctuation	–	Intermittent	Protection malfunction, light intensity changes
Voltage unbalance	–	Symmetrical components	Shortened lifespan, increased losses, the degraded performance of three-phase loads
Power frequency variation	–	Frequency	Reduced motor efficiency, light intensity changes, damage to generator and turbine shaft
Voltage distortion	Harmonic	THD, harmonic spectrum	Increased losses, poor power factor
	Notching	THD, harmonic spectrum	Damage to capacitive components
	Inter-harmonics	THD, harmonic spectrum	Acoustic noise in power equipment
	DC offset	Volts, amperes	Transformer saturation
	Noise	THD, harmonic spectrum	Capacitor overloading, disturbances to loads
Voltage flicker		Occurrence frequency, modulating frequency	Nuisance tripping, malfunction of loads

Source: Adapted from IEEE.

Table 7.2 Standards for common power quality problems.

Power quality problems	Standards
Short duration voltage variation	IEEE 1159-2009 IEC 61000-2-8
Steady-state long duration voltage variation	TSC Appendix 2 IESO market rules Appendix 4.1 CSA Standard CAN3-C235-83
Over-voltage	IEEE 1159-2009 IEC 61000-2-8
Under-voltage	IEEE 1159-2009 IEC 61000-2-8
Voltage unbalance	NOP-32 CAN/CSA-C61000-12-2:04 CAN/CSA-C61000-2-2:04
Voltage harmonic	IEEE 519 2014 CAN/CSA-C61000-2-2:04
Power frequency variation	NPCC Directory 12
Voltage flicker	CAN/CSA-C61000-3-7 CAN/CSA-C61000-2-2 CAN/CSA 61000-4-15

Table 7.3 Recommended voltage variation (up to 1000 V).

	Nominal system voltage (V)	Normal operating conditions		Extreme operating conditions	
		Min (V)	Max (V)	Min (V)	Max (V)
Single phase system	120/240	110/220	125/250	106/212	127/254
	600	550	625	530	635
Three phase system (4 wire)	120/208	112/194	125/216	110/190	217/220
	347/600	318/550	360/625	306/530	367/635
Three phase system (3 wire)	240	220	250	212	254
	600	550	625	530	635

Source: CSA standard CAN3-C235-83.

Table 7.4 IEEE Standard 519-2014: current distortion limits for general distribution systems [3].

	Maximum harmonic current distortion (in percent of I_L)					
	Individual harmonic order (odd harmonics)					
I_{sc}/I_L	$3 \leq h < 11$	$11 \leq h < 17$	$17 \leq h < 23$	$23 \leq h < 35$	$35 \leq h \leq 50$	TDD (%)
<20	4.0	2.0	1.5	0.6	0.3	5.0
20–50	7.0	3.5	2.5	1.0	0.5	8.0
50–100	10.0	4.5	4.0	1.5	0.7	12.0
100–1000	12.0	5.5	5.0	2.0	1.0	15.0
>1000	15.0	7.0	6.0	2.5	1.4	20.0

I_{sc}: maximum short-circuit current at PCC.
I_L: maximum demand load current (fundamental frequency component) at the PC under normal operation conditions.

Table 7.5 Transmission level voltage distortion limits.

Bus voltage at PCC	Individual harmonic (%)	THD (%)
<1.0 kV	5.0	8
1 kV–69 kV	3.0	5
69 kV–161 kV	1.5	2.5
>161 kV	1.0	1.5

Source: IEEE, 2014.

In this standard, the total demand distortion (TDD) is used to measure the harmonic distortions produced by converters. The definition of TDD and THD can be expressed as

$$\text{TDD} = \frac{1}{I_L}\sqrt{\sum_{n=2}^{\infty} I_n^2} \times 100\% \tag{7.1}$$

$$\text{THD} = \frac{1}{I_1}\sqrt{\sum_{n=2}^{\infty} I_n^2} \times 100\% \tag{7.2}$$

where I_L is the max RMS current at a fundamental frequency while I_1 is the fundamental component of the analyzed current. Therefore, different from THD, TDD only describes the distortion under full load conditions.

For a system that only consumes power from the grid, relaxed TDD limits can be applied. For example, when the short circuit ratio (I_{SC}/I_L) is in the range of 20–50, the TDD limit is 8%. However, these relaxed limits are not applicable for generation systems. For example, for a converter system that injects active power into the AC system, the TDD shall always be lower than 5% and fulfill the constraints on individual harmonic defined for the weak grids ($I_{sc}/I_L < 20$). Besides, the voltage distortion is also limited in IEEE 519-2014, which shall be considered in the weak grid and islanded systems and the limit on voltage THD is 8% for low voltage systems.

In a three-phase system, voltage unbalance occurs when the magnitudes of phase or line voltages are different, or the phase angles differ from the balanced conditions or both. The voltage unbalance definition and calculation are given below.

1. The ANSI definition of voltage unbalance, also known as the line voltage unbalance, is calculated as

$$\frac{\text{maximum derivation from average line voltage}}{\text{average line voltage}} \times 100\%. \tag{7.3}$$

The ANSI definition (C84.1-2016) [4] only considers the voltage magnitude, and phase angles are not included. Based on this standard, electric supply systems should be designed and operated to limit the maximum voltage unbalance to 3% when measured at the electric utility revenue meter under no-load conditions.

2. Besides the ANSI definition, IEC 61000-2-2 defines voltage unbalance as a ratio of the negative sequence component of three-phase voltage to the positive sequence component, as follows:

$$\frac{\text{negative sequence voltage}}{\text{positive sequence voltage}} \times 100\%. \tag{7.4}$$

The above definition involves both the magnitude and angles when calculating the positive and negative sequence voltage components. An approximation method is given in (7.5) to avoid complexity:

$$\sqrt{\frac{6\left(U_{ab}^2 + U_{bc}^2 + U_{ca}^2\right)}{(U_{ab} + U_{bc} + U_{ca})^2} - 2} \times 100\% \tag{7.5}$$

where U_{ab}, U_{bc}, U_{ca} are three line-to-line voltages.

In this standard, voltage unbalance is considered in relation to long-term effects, i.e. for 10 min or longer. The compatibility level for unbalance is the negative sequence component of 2% of the positive sequence component. In some areas, especially where it is the practice to connect the large single-phase load, values up to 3% may occur.

3. In IEEE 1159-2019 [2], the voltage unbalance can be calculated using only phase-to-phase RMS measurement without angle with the following equations:

$$\sqrt{\frac{1 - \sqrt{3 - 6\beta}}{1 + \sqrt{3 - 6\beta}}} \times 100\% \tag{7.6}$$

where

$$\beta = \frac{|U_{ab}|^4 + |U_{bc}|^4 + |U_{ca}|^4}{\left(|U_{ab}|^2 + |U_{2c}|^2 + |U_{ca}|^2\right)^2}. \tag{7.7}$$

Equation (7.4) is always valid using either phase-to-neutral or phase-to-phase measurements, and Eq. (7.6) is valid only if the system zero sequence component is zero, and the voltage measurements are phase-to-phase. The voltage unbalances of a three-phase service should be less than 5%.

7.4 Mitigation Techniques of Power Quality Problems

As discussed above, various power quality disturbances can impact grid-connected equipment. There are various industry standards to regulate power quality. Power conditioning for mitigating common power quality problems includes various alternatives for utilities, customers, and equipment manufacturers. Given increased problems due to power quality in terms of financial loss and loss of production, a wide variety of mitigation techniques for improving the power quality have been developed. This section provides an overview of the widespread power quality mitigation techniques. A summary figure of different power quality compensation methods is shown in Figure 7.3.

7.4.1 Passive Mitigation Solutions

Tap Changer

As discussed in Section 7.2.3, the distribution grid can suffer from long-term voltage variations, mainly caused by line impedance voltage drop under high power flow. To address these issues, a popular practice is to install transformers with tap changers. Adjusting the tap position, the transformer can have different turn ratios within a range, therefore compensating for the voltage variations. According to

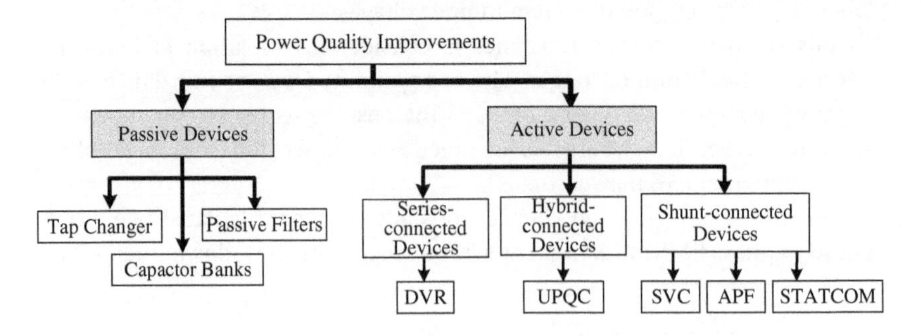

Figure 7.3 Traditional power quality compensation methods.

the different functions, tap changers can be classified as off-load tap changers or on-load tap changers (OLTCs).

An off-load tap changer can only change its tap position when the loads are cut off. Transformers with an off-load tap changer are widely used in the utility grid. The transformer is generally with +5%, 0, and −5% tap position, and it can be adjusted to resist the seasonal load change. The adjustment is generally made once or twice a year due to the necessity to cut off the loads.

By contrast, an OLTC can change its voltage ratio without disconnecting the loads, providing higher flexibility. Since the adjustment can be made without causing power outages, the tap position can be changed many times a day with automatic controller help. With tap selector development, the arcing during the transition between two taps is no longer a significant challenge, and the lifetime can be ensured. As a result, the OLTC provides a solution to address the hourly change in load demand and ensures a good regulation of distribution lines.

However, the tap changers cannot provide reactive power, and, as a result, it requires the upstream power lines to provide the reactive power. During extreme conditions, this can lead to stability issues.

Capacitor Banks

Capacitor banks, particularly in shunt capacitor compensation, are widely used in power systems to provide reactive power support. They can provide leading current and help correct the lagging current absorbed by the loads. In practice, they can be directly installed for long-term low power factor loads or use mechanical switches to connect them to the grid when needed.

Since the capacitance is fixed while the load conditions vary, the capacitor banks may lead to over-compensation or under-compensation. Adjustable levels of compensation can be achieved through parallel capacitor branches with individual switch control. For shunt capacitor compensation, the generated reactive power is dependent on the grid voltage. As a result, when the grid voltage is low and requires more leading current compensation, the capacitor banks can only reduce capacitive power. On the other hand, when the grid voltage is high, capacitor banks will aggravate the over-voltage as more reactive power will be injected.

It is well known that capacitor bank switching will introduce transients into the power system, with oscillatory over-voltages that may cause stress to the power system equipment. Moreover, capacitor banks can be sensitive to harmonics, especially when the line impedance and the capacitor banks can create resonance and provide a low impedance path for harmonics, leading to harmonic amplification.

Passive Power Filter (PPF)

Passive power filters (PPFs) are traditionally used to reduce the harmonics of the AC loads. Those filters are used even nowadays in high-power rating systems due

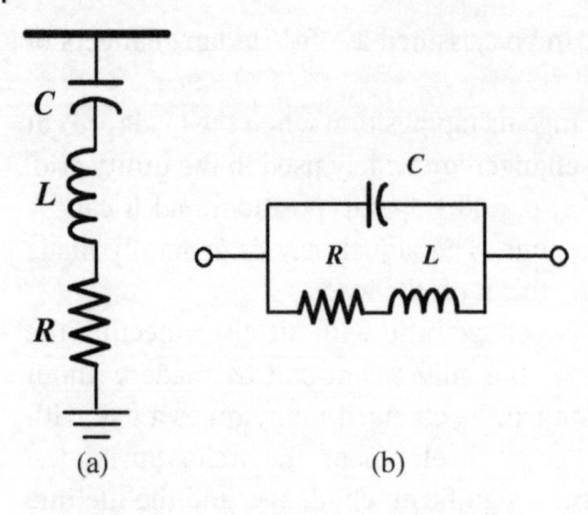

Figure 7.4 Passive single-tuned filters: (a) shunt passive filter, (b) series passive filter.

to their low cost, robustness, simplicity, and ability to provide reactive power at the fundamental frequency. However, in medium- and low-power rated systems, the design of PPFs is quite challenging. The reactive power requirement at the fundamental frequency is relatively low in such a system, but the harmonic currents are with high amplitude and broad-spectrum. Moreover, due to the low value of power capacitors in such filters, these filters are prone to parallel resonance between PPF capacitors and grid impedance, leading to a severe voltage distortion and amplification of harmonic currents. This can be avoided by combining passive filters with active filters that block or avoid such parallel resonance.

PPFs can be classified as tuned filters, damped filters, or a combination of both. The typical topologies of shunt and series single-tuned passive filters are shown in Figure 7.4. In practice, the passive tuned filters are mainly placed at the load end, injecting equal compensating currents with opposite phase angle to cancel harmonics and/or reactive currents caused by nonlinear loads. On the other hand, passive damped filters can eliminate all higher-order harmonics flowing toward the AC grids. In this case, the filters are placed near the PCC. The typical structures of damped filters, shunt or series, are shown in Figure 7.5.

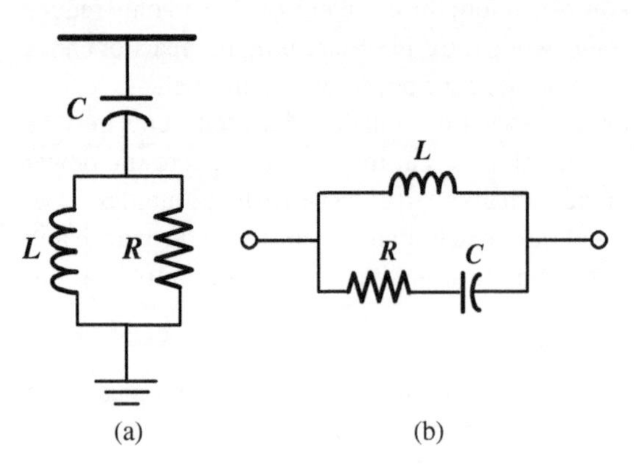

Figure 7.5 Passive damped filters: (a) shunt passive filter (second-order), (b) series passive filter.

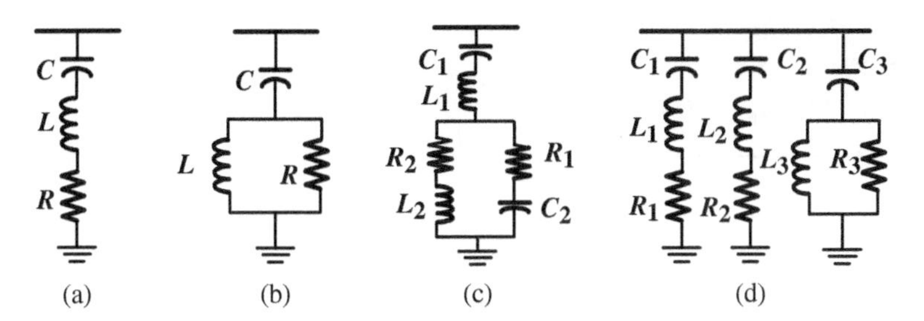

Figure 7.6 Shunt passive filter with impedance frequency plots: (a) band-pass, (b) high-pass, (c) double band-pass, (d) composite.

As discussed, both tuned or damped PPFs can be connected as shunt filters, series filters, or a combination of both (also known as hybrid filters). The passive shunt filters are connected in parallel to the harmonic source. This filter will provide a parallel low-impedance path for harmonic currents, which leads the harmonic currents to the filter and prevent their injection into the supply power system. Typically, these filters consist of lossless passive elements, including inductors (L) and capacitors (C). In Figure 7.6, various types of passive shunt filters are shown [5].

Alternatively, PPFs can also be connected in series with harmonic sources to block the harmonic current's propagation with a high impedance at harmonic frequencies. These filters provide very low impedance at the fundamental frequency but very high impedance at single specific harmonic order, such as the third order, to block harmonics at the same frequency. It is also possible to connect multiple filters in series to block harmonics at multiple harmonic orders. In practice, series filters are not popular compared to shunt filters due to the significant voltage drop and power losses (especially for multiple series-connected filters). One exception is the third-order series filters, which have wide applications to address harmonics of small-power, single-phase nonlinear loads. Typically, the passive series filters consist of parallel lossless passive elements such as inductors (L) and capacitors (C), shown in Figure 7.7.

To overcome the drawbacks of individual passive shunt and passive series filters, passive hybrid filters, which consist of both series and shunt passive filters, can be used. Different types of passive hybrid filter are shown in Figure 7.8 [1]. Typically, passive hybrid filters consist of a single tuned passive series filter, a single tuned

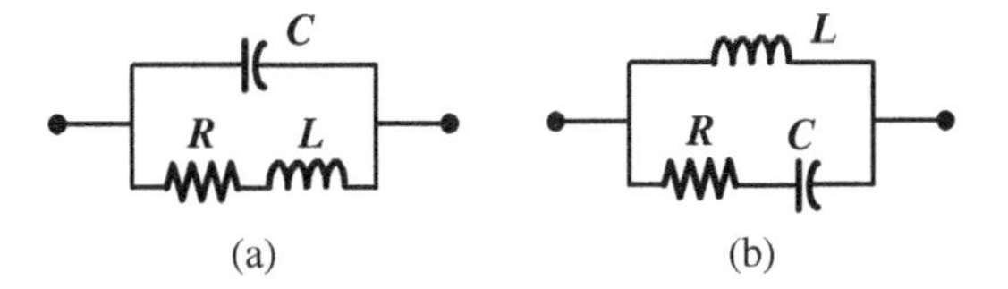

Figure 7.7 Series passive filters: (a) single-tuned filter, (b) damped filter.

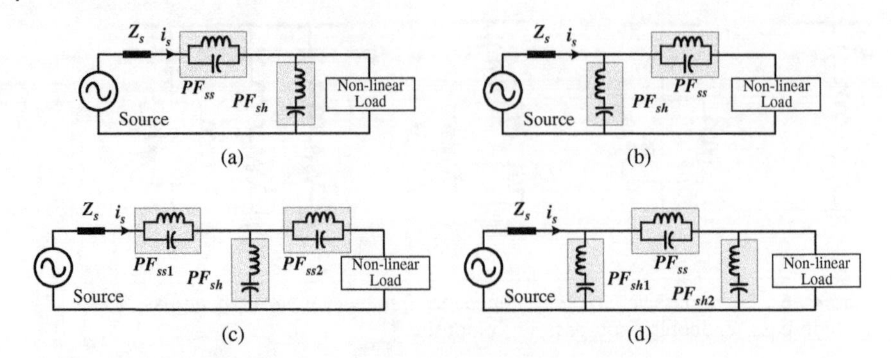

Figure 7.8 Different combinations of passive series and shunt filters for hybrid filter construction: (a) passive series and passive shunt filters, (b) passive shunt and passive series filters, (c) passive series, passive shunt, and passive series filters, (d) passive shunt, passive series, and passive shunt filters.

passive shunt filter, and a high-pass passive shunt filter. In such filters, the single tuned passive series filter could block the resonance between the supply and the shunt filter as well as absorb the passive shunt filter's excess reactive power. For large power rating applications, a passive shunt filter and a small series active filter is used as an ideal hybrid filter.

7.4.2 Active Mitigation Solutions

The passive component is generally easy to implement but lacks flexibility. Facing the utility grid's uncertainty in terms of various faults, incidents, and adjustments, the passive components can only provide limited mitigation performance. Under some circumstances, they may add inverse effects such as resonance and over-voltages, which could potentially affect the system stability.

To combat the power quality disturbances more effectively, active compensation devices utilizing power converters have been developed. Combining various power electronic topologies and flexible control schemes, the power quality disturbances can be well mitigated with high mitigation performance and fast dynamic response. A few popular active power quality compensation schemes are introduced in this section.

Static VAR Compensator

Static var compensators (SVCs) have been a popular choice to enable flexible and dynamic reactive power compensation. SVCs are shunt components to provide reactive power. An SVC can provide inductive reactive power to reduce AC voltage and provide capacitive reactive power to mitigate under-voltage issues. The most popular configurations are thyristor-switched reactors (TCRs), thyristor-switched capacitors (TSCs), as well as fixed and switched capacitors.

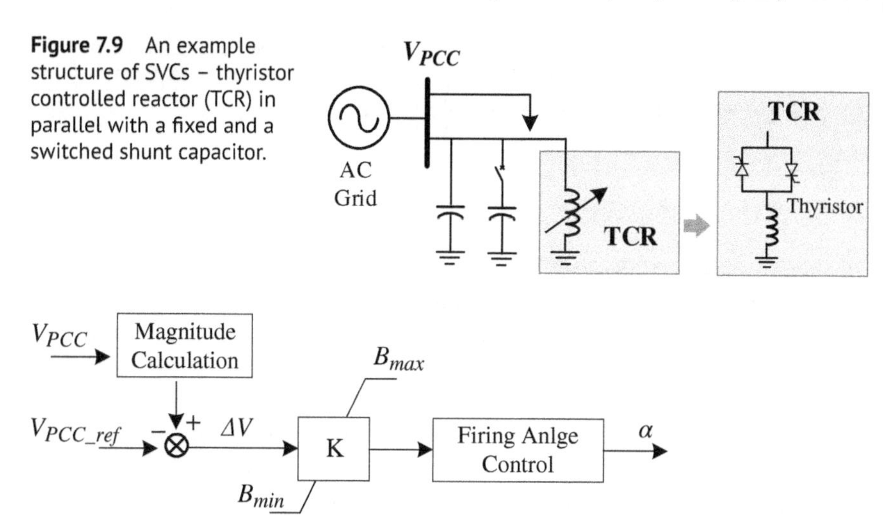

Figure 7.9 An example structure of SVCs – thyristor controlled reactor (TCR) in parallel with a fixed and a switched shunt capacitor.

Figure 7.10 The control scheme of an SVC.

Figure 7.9 shows an example SVC configuration with TCR in parallel with fixed and switched capacitor banks. With multiple branches, the SVC can adjust the reactive power flexibly. For example, the TCR in Figure 7.9 can dynamically adjust the reactive power through the thyristor firing angle control. Combined with the fixed and switched capacitor bands, a wider range of compensation can be achieved. To properly control the SVC, suitable control schemes, particularly for the TCR should be designed. An example of TCR control is given in Figure 7.10. As can be seen, the control scheme detects the voltage of the AC grid, and a feedback loop is applied to reduce the error between the reference voltage and measured voltage. By changing the applied number of units as well as the firing angle of the TCR, the injected reactive power is adjusted until the error is eliminated or the device reaches its maximum capacity. In this example, the constraints on susceptance (B_{max} and B_{min}) are related to the limit of the reactive power of the TCR.

SVCs have low cost among different converter-based compensation schemes. However, it may not smoothly change its reactive power when switched shunt compensation is needed. Also, the TCR produces significant harmonics into the system, which may cause other power quality problems.

Static Synchronous Compensator

Static synchronous compensators, also known as static synchronous compensators (STATCOMs), are shunt devices that can provide flexible reactive power compensation. With the help of fully controllable power electronic devices such as GTO and IGBT, STATCOM can dynamically control its output reactive power effectively supporting the grid and stabilizing the grid voltage [6].

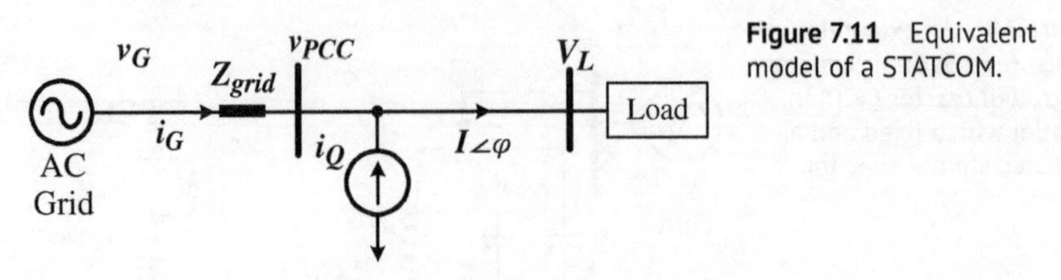

Figure 7.11 Equivalent model of a STATCOM.

Different from SVC, STATCOM does not rely on passive components to create reactive power. Instead, it adopts fully controlled power semiconductors to actively regulate the amplitude and phase angle of output currents. As shown in Figure 7.11, the STATCOM acts as a current source that produces only inductive or capacitive current (with reference to the grid voltage). This will provide reactive power compensation for voltage regulation. To properly control the injected current, proper converter structure and control schemes should be used.

Generally, high-power STATCOMs are connected to medium voltage distribution lines and perform compensation for several mega VAR. Multilevel converters are now widely adopted for medium voltage STATCOMs. A popular topology is the cascaded H-bridge converter without coupling transformer [7], shown in Figure 7.12. The cascaded H-bridge converter has a modular structure and is easy to extend to fit different voltage levels of the distribution lines by configuring different modules. As STATCOM mainly produces reactive power, it does not need energy sources at its DC side. Instead, it can absorb a minimal amount of active current components to keep a constant DC voltage to ensure the proper operation of the voltage source converters (VSCs).

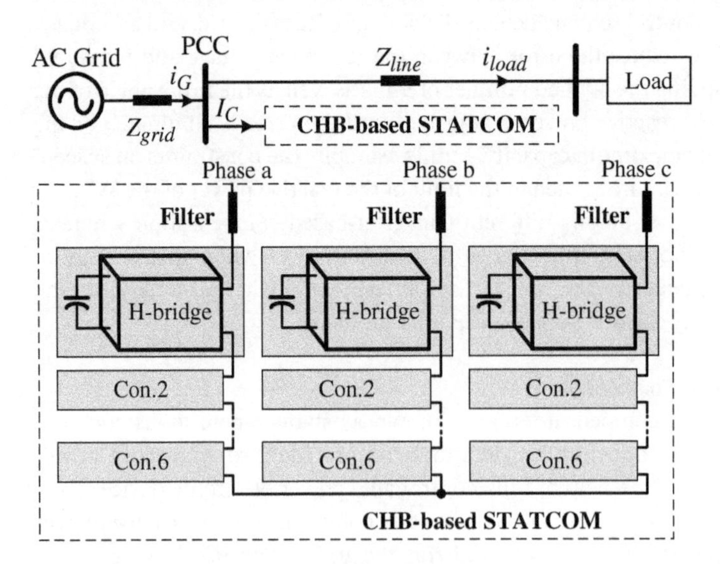

Figure 7.12 Cascaded H-bridge converter based high-power medium-voltage STATCOM.

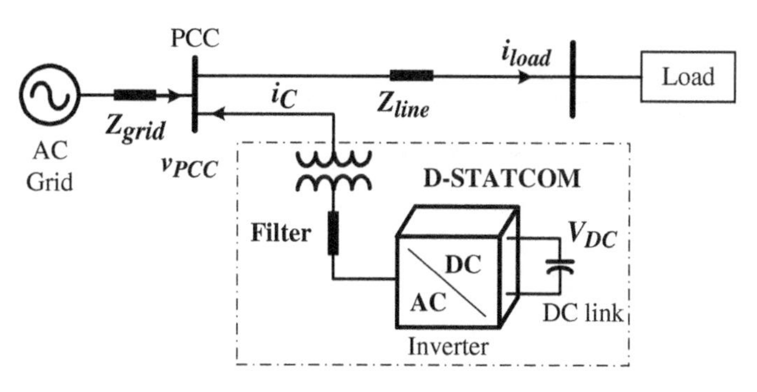

Figure 7.13 Structure of a distributed STATCOM (D-STATCOM).

In addition to high power STATCOMs, low power distributed STATCOMs (D-STATCOMs) are also increasingly used in low voltage systems. In general, two-level converters can be used to construct D-STATCOMs. A typical structure of a D-STATCOM is shown in Figure 7.13. Again, the primary function of D-STATCOM is to provide reactive power, its DC side does not need energy sources. Instead, capacitors are used on the DC link with contact DC voltage control to ensure the proper operation of the VSC.

Figure 7.14 shows an example of the control scheme of a D-STATCOM. The control scheme is implemented in the *dq* frame. The *q*-axis component is related to reactive power, where the PCC voltage v_{PCC} is regulated by the voltage controller. The generated reactive current control signal I_{q_ref} will be fed to the inner current loop and force reactive current competent to follow the reference. As a result, proper reactive power will be injected into the grid to eliminate the voltage between the expected voltage amplitude and actual AC voltage amplitude. On the other hand, the DC voltage is regulated by the *d*-axis controllers, enabling a small amount of active current flow to ensure a constant DC link voltage.

Dynamic Voltage Restorer

A dynamic voltage restorer (DVR) is a kind of series compensation device to mainly address AC voltage disturbances. As it directly applies a series

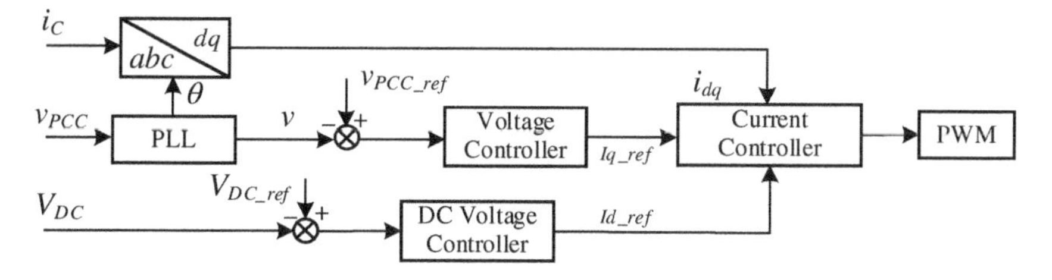

Figure 7.14 Example control scheme of a D-STATCOM.

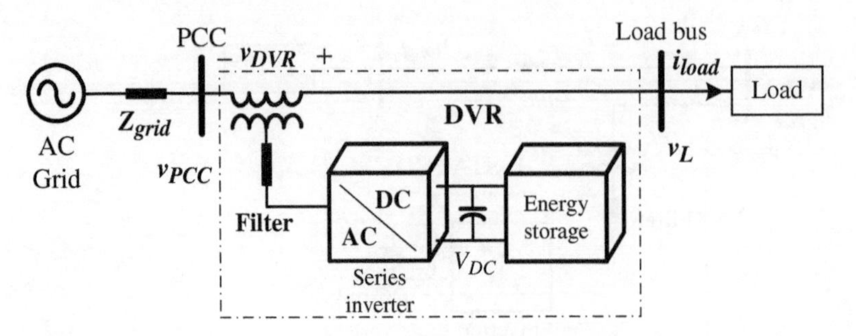

Figure 7.15 Structure of a dynamic voltage restorer (DVR).

compensation voltage to the feeder, it can compensate for short-duration issues like voltage swell and voltage sag with fast dynamic response. As the DVR only generates the compensating voltage, and most voltage sags and swells in the grid are not severe, the DVR voltage rating requirement is usually a fraction (e.g. 50%) of the system voltage.

As shown in Figure 7.15, a typical DVR consists of an energy source, a power converter, and a transformer whose primary winding is series connected to feeder lines. When the power converter applies a voltage to the transformer's secondary winding, a corresponding voltage will be added to the feeder line, and the loads will see a steady voltage even if the PCC voltage suffers from significant disturbance. As the power converter will provide active power to create the compensating voltage on the transformer, energy storage is generally employed. Alternatively, it can also absorb active power from the feeder line, either the grid side or load side [8]. In this case, the DVR can operate for a longer duration without worrying about the capacity of the energy storage component.

The principle of DVR compensation can be explained by the phasor diagram, as shown in Figure 7.16. Three different compensation schemes can be used: pre-sag compensation, in-phase compensation, and phase advanced compensation. Pre-sag compensation aims at regulating the load voltage ($V_{\text{Lpre-fault}}$) to be the same as pre-sag normal voltages, which ensure minimum disturbance of the voltage magnitude and phase angle to the load, as shown in Figure 7.16 (a). In-phase compensation, shown in Figure 7.16 (b), focuses on amplitude compensation, and once a fault happens, the DVR only compensates the voltage amplitude back to the normal value. As a result, the load will see a different voltage phase angle. However, this method has the smallest DVR output voltage requirement. In Figure 7.16 (c), the phase advanced compensation control generated the DVR voltage 90° leading to the load current, and as a result, this method has the minimum energy storage requirement on the DC link, but it has the highest voltage requirement for DVR. In practice, pre-sag and in-phase compensations are more popular.

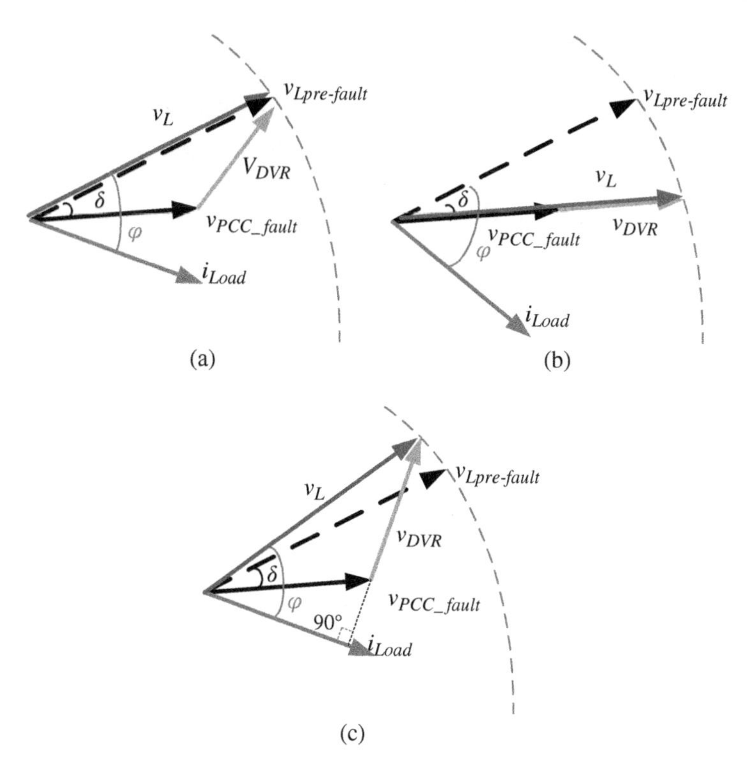

Figure 7.16 Phasor diagram for different DVR operating schemes: (a) pre-sag compensation technique, (b) in-phase compensation technique, (c) phase advanced compensation techniques.

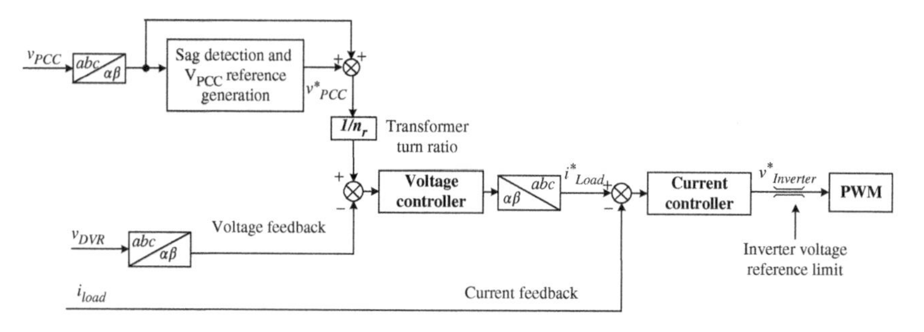

Figure 7.17 Example pre-sag control scheme for a DVR.

A typical control scheme for pre-sag control of DVR from [9] is shown in Figure 7.17. As can be seen, the inner loops of the DVR employ the current feedback loop and AC voltage outer loop to regulate the voltage that is applied to transformers' secondary windings. To correctly control the secondary voltage,

The PCC voltage under nominal condition, including initial phase angle and amplitude, are kept as voltage reference so that the load will not experience any change of voltage. The transfer turns ratio must be considered as the power converters regulate the secondary-side voltage, not primary voltage.

A DVR is a low-cost device to compensate for voltage disturbance for critical loads. Faults typically cause those voltage disturbances on other feeders. In case a fault on a feeder with a DVR installed, recent work has been developed to control the DVR as a fault current limiting device to limit the fault current and maintain the PCC voltage for other feeders [10].

Active Power Filter (APF)

In recent years, active power filters (APFs) have become a good option for harmonics dynamic compensation. It can be built by VSCs and current source converters (CSCs) [11]. The VSC-based APFs, which have a large DC capacitor to provide a self-supporting DC voltage bus, is widely used in power systems due to lower weight, lower cost, and the capability to be expanded to the multilevel operation mode. Mainly, those filters are commonly used in UPS-based applications. In Figure 7.18, a block diagram of the VSC-based APF is shown.

According to the application requirements, the APFs can be installed into a single-phase system or three-phase system. For single-phase systems, half-bridge converters or full-bridge converters can be used in APFs. In a three-phase system, APFs can be designed for three-wire or four-wire connections.

In Figure 7.19, a typical control strategy example of a shunt three-phase VSC-based APF is shown. In this control strategy, the three-phase load current (i_{Load}) and the DC bus voltage (V_{DC}) are sensed as the feedback signal. The harmonic components in load current are extracted and then be regulated to zero. As a result, the utility grid will see a sinusoidal current. Besides, an alternative way is R-APF control [12], which senses the PCC voltage (v_{PCC}) and uses a virtual resistor to generate expected harmonic control reference. More detailed

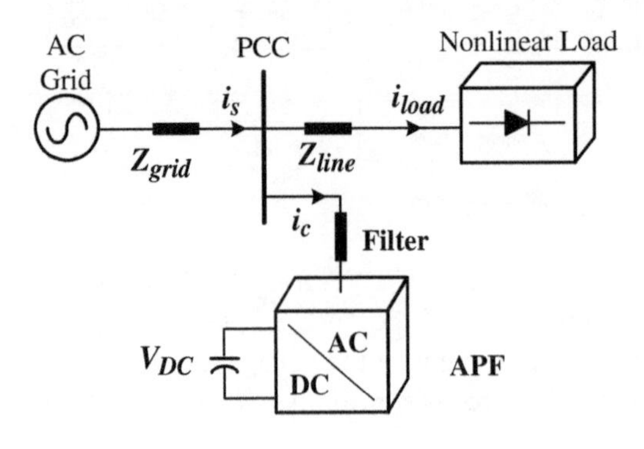

Figure 7.18 Structure of shunt active power filters.

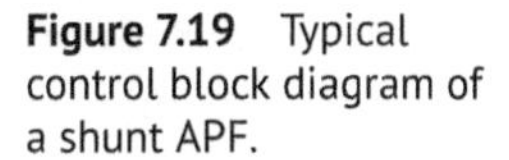

Figure 7.19 Typical control block diagram of a shunt APF.

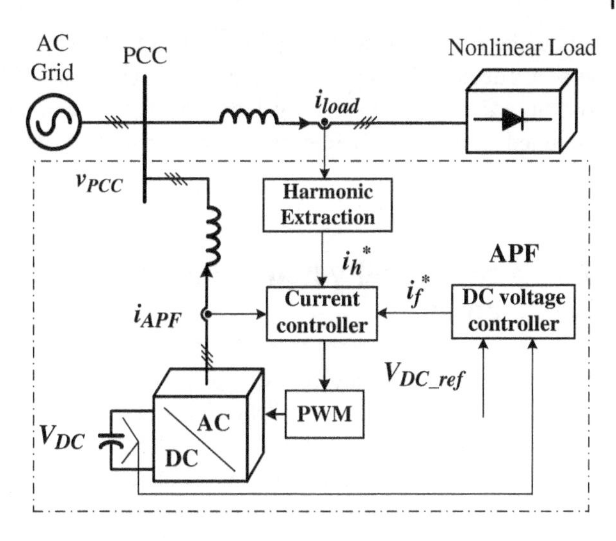

examples of using virtual impedance-based harmonics compensation using smart converters can be found in Chapter 10.

Unified Power Quality Conditioner

As discussed above, both shunt device and series devices have been used in power quality compensation. When shunt and series compensation are combined into one compensation strategy, it can be called a unified power quality conditioner (UPQC). The structure is similar to the unified power flow control (UPFC) [13] in the transmission system but is dedicated to distribution system power quality management.

A typical structure of a UPQC is shown in Figure 7.20. As can be seen, it has two converters back-to-back and they are coupled with the grid by a series transformer and a shunt device. Shunt transformers can be adopted or omitted, and various

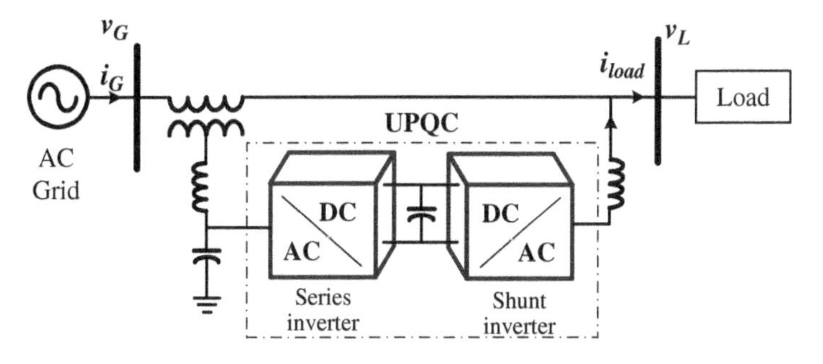

Figure 7.20 Unified power quality conditioner (UPQC) system configuration.

topologies can be adapted to fit the needs [14]. As a result, the UPQC can perform as series devices like a DVR, and shunt devices like a STATCOM and a shunt APF. As long as the DC-link can be well regulated and the necessary active power can be obtained, the other active device control methods can be integrated or transplanted into the UPQC.

7.5 Power Quality Issues and Compensation in Microgrids

Many power quality disturbances affect the grid, and therefore the power quality is an essential topic for the distribution system. The power quality disturbances in microgrids have some different features from the traditional distribution grid. The traditional distribution grid has a radial structure and unidirectional power flow. Microgrids can have more complex structures, and power flows can be bidirectional due to DG systems. In particular, an islanded microgrid may have more challenges as the system frequency and voltage amplitude can change in a larger range. Additionally, the increasing penetration of single-phase and unbalanced loads, nonlinear loads, and single-phase and three-phase power electronics-based DG and ESSs further contributes to the microgrid power quality issues. In addition to AC microgrids, the power electronics-based DC DG and ESSs, and DC loads produce power quality issues in DC microgrids. In the hybrid AC/DC microgrids, the interfacing converters for the AC and DC subgrids can also transfer the power quality issues from one side to another.

In the following sections, the power quality issues of microgrids and the solutions are discussed.

7.5.1 Power Quality Issues in an AC Microgrid

As microgrids can operate in islanded or grid-connected mode, where the grid connection can be considered weak compared to the traditional power system, the voltage and frequency variations can be a major challenge as the voltage is no longer strictly bonded to the utility systems. As a result, AC microgrid voltage and frequency are vulnerable to abrupt changes in loads, faults, and power unbalance.

In an islanded microgrid with diesel/gas generators, the diesel/gas generators are usually considered the primary suppliers while renewable energy sources and energy storage systems generate additional power. In this case, the microgrid voltage is mainly determined by the generator's excitation system, and frequency is regulated by the speed governor embedded in the generator set (gensets). Variations in the fuel supply or abrupt changes in loads can lead to voltage swell or voltage sag, and the frequency can have a long-time deviation from the rated frequency. In this case, the power converters, which have faster responses than gensets, can participate in the voltage and frequency regulation.

Besides the islanded microgrid, the grid-connected microgrid also faces voltage and frequency disturbances from the grid side. New grid codes are being developed to require the DGs and microgrids to remain connected to the grid in a disturbance to improve system stability. However, maintaining microgrid power quality while connected to a grid during a grid disturbance can be challenging [15]. More discussion on the disturbance ride-through control of DGs and microgrids can be found in Chapter 8.

For the critical loads in a microgrid with a high requirement for energy availability, it is necessary to deploy solutions to increase the reliability of the microgrid power supply. A popular solution is to install backup power supplies. For example, AC data centers (see Chapter 1 for more information) generally have a utility grid as the primary power supply. However, voltage sags and momentary interruptions are the primary concern for data centers as the data centers' availability is critical; otherwise, loss of data and denial of access can lead to unacceptable consequences. Therefore, uninterruptable power supplies (UPSs) with ESSs and backup generators are typically used to deal with voltage sags and momentary interruptions.

Another typical quality issue in a microgrid is the voltage unbalance, caused by single-phase/unbalanced loads, single-phase/unbalanced DG, and remote grid faults. One example of the unbalanced microgrid is the single-phase railway power system, which creates significant unbalance to the grid. There are various solutions, such as booster transformers or autotransformers. Also, the previously introduced compensation method such as STACOM can compensate for the unbalance [16]. Another typical example is the residential/community microgrid, where the household loads are generally designed to be distributed evenly for the three phases. However, the load unbalance is unavoidable. The single-phase residential DG, such as PV generation and tidal loads like electric vehicle (EV) chargers, can also cause severe unbalance in the microgrid. It should be noted that in addition to the three-phase voltage amplitudes, their corresponding phase angles should be maintained balanced as well. For example, due to the high penetration of single-phase PV power systems in the Hawaii power system, the system had a voltage unbalance issue. In that system, although the voltage amplitudes in different phases were regulated, the phase angle between every two phases was not 120° [17]. An effective solution for the unbalanced compensation is deploying energy storage systems to balance the active powers in different phases. Moreover, the reactive power compensation can also help for unbalanced voltage reduction/mitigation.

While the voltage quality issues in a microgrid are challenging, the current harmonics in such systems also increase rapidly. For example, increasing adoption of modern high-efficient loads such as compact fluorescent lamps (CFLs),

LEDs, adjustable speed drive fridges, high-efficiency washers, etc. in residential distribution microgrids can produce high harmonics due to their nonlinear characteristics. Depending on the manufacturers, both CFLs and LEDs can produce high low order harmonics and high THD in the current. Also, the power electronics interfaces of many DG, ESSs, and loads can contribute to harmonic issues in microgrids. For example, current harmonics are the main power quality issue in AC charging stations. In an onboard charger, where PFC and pulse width modulation (PWM) control techniques are used, the generated harmonics are small. However, in the presence of a 12-pulse diode rectifier in DC stations, low order harmonics are produced. The power quality issues in the EV stations can be compensated for using passive and active power filters, PWM rectifiers, and PFC stages. It should be mentioned that the power quality control in the EV charging station is a relatively new topic, and more studies are being done in this field.

Traditional power quality mitigation techniques rely on installing dedicated devices for power quality improvement purposes. Those compensation methods increase the investment costs and are typically centralized compensation, which may not be effective for a microgrid where the source of power quality distortions is highly distributed in a microgrid. As a result, the distributed compensation will be more effective in this case. Further considering there are many power converters in a microgrid due to the DG and ESSs, utilizing the existing power electronics converters in the microgrid for power quality compensation as a smart converter with ancillary functions can be a very promising concept. Some main benefits can be summarized as below.

- Due to the increasing penetration of smart IFCs from DG and ESSs, they can be great candidates to help address the power quality issues in addition to their primary power management targets.
- The power quality compensation using smart IFCs is distributed compensation and, therefore, more effective than the traditional centralized compensation.
- Most of DG and ESSs do not operate at their full rating due to the intermittent nature of renewable generation. So, available ratings can be used to improve the power quality issues free of cost.
- Smart IFCs already have control and communication capabilities and can be easily coordinated to provide valuable ancillary functions. These ancillary functions can further help to improve the cost-effectiveness of DG and ESSs.

An example control scheme for a DG system to participate in harmonic compensation is shown in Figure 7.21. A multiloop control system is adopted for the fundamental frequency control, whose primary task is to produce expected active power. The power control references can be dispatched from supervisory control or automatically generated by MPPT control. In addition to the fundamental frequency control, the harmonic control loop can be added. The available rating

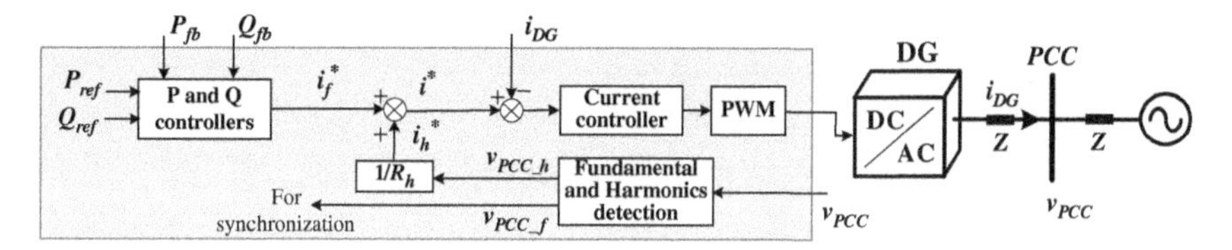

Figure 7.21 An example control scheme for distributed generation systems with harmonic control as an ancillary function.

for harmonic control is calculated according to the actual fundamental power and converter capacity, avoiding the converter's overcurrent. In this case, widely installed DG can participate in the harmonic compensation, and DG converters can be better utilized.

7.5.2 Power Quality in a Hybrid AC/DC Microgrid

A hybrid AC/DC microgrid can facilitate installing DC components like battery storage, PV generation, and emerging modern DC loads and reduce the number of power conversion stages. As mentioned, in AC subgrids, voltage variations and current harmonics are major power quality concerns. However, the power quality issues in the DC subgrid should also be considered. In hybrid AC/DC microgrids, power quality issues in the AC and DC subgrids could transfer to the other side through the IFCs between the AC and DC subgrids. For example, the DC subgrids harmonics, particularly the low-order harmonics, can be introduced by the AC-side harmonics or unbalanced three-phase AC voltage.

Figure 7.22 shows harmonics interactions between AC and DC subgrids. As can be seen, the nth order harmonics can lead to $(n-1)$ negative sequence harmonics and $(n+1)$ positive sequence harmonics at the DC side. Similarly, the DC ripples can also lead to distortions on the AC side. To be specific, the nth order ripples at the DC side can cause $(n-1)$th and $(n+1)$th order harmonics on the AC side. In return, the AC harmonics will again drive harmonics on the DC side, leading to $(n-2)$th and $(n+2)$th order ripples at the DC side. The interactions will continue until the transferred harmonics have neglectable amplitudes. A more detailed analysis of AC and DC subgrid interaction will be explained in Chapter 10.

Also, the DC harmonics can be caused by DC/DC converters in the loads and DG connected to the DC subgrid, such as maximum power point tracking control of PV systems or LC oscillation due to parasitic DC line impedance.

Like an AC microgrid, two options are available in hybrid AC/DC microgrids to address power quality issues: the first option is installing additional power quality compensation devices. The second option is controlling the existing smart power electronics converters for power quality compensation. In the first option,

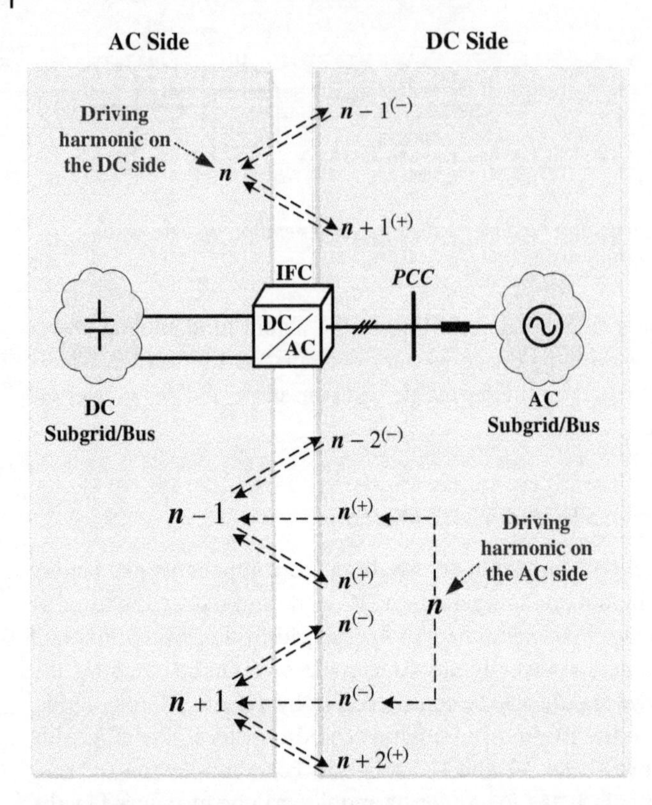

Figure 7.22 Harmonic interactions between AC and DC subgrids.

proper compensation devices are installed for each type of power quality issue. For example, the AC-side three-phase unbalance condition can be compensated for by installing a UPQC and a STATCOM. The harmonics in the AC subgrid can be mitigated by using PPFs and APFs, while installed large capacitors and tuned filters can easily filter the DC-side harmonics and variations.

Like AC microgrids, the hybrid AC/DC microgrids also have power electronic converters from DG and ESSs to utilize for power quality compensation purposes. Also, AC/DC microgrids have one or more AC/DC subgrid IFCs to realize power exchange between the two subgrids, which can be a good option for power quality conditioning as the AC/DC subgrid IFCs can obtain information on both subgrids without relying on communications. However, it should be considered that when the AC/DC subgrid IFC is used for power quality compensation in an AC or DC subgrid, it may affect the other side's performance. For example, if the IFC is used for unbalanced voltage compensation in the AC side, it will magnify the DC side's voltage ripples due to the oscillating power flow under unbalanced AC voltage. In this case, the unbalanced voltage compensation, while attenuating the DC voltage ripples and ensuring the IFC acceptable current quality, is challenging. One

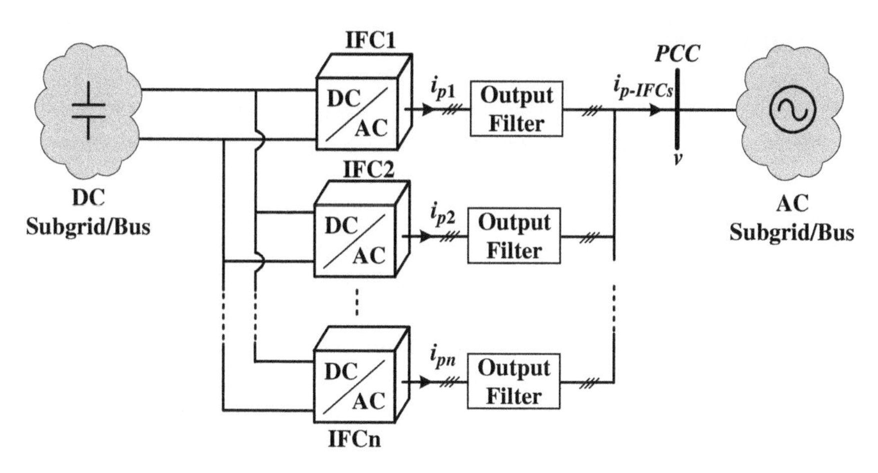

Figure 7.23 Parallel IFCs for DC ripple cancelation.

solution is to use parallel IFCs. In this case, when the unbalanced voltage compensation results in active power oscillation in the AC side and leads to second order ripples in the DC subgrid, the parallel IFCs, as shown in Figure 7.23, can potentially mitigate the situation. Although the power oscillations will still exist for each converter, the parallel converters' output power and the DC link voltage can be oscillation-free if the parallel IFCs are controlled properly. The details of this control scheme can be found in Chapter 9.

Another way to mitigate ripples in the DC subgrid is to install dedicated DC/DC converters or add ripple mitigation functions as ancillary functions in DC IFCs. As shown in Figure 7.24, ripple mitigators can be distributed in the DC subgrid. This solution does not have concerns of deteriorating AC power quality and is thus more straightforward and easier to implement than DC/AC IFCs. The details on this solution can be found in Chapter 10.

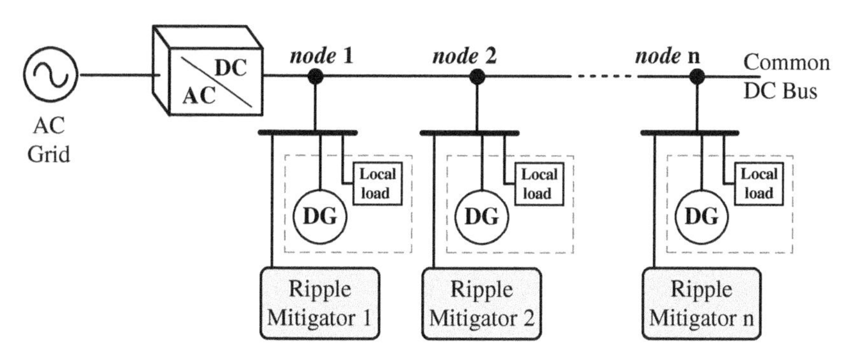

Figure 7.24 DC ripple mitigators in a DC subsystem.

7.6 Summary

This chapter provides an overview of common power quality problems in power systems, focusing on microgrid power quality. The popular industry standards on power quality are briefly reviewed. The traditional methods to compensate for the power quality issues are reviewed, including the passive approach as well as active power electronics-based methods. For microgrids and hybrid AC/DC microgrids, some power quality problems such as voltage and frequency quality, unbalance, and harmonics are more obvious. The AC subgrid power quality and DC subgrid power quality may also affect each other through the interfacing converters between the AC and DC subgrids. Compared to traditional centralized compensation through dedicated compensation devices, the smart converter concept is a promising approach to provide ancillary functions for the microgrid to address power quality problems more effectively. Therefore, the next few chapters focus on controlling smart IFCs under grid disturbances, unbalance, and harmonics.

References

1 Dugan, R.C., McGranaghan, M.F., and Beaty, H.W. (2002). *Electrical Power Systems Quality*. McGraw Hill.

2 (2019). IEEE Recommended Practice for Monitoring Electric Power Quality. In: *IEEE Std 1159-2019* (Revision of IEEE Std 1159-2009), 1–98.

3 (2014). IEEE recommended practice and requirements for harmonic control in electric power systems. In: *IEEE Std 519-2014* (Revision of IEEE Std 519-1992), 1–29.

4 American National Standard for Electric Power Systems and Equipment—Voltage Ratings (60 Hz), ANSI C84.1-2016, NEMA.

5 Singh, B., Chandra, A., and Ai-Haddad, K. (2015). *Power Quality: Problems and Mitigation Techniques*, 1–596. Wiley.

6 Dixon, J., Moran, L., Rodriguez, J., and Domke, R. (2005). Reactive power compensation technologies: state-of-the-art review. *Proceedings of the IEEE* 93 (12): 2144–2164.

7 Peng, F.Z. and Lai, J.-S. (1997). Dynamic performance and control of a static VAr generator using cascade multilevel inverters. *IEEE Transactions on Industry Applications* 33 (3): 748–755.

8 Nielsen, J.G. and Blaabjerg, F. (2005). A detailed comparison of system topologies for dynamic voltage restorers. *IEEE Transactions on Industry Applications* 41 (5): 1272–1280.

9 Li, Y.W., Blaabjerg, F., Vilathgamuwa, D.M., and Loh, P.C. (2007). Design and comparison of high performance stationary-frame controllers for DVR implementation. *IEEE Transactions on Power Electronics* 22 (2): 602–612.

10 Li, Y.W., Vilathgamuwa, D.M., Loh, P.C., and Blaabjerg, F. (2007). A dual-functional medium voltage level DVR to limit downstream fault currents. *IEEE Transactions on Power Electronics* 22: 1330–1340.

11 Singh, B., Al-Haddad, K., and Chandra, A. (1999). A review of active filters for power quality improvement. *IEEE Transactions on Industrial Electronics* 46 (5): 960–971.

12 Akagi, H., Fujita, H., and Wada, K. (1999). A shunt active filter based on voltage detection for harmonic termination of a radial power distribution line. *IEEE Transactions on Industry Applications* 35 (3): 638–645.

13 Gyugyi, L., Schauder, C.D., Williams, S.L. et al. (1995). The unified power flow controller: a new approach to power transmission control. *IEEE Transactions on Power Delivery* 10 (2): 1085–1097.

14 Fujita, H. and Akagi, H. (1998). The unified power quality conditioner: the integration of series- and shunt-active filters. *IEEE Transactions on Power Electronics* 13 (2): 315–322.

15 Li, Y.W., Vilathgamuwa, D.M., and Loh, P.C. (2005). Microgrid power quality enhancement using a three-phase four-wire grid-interfacing compensator. *IEEE Transactions on Industry Applications* 41: 1707–1719.

16 ABB (2019). FACTS Solutions for Railways. [Online]. https://new.abb.com/facts/solutions-for-railways.

17 Stewart, E., MacPherson, J., Vasilic, S., et al. (2013). Analysis of high-penetration levels of photovoltaics into the distribution grid on Oahu, Hawaii: Detailed analysis of HECO feeder WF1 (No. NREL/SR-5500-54494). National Renewable Energy Lab.(NREL), Golden, CO (United States).

8

Smart Microgrid Control During Grid Disturbances

8.1 Introduction

During grid disturbances, the control of AC interfacing converters (IFCs), distributed generation (DG), energy storage systems (ESSs), and AC/DC grids, is critical to providing a reliable and stable operation for smart microgrids. The converters' operations under grid disturbances can be classified into two different groups when they are islanded from the upstream part of the electric power systems and interconnected to the power system upstream.

The islanding condition can be intentional and unintentional. In intentional islanding operation, a planned electrical island of microgrids capable of being energized by one or more DG sources is formed, and the transition to and from these islanded conditions and how different power sources should operate is planned. During unintentional islanding, IFCs should detect the island and cease to energize the system. This operation is now common practice for most grid-connected DG under grid disturbances.

More recently, with the increasing connection of DG and ESSs and the consideration of their impact on the grid, grid disturbance ride-through operation, where the IFCs are not disconnected from the grid under a short term disturbance, is getting more critical and is being considered for newer versions of grid codes. Unlike the intentional islanding microgrid operation, the power converters can ride-through or trip (anti-islanding) depending on the disturbance's duration and effects on the power system (IEEE 1547 standard, 2018). So it is essential to control the converters during the ride-through region with specific performance, providing predictable performance before a disturbance recovery or eventual anti-islanding trip.

Finally, it is essential to understand that when the converter operates under the ride-through mode, its fault current contribution to the grid should be studied carefully to avoid confliction with the distribution system protection such as overcurrent settings and fuse-recloser coordination.

Smart Hybrid AC/DC Microgrids: Power Management, Energy Management, and Power Quality Control, First Edition. Yunwei Ryan Li, Farzam Nejabatkhah, and Hao Tian.

In this chapter, the operation of IFCs under grid disturbance is discussed. It mainly includes three topics: (i) different islanding detection algorithms for either traditional anti-islanding DG operation or intentional microgrid islanding operation, (ii) the fault ride-through (FRT) capability of IFCs when connected to the upstream power system and under short-duration grid disturbances; and (iii) fault current contribution of DG and control of the IFCs focusing on their impacts on fault current and fuse-recloser coordination.

8.2 Islanding Detection

Islanding detection is essential for both the traditional anti-islanding requirement of DGs and the intentional islanding microgrid operation. The islanding condition in a distribution system happens when a portion of the distribution system is energized solely by DG or ESSs while that portion of the distribution circuit is disconnected from the main grid. For islanding microgrid operation, islanding detection is needed so that the DG controllers can smoothly transition from grid-connected control to islanding microgrid control unless some uniform control scheme suitable for both scenarios is adopted. If properly planned, protected, and controlled, intentional islanding microgrids can improve service reliability to end-users and be highly beneficial. The topic of microgrid power management in different operation modes is provided in Chapter 5. On the other hand, unintentional islands can introduce significant risks to equipment or personnel due to frequency, voltage quality, and line energization concerns, and this is the main reason for the traditional anti-islanding requirement of DGs, where a DG must be disconnected from the distribution system when an island is formed.

There are many grid codes and standards regarding the islanding detection requirement. A review of the microgrid-related standards is provided in Chapter 3. Among the different standards, the most well-known one is IEEE 1547, which requires that DG detects an unintentional island operation condition and ceases operation within two seconds [1, 2]. Many countries and utilities have adopted grid codes which are mostly consistent with IEEE 1547. For example, UL 1741 is relatively consistent with IEEE 1547 and includes test procedures to verify the islanding detection method. The converters certified under those standards have undergone extensive testing to meet the anti-islanding requirements and can be installed.

There are various islanding detection methods, which can be generally classified into local methods and remote methods. As the name suggests, the local methods are based on DG local measurements and control and communications are not required. The local method can be further clarified into passive methods and active methods, where the passive method is based on measurement (e.g. voltage

Table 8.1 Islanding detection methods.

Islanding detection methods		
Local methods		**Remote methods**
Passive methods: Examples: - Under/over voltage and under/over frequency - Rate of change of frequency or frequency/power - Impedance change detection - Phase jump detection	Active methods: Examples: - Frequency shift based methods - Voltage shift based methods - Active impedance measurement	Examples: - Power line communication-based method - Transfer trip detection method

and frequency) only, and the active method requires the DG to generate perturbing signals to the system to facilitate the islanding condition detection. The remote methods rely on communications to detect the islanding condition. In Table 8.1, a general summary of islanding detection methods is provided, and they are discussed in detail in the following section.

8.2.1 Local Islanding Detection Methods

Passive Islanding Detection Methods

For the passive islanding detection methods, the system parameters such as voltage, current, impedance, and frequency are measured or calculated at the DG terminals. Typically, those parameter values will undergo noticeable change when the grid is disconnected, and DG islanding operation is formed. After sensing a significant change in one or more of those parameters, the islanding condition is detected, and the DG should be disconnected from the system. The popular passive islanding detection methods include frequency and voltage-based relays. There are also passive detection methods based on phase jump detection, harmonics measurement, grid impedance measurement, and power variations. The passive approaches do not disturb the system during normal operation and therefore do not affect the power quality or grid operations. The main drawbacks of passive methods are the relatively large non-detection zone (NDZ) and detection speed. An NDZ is typically defined as a range of operation region that islanding cannot be detected and anti-islanding protection cannot be provided in a timely manner. The NDZ is related to how the measured system parameters do not change obviously after an islanding condition is formed. This usually happens when the microgrid loads are close to the DG output power during grid-connected operation, and therefore the voltage and frequency of the islanding microgrid do not noticeably change. A small NDZ means a method is more effective. To reduce the

NDZ, a combination of multiple parameters can be considered. A brief description of a few popular passive islanding detection methods is given below.

Under/Over Voltage and Under/Over Frequency Detection

The conventional protection relays usually used in the under/over voltage protection or under/over frequency protection can be installed in a distribution feeder to detect the abnormal conditions related to islanding operation. Unusually, when islanding operation in a segment of the distribution feeder happens, the DG will see a change in real and reactive power outputs. This change in output power will typically affect the islanding microgrid voltage and frequency. For example, a real power change can lead to voltage variation at the connection point, and a reactive power change will affect the phase angle of the DG output current, which affects the current frequency. As a result, in islanding operation, those relays can detect if the voltage and frequency are beyond the pre-set thresholds and then trip the DG from the feeder. These methods are simple with low cost and without complex control in the DG. But they have an NDZ when the DG output power does not change obviously when the transition from grid-connected mode to islanding mode happens.

Rate of Change of Frequency or Frequency/Power

As the name suggests, the rate of change of frequency method monitors DG output frequency change rates ($\mathrm{d}f/\mathrm{d}t$). The rate of change of frequency/power uses the value of $\mathrm{d}f/\mathrm{d}P$, which is typically larger when the main grid is disconnected. These methods also rely on the DG output power and frequency variation when islanding occurs. But compared to the under/over voltage and under/over frequency detection, these methods can be faster. Moreover, by focusing on the rate of change, this method is not affected as much by the small power mismatches between the DG and local loads, and therefore have a smaller NDZ.

Impedance Change Detection

When an islanding operation occurs, a change in the grid side impedance can usually be detected, considering that the utility impedance is considerably smaller than an islanding microgrid's impedance. Therefore, monitoring the source impedance can be an effective method for islanding detection, where the DG can be disconnected when a sudden increase in impedance is detected.

Phase Jump Detection

Similar to the frequency various based detection, the phase jump detection is based on the fact that when DG output power changes from grid-connected operation to islanding operation, the microgrid voltage will change (due to real power variation) and a phase angle jump of the DG output current will happen (due to reactive power variation). Therefore, by detecting the phase jump, islanding detection can be conducted. This method is suitable for multiple inverter-based DGs using phase-lock loops. However, this method suffers from potential malfunctions when a grid disturbance happens as well as NDZ if the DG output power does not change much during the islanding transition

Active Islanding Detection Methods

In the active islanding techniques, a small perturbing signal is injected into the DG output current or voltage. When the DG is operating under grid-connected operation, the injected signal should not affect the operation. In contrast, the injected signal modifies the system parameters under islanding operation, facilitating islanding detection. Many active islanding detection methods rely on positive feedback that usually does not affect the system stability during a grid-connected operation but will make the system unstable in unintentional islanding operation. As a result, the active techniques can effectively reduce the NDZ, but they may introduce disturbances and power quality issues to the power system. Typically the active methods also take a longer time for islanding detection compared to the passive methods. The active islanding detection techniques can be easily used with inverter-based DG systems, where flexible control of the DG output is possible. In this subsection, a few well-known active islanding detection methods are reviewed.

Frequency Drift-based Methods

Frequency drift-based islanding detection relies on either a slightly distorted current or positive feedback in the DG control. An active frequency drift method based on the slightly distorted DG output current is shown in Figure 8.1. It can be seen that there is a small dead time or zero time T_Z created between the voltage (with a period time of T_V) and current (with a period of T_I) for half of a fundamental cycle. During this dead time, the DG output current remains zero. In grid-connected operation, the frequency and voltage waveform changes will not be shown at the DG terminals due to the main grid presence. While the main grid is disconnected in islanding operation, a zero crossing of the voltage occurs due to the injected current with a small disturbance signal. This will introduce a phase error in each cycle's voltage and, therefore, a frequency drift (increase in this example). This frequency drift continues until the voltage frequency exceeds the predefined threshold, and the islanding condition can be detected.

Another method to introduce the active frequency drift method is positive frequency feedback added to the inverter output current or voltage. This positive feedback of frequency is usually called the Sandia frequency shift method. Similarly, when the main grid is disconnected, this method will introduce a change in converter output frequency due to the positive feedback to destabilize

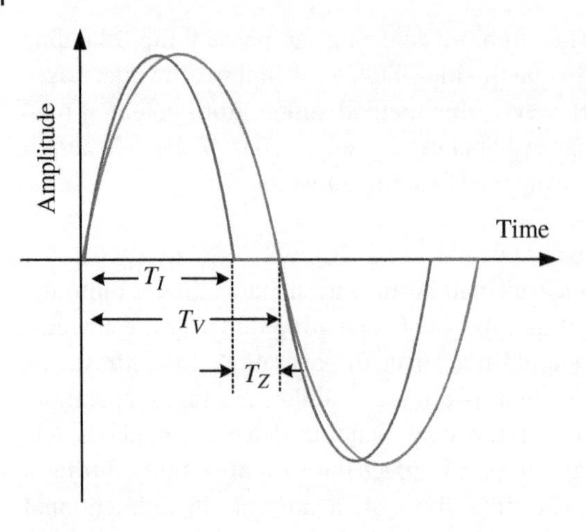

Figure 8.1 Distorted current waveform for the active frequency drift method.

the system, and this change continues until the frequency exceeds a predefined threshold and islanding is detected.

The active frequency drift can also be introduced by positive feedback control of the converter output current. An inverter output current has three parameters that can be controlled (current amplitude, frequency, and phase angle). When positive feedback of the inverter output current phase angle is added using the measured DG terminal voltage frequency, the DG terminal voltage frequency will deviate away from its nominal value when islanding operation occurs. Therefore, islanding detection can be performed.

Voltage Shift-based Methods

Voltage shift-based methods rely on the voltage amplitude's positive feedback to disturb the system in islanding operation. A well-known example is the Sandia voltage shift (SVS) method, where a positive feedback is applied to the voltage amplitude at the DG unit's output terminals. In the grid-connected operation, this method has a negligible effect on the system stability. When the grid is disconnected, a quick reduction in the DG terminal voltage is introduced by this method, facilitating islanding detection. The SVS method is one of the most efficient islanding detection methods among other feedback-based detection methods.

Active Impedance Measurement

Unlike passive impedance change detection, the active impedance measurement is based on signal injection from the DG side and then measurement of impedance variations for islanding detection. For example, with a high-frequency signal added to the output of the DG, the calculation of high-frequency impedance using dV/dI can be performed. When islanding operation occurs, this high-frequency impedance becomes more significant, and therefore islanding operation can be detected. Compared to the passive impedance change detection, the active impedance detection method will have a smaller NDZ. However, like other active detection methods, the intrusive nature can potentially produce voltage and current harmonics to the system.

8.2.2 Remote Islanding Detection Methods

Remote islanding detection methods are based on the communications between the utility grid and each DG. These methods perform quickly, effectively, and have negligible NDZ. However, the remote detection method may have higher cost, high computational burden, and complexity due to the communication requirement. Many remote islanding detection methods work for multiple DGs together or can be integrated into the supervisory control and data acquisition (SCADA) system and therefore considered utility-level islanding detection methods.

A typical remote islanding detection scenario is shown in Figure 8.2, where the utility grid and DG unit have a communication link. The islanding detection is typically performed upstream, and a tripping signal is sent to each DG when islanding happens. Two common examples of remote islanding detection schemes are power line communication (PLC)-based detection and transfer trip-based methods.

Power Line Communication (PLC)-based Detection Method

The PLC-based islanding detection method uses power lines as the communication carrier. Typically, the transmitter device produces continuous small voltage distortions, which are then propagated downstream to the distribution feeders. The DG on the feeders will each have a receiver device that can detect the feeders' voltage distortions. If the distortion cannot be detected, meaning islanding occurs, the DGs will be disconnected from the feeder. This method is suitable for a radial system with many downstream DGs. Interference of the signal could happen due to transformers or other load-related distortions to the voltage.

Transfer Trip Detection Method

The transfer trip method is usually integrated with the SCADA system. This method monitors the breaker's status and other switching and control circuits

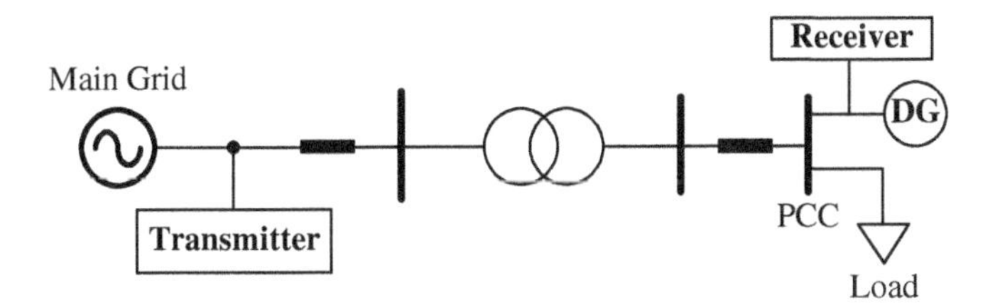

Figure 8.2 Remote islanding detection scenario.

continuously as part of the SCADA system. When the SCADA system detects a system loss, the relays operate to disconnect the DG. Such a method has fast islanding detection with negligible NDZ. However, hardware limitations, high cost, and risk of communication failure are among the disadvantages of this method.

8.2.3 Signal Processing Techniques Used in Islanding Detection

Facilitated by the widely implemented time and frequency-domain signal processing techniques, e.g. Kalman filter, wavelet transform, autocorrelation function, S-transform, Fourier transform, etc., the state-of-the-art islanding detection methods for microgrids have improved in recent years, with faster detection speed and smaller NDZ [3]. The primary function of above-mentioned advanced signal processing techniques is to capture and determine the measured signal key characteristics to better facilitate power system operations.

The Fourier transform's primary function is to capture the targeted signal characteristics with specific frequencies, thus being classified as a frequency-domain method. To deal with time domain-related analysis in islanding detection, the short-time Fourier transform (STFT) tool is then developed, which could incorporate multiple frames of the measured signal in a moving window. Besides, some other Fourier transform-based methods are also popular in industrial applications, such as the discrete Fourier transform (DFT) and fast Fourier transform (FFT), which can convert the discrete-time sequence with finite length to a discrete-frequency sequence effectively.

Wavelet transform and S-transform are also efficient tools in capturing the key characteristics of the measured signals under distorted situations. Wavelet transform is often implemented together with the STFT and multi-resolution tools [4]. For the islanding detection methods based on wavelet transform, a proper threshold should be defined, which will be compared with the wavelet coefficients, i.e. a larger value of the wavelet coefficient than the threshold indicates that the islanding situation occurs. However, the wavelet transform-based methods have certain limitations, e.g. the proper selection of the threshold value, the various sampling frequencies, the mother wavelet's determination, etc. Besides, only the low-frequency components can be processed by the wavelet transform. If the high-frequency band signals should be processed, the wavelet packet transform (WPT) can be adopted. The WPT is highly related to the discrete wavelet transform (DWT) but with the same resolution for low-frequency and high-frequency signals.

The S-transform-based method can be considered as an extension of the wavelet transform-based tools. The detection of the islanding situations in microgrids using the S-transform is also based on capturing the measured signal key characteristics. The process is typically as the following: firstly, the S-transform analyzes the measured electrical signals at the DG terminals, e.g. voltage or current, based on which the S-matrix and its equivalent time-frequency contours are obtained. Next, a calculation of the spectral energy content of these contours is made, including the variation of frequency and magnitude that can be used with the islanding detection.

Another widely implemented time and frequency-domain-based signal processing method is the Kalman filter, which is widely adopted in microgrids to capture and filter the harmonic characteristics based on the measured voltage and current signals (such as in phase-locked loops). The Kalman filter's islanding detection methods are typically based on evaluating voltage harmonic components and the selected harmonic distortion.

8.2.4 Intelligent Techniques Used in Islanding Detection

With the development of artificial intelligence (AI) and learning algorithms in recent years, different intelligent techniques and data mining tools are also adopted in detecting islanding operations [3]. The well-known algorithms include fuzzy logic, artificial neural networks (ANNs), probabilistic neural networks (PNNs), a decision tree, and a support vector machine (SVM). Aided by the intelligent tools, the multi-objective model in islanding detection can be well addressed.

The diagram of typical intelligent algorithm-based islanding detection is presented in Figure 8.3. Firstly, the measured voltage or current signals at the PCC should be incorporated into the detection system for model training and feature

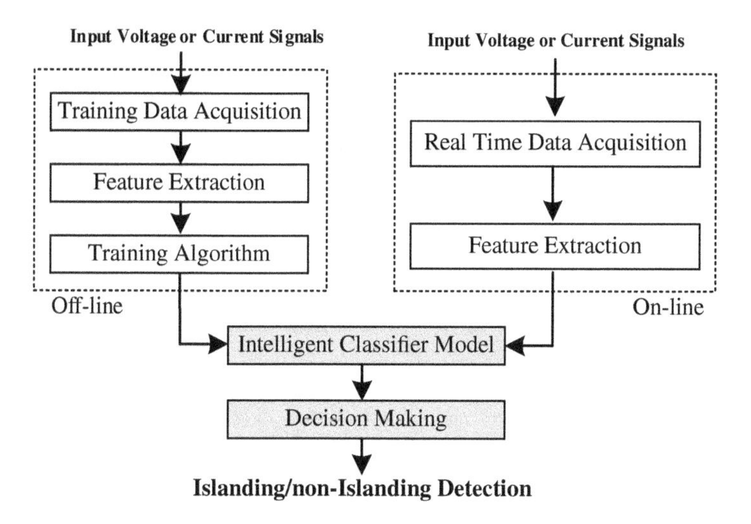

Figure 8.3 Schematic of intelligent process in islanding detection.

capturing purposes, while such steps are conducted off-line to avoid the islanding detection delay. Based on the off-line process results, an intelligent classifier is used in the online process for real-time detection of the islanding. Brief descriptions of each intelligent detection tool are presented as follows.

The primary function of the ANN-based methods is to capture the key characteristics based on the measured data to closely monitor the system's parameter variations. An ANN is a nonlinear and adaptive information processing system composed of many interconnected processing units. The information processing is based on simulating brain neural network processing and memorizing information.

Another possible intelligent DG islanding detection method is the decision tree, a commonly used data mining method for establishing classification systems based on multiple covariates or developing prediction algorithms for a target variable. The decision tree method can also be combined with a weighted pattern tree (WPT) or DWT for signal processing, which can be used for feature extractions based on the measured voltage or current signals for DG islanding detection.

DG islanding detection using the fuzzy logic tools is based on the fuzzy-rule classifier. The fuzzy logic techniques can be based on the decision tree. It can also be improved by combining the rule-based formulations and fuzzy membership functions. As a result, high detection accuracy can be achieved with fuzzy logic-based DG islanding detection.

SVM is a kind of generalized linear classifier that classifies data in the way of supervised learning. Its decision boundary is the maximum margin hyperplane of learning samples. For DG islanding detection issues, the SVM classifier is often adopted together with autoregressive modeling in extracting the key characteristics of the targeted voltage or current signals at the PCC, which can give a short detection time and high detection accuracy.

Much of the above intelligent algorithm-based islanding detection has the potential challenges of reliable training data (islanding does not happen very often in a distribution system), relatively high computational burden, innate complexity, and is influenced by noises when compared to the traditional detection methods, especially for the system with a lot of DG.

8.3 Fault Ride-through Capability

FRT ability is defined as a grid-connected converter's capability to remain connected to the grid under specific voltage or frequency disturbances for a specific duration. This allows the system to recover from short-duration faults and turn back to normal operation. The FRT capability initially applies for centralized PV or wind-turbine generation plants as they play an essential role as power sources in the power system, and their sudden disconnections can lead to significant disturbances.

With the increasing penetration level of DG or other converter-based energy sources in distribution systems, the DG and microgrid's FRT capability becomes an important solution for improving the weak grid's fault-tolerance capability. In the systems with a high penetration level of DG, triggering anti-islanding protection at the beginning of fault conditions can lead to sudden losses of a significant portion of power sources, threatening the grid's stability and aggravating the faults. For example, under some faults that reclosers can solve, DG loss can lead to a severe mismatch between sources and loads, leading to power outage for the whole system.

The ride-through capabilities can be classified into voltage ride-through and frequency ride-through. In this section, the voltage ride-through capability is discussed.

8.3.1 Fault Ride-through Requirement

With the increasing penetration of DG into power systems, many utility operators have increased the demand for DG interconnection and operation during grid disturbances. However, their operation should be in line with standards and grid codes. The grid codes vary from country to country because they are directly related to the generation characteristics and network operating requirements. For example, the frequency related requirements are generally more stringent in countries with relatively isolated (i.e. weakly interconnected) systems than those with large, highly interconnected systems. The low-voltage ride-through (LVRT) and high-voltage ride-through (HVRT) capabilities are discussed in this section.

Low Voltage Ride-through (LVRT) and High Voltage Ride-through (HVRT)

LVRT is the capability of DG to withstand low voltage conditions, i.e. voltage sags. The voltage sags are usually caused by grid faults, which refers to a sudden drop in the voltage of a node in the grid, exceeding the allowable range of normal voltage deviation, and finally returns to the normal allowable range in a short time. Please refer to Chapter 7 for more discussion about grid abnormal conditions and voltage sags. The voltage drop depth is the voltage minimum value ratio to the rated voltage at the grid connection point during the voltage sag. The DG LVRT operation requires it to be connected to the grid safely and reliably during the voltage sag and helps recover the grid connection point voltage.

Various countries have different grid connection guidelines, including specific requirements for LVRT. Most of them are for large-scale renewable power generation, such as wind farms connected to the transmission system, but there is also more demand for the LVRT of DG systems. Typically, grid-connected DG allows the voltage of the grid-connected point to drop to a certain limit according to the requirements. When the voltage drops under the graph, the DG will trip.

The HVRT is the DG's capability to stay connected to the power grid when a voltage swell happens. The voltage swells are usually associated with system fault conditions; however, they are not as common as voltage sags. The temporary voltage rise on the unfaulted phases during a single line-to-ground fault is a common cause of swells. Also, switching off a large load or energizing a large capacitor bank may cause voltage swells. Chapter 7 provides more detail on the grid abnormal conditions and voltage swells. In the next section, details of LVRT and HVRT requirement of DG in the most recent IEEE Standard 1547 is discussed.

LVRT and HVRT Requirements in IEEE Standard 1547™-2018

The IEEE Standard 1547-2018 is one of the most well-known and practical standards in DG interconnection. In this section, a brief discussion of the LVRT and HVRT requirements in this standard is provided. From this standard, the grid-connected DG operation can be summarized [2, 5] and is shown in Figure 8.4.

In this figure, ride-through is the ability of the DG to withstand frequency or voltage disturbances. In case the DG cannot ride through the fault, it will trip. The trip means that the DG ceases output power production without an immediate return to service. In this operation condition, the DG is not necessarily disconnected from the grid. When the DG trips, it will return to service following a trip, and this operation is equivalent to DG start-up. For DG ride-through mode under the permissive operation condition, the DG may cease to energize (called momentary cessation) or continue to operate (called mandatory operation). In the momentary cessation period, the DG ceases to energize the grid during the disturbances and should rapidly recover when the voltage or frequency return to a defined range. The DG should restore output to normal values after the disturbance if the disturbance does not cause a trip.

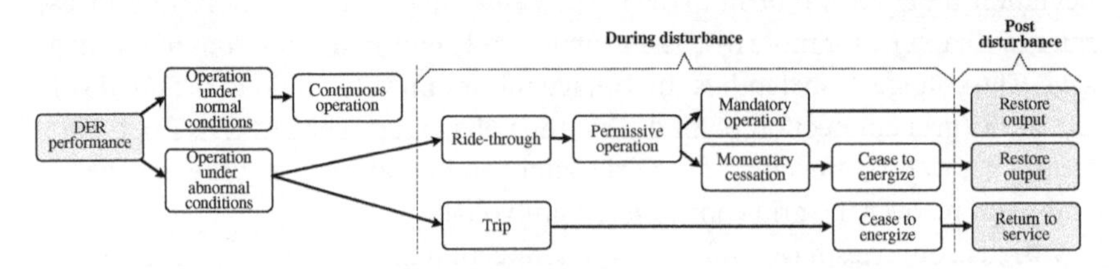

Figure 8.4 Operation of grid-connected DG based on IEEE Standard 1547-2018 adapted from [5].

Based on IEEE 1547-2018, particular specifications and requirements are dependent on application considerations. The requirements can be classified into three categories [6]:

- Category I: The power system's *essential* stability and reliability needs are considered for the abnormal operation performance.
- Category II: *All* stability and reliability needs of the power system are considered for the abnormal operation performance. This is also in coordination with existing reliability standards. As a result, DG tripping should be avoided for a wider range of disturbances.
- Category III: Both the needs of distribution system reliability/power quality and the bulk power system's stability/reliability are considered for the abnormal operation performance. It is also coordinated with existing interconnection requirements for very high DER penetration.

IEEE Standard 1547-2018 has specified DG operation regions for each category differently. The DG ride-through is an inherent capability that may not be adjustable. However, the trip settings can be adjustable to satisfy the operation regions. As described earlier, DG operation regions include "continuous operation capability," "mandatory operation capability," "momentary cessation capability," and "shall trip" region. The "continuous operation capability" is the same for all three categories. Category I has the smallest "mandatory operation capability" requirement, and therefore DG with limited RT capabilities is allowed in Category I. With increasing RT demands of DG, Category II and Category III have larger regions for "mandatory operation capability," particularly for LVRT. Category II and III are achievable by UL 1741 SA certified inverters.

In Figure 8.5, voltage ride-through requirements for Category III are shown [2]. It can be seen that continuous DG operation is required for voltage within 0.88–1.1 p.u., and LVTR is mandatory for voltage as low as 0.5 p.u. up to 10 s. For an even lower voltage (0–0.5 p.u.), the DG should remain connected but cease the operation momentarily and not energize the feeder for up to 1 s (including zero voltage ride through). This momentary cessation capability can ensure that power quality and the stability issues associated with DG resynchronization can be minimized. In case the grid voltage does not return after 1 s, the DG can trip off or continue the ride through operation up to 2 s when it must trip.

After DG ride-through operation, according to IEEE Standard 1547-2018, DG output should be restored to 80% of pre-disturbance active current within 0.4 s. The 0.4 s time starts when the voltage range returns to continuous operation or mandatory operation ranges. If the output power is oscillating, but the oscillation is damping, it is acceptable. In case the DG provides dynamic reactive power support, it must continue its operation for 5 s before it returns to pre-disturbance reactive control mode.

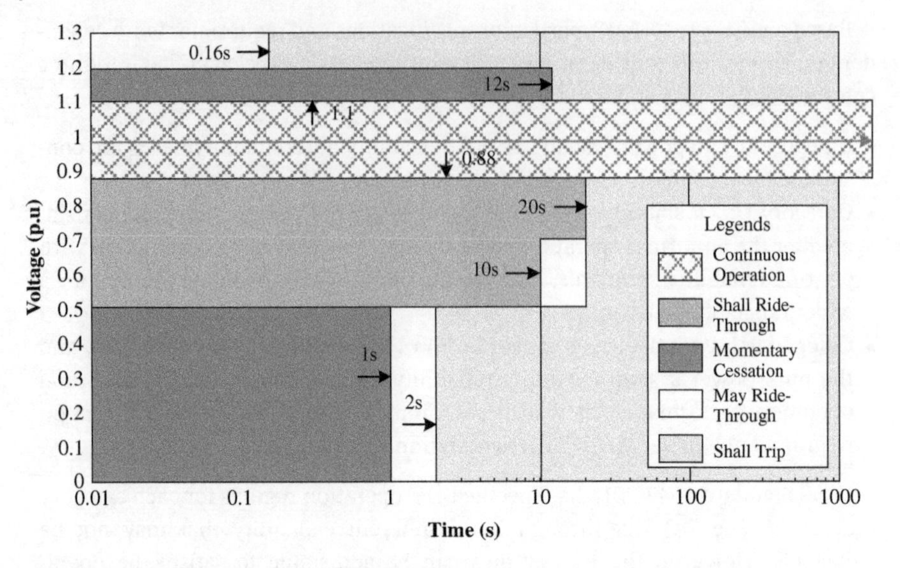

Figure 8.5 DG response to abnormal voltage and high voltage and low voltage ride-through requirements based on IEEE Standard 1547-2018 for Category III. Adapted from [2].

8.3.2 Ride-through Enhancement

The general goals of FRT enhancement strategies can be summarized as follows:

- At the moment of grid failure, according to the grid connection requirements, the power system needs to be compensated for reactive power to support the voltage of the grid connection point to ensure that the DGs do not run off the grid.
- At the moment of grid failure, the DG grid interface inverter output current should be controlled so as not to exhibit current overshoot to avoid harmonics from polluting the power system.
- DG grid interface inverter shutdown at the moment of grid failure due to instantaneous overcurrent or DC side bus overvoltage should be avoided.
- After the grid fault is restored, the system should be quickly restored to the state before the fault to ensure that all devices continue to operate in the normal operation mode and the grid interfacing inverter of the DG is still connected to the grid with a unity power factor.

Commonly, the ride-through of DG can be enhanced by installing auxiliary devices or/and by grid-interfacing converter control modifications (called active control methods).

Auxiliary Devices for Ride-through Enhancement

DG, particularly the rotating machine-based DG, usually requires auxiliary device support to avoid tripping in voltage sags and other grid fault situations. Different auxiliary devices can offer sufficient dynamic voltage support and improve the DG and microgrid FRT capability. The auxiliary devices can be categorized into series-connected devices, shunt-connected devices, and hybrid-connected devices. More details on power quality compensation device is in Chapter 7.

Series-connected auxiliary technologies, including thyristor-controlled series compensation, dynamic voltage restorer, series dynamic braking resistor, fault current limiter (FCL), are essentially implemented in controlling the voltage or limiting fault current and improving voltage stability in distribution systems. The most widely adopted solution among the series-connected auxiliary devices is the FCL with a more straightforward control strategy. In detail, the FCL is a variable impedance device connected in series with the feeders. In case of a failure, its impedance increases to the value that the system can deal with. After clearing the fault, the device returns to its original low impedance mode within several cycles.

One candidate of FCL is the superconducting fault current limiter (SFCL). The working principle of the SFCL is based on the quench phenomenon of the superconducting material: when the current density, temperature, and magnetic field strength in the superconductor exceed the critical value, the superconductivity is lost immediately. This characteristic makes the superconductor a self-triggered fast transfer switch, transferring its large current to its parallel branch. One typical type of SFCL is the resistive SFCL, in which the superconducting element is connected in parallel with the current limiting resistor or reactance [7]. A typical example of the SFCL is shown in Figure 8.6.

The technical performance of the superconducting current limiter is close to the ideal current limiter, and its advantages include: (i) zero resistance in the normal operation, (ii) self-triggering without an external control system, (iii) fast action speed (the current limiting action can be completed within 1 ms), and (iv) the current limiting depth can generally reach 50% of the expected fault current peak value.

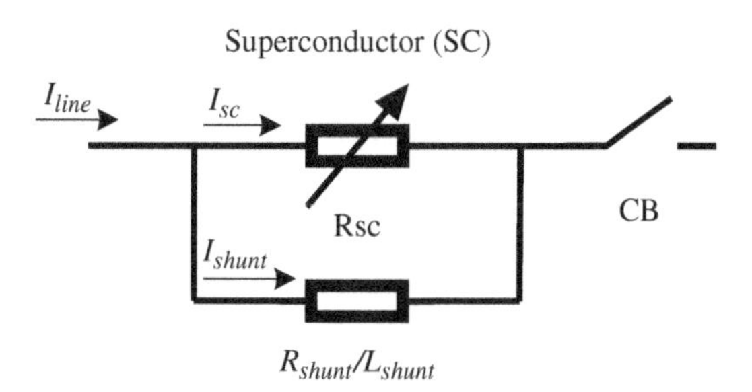

Figure 8.6 A typical example of a resistive type SFCL.

Recently, a power electronics-based FCL with virtual-impedance-based control is also considered to suppress the fault current (please refer to Chapter 4 for more information about virtual impedances) [8, 9]. Low capital and maintenance costs are the key benefits over the other FCLs because additional devices are not needed. In the virtual impedance-based FCL, an additional control loop is added to the inverter control system of the DG grid interface converter to implement large impedance during fault conditions and keep it zero in the normal operating conditions to avoid unnecessary voltage drops. An example control strategy of DG FRT using virtual impedance is provided in Section 8.3.2.0.

In addition to series-connected auxiliary devices for FRT enhancement, some hybrid-connected and shunt-connected devices can also be used, such as a unified power quality conditioner (UPQC), a static synchronous compensator (STATCOM), and a static VAR compensator (SVC). These devices can regulate the reactive power, suppress the voltage variations, and provide enough time for the system to reduce the load and achieve stability.

Another important shunt auxiliary device for fulfilling the FRT requirement is energy storage on the DC side of the DG or DC microgrids, which can control the DC-link voltage under the grid faults. Hence, the disconnection of the grid-interfacing converter of the DG or microgrid can be avoided. The energy storage can also enhance the microgrid stability by injecting power into the system during power shortages and dynamic load situations.

IFC Control Examples for Fault Ride-through

In this section, two examples of IFC control for FRT operation are introduced. Considering the two popular IFC primary control strategies, one example is based on a current-controlled IFC or grid following PV inverter system operation. The other is for a voltage-controlled IFC for use in droop control or virtual synchronous generator control of DG for grid forming operation. A more detailed introduction of the converter's primary control can be found in Chapter 4.

Ride-through Control Example for Current Control Mode of DG The current control method refers to the active and reactive power control of IFCs through current regulation and grid voltage synchronization. This control is also referred as a grid following mode of operation. When the voltage sag occurs on the grid side, the inverter's output current should be appropriately controlled. To ensure that the active power is sufficiently supplied to the load, a certain reactive current should be sent to the grid to support the grid voltage recovery. Thus, the active and reactive output currents must be redistributed according to the voltage drop depth.

Here, a PV system is considered as an example. The PV grid-connected inverter's FRT ability denotes that when the grid voltage momentarily drops, the grid-connected inverter can withstand a certain voltage drop depth and cannot

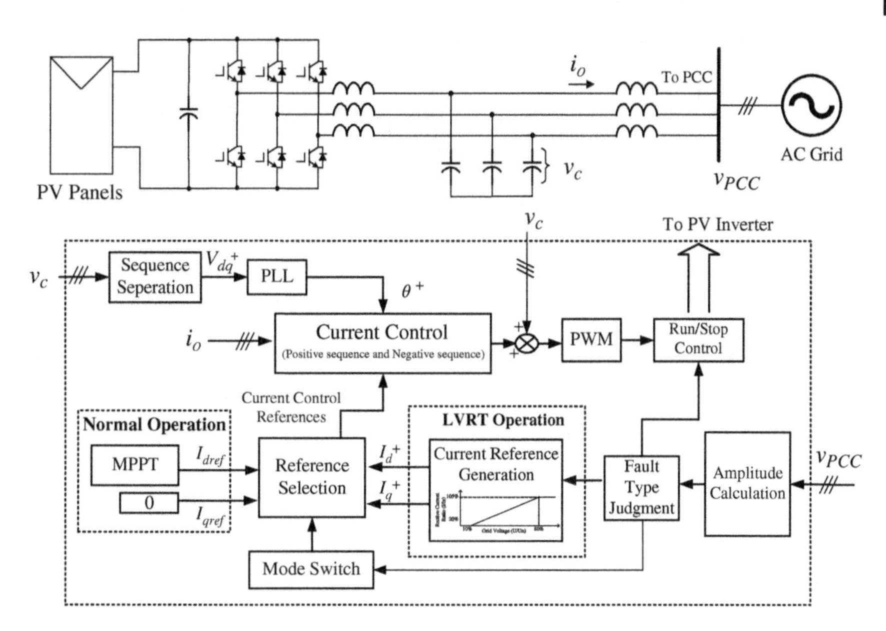

Figure 8.7 Example of LVRT control for a grid-connected PV system.

enter the islanded operating state before a specific time. At the same time, it would produce reactive power to the grid, which helps to support the recovery of grid-side voltage.

A typical example of a control flowchart of FRT enhancement for PV systems is shown in Figure 8.7. For FRT, the new reference values of active and reactive powers are provided according to the predefined voltage-reactive power curve in FRT standards. The generation of reactive power inevitably affects the output of active power. At this time, the active power's reference value should be appropriately reduced to avoid IFC overload.

The converter will focus on power control during normal operation and ensure the PV panel work at its maximum power point (MPP). The control system will keep monitoring the grid voltage amplitude of each phase. Once the voltage exceeds the normal operation range, the IFC will switch its control modes to FRT control. In this case, the reactive power current control references will be generated according to the fault ride-through standards. At the same time, the active power component will be limited to avoid overcurrent. In this case, the MPPT control under normal conditions can no longer be performed. Instead, the highest priority of active power control will turn to keep the DC-link voltage of the converter to be stable. To ensure proper current regulation under disturbances of PCC voltage, PCC voltage feed-forward control is added to the system. As a result, the modulation reference will contain grid voltage amplitude information, and the voltage drops will be immediately compensated.

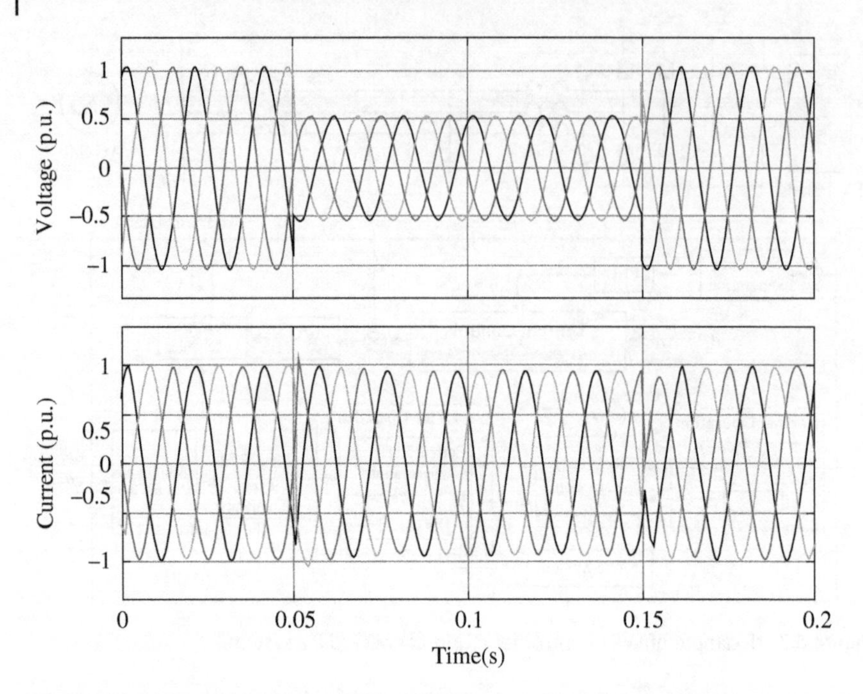

Figure 8.8 Simulation results of FRT of PV inverter under balanced faults.

Figure 8.8 shows the simulation results of the PV inverter's LVRT. As expected, when grid voltage drops to 0.5 p.u., the output current responds quickly with a smooth transient while the current phase angle is adjusted to inject reactive power. As a result, the PV IFC will not perform MPPT. Instead, the active power is reduced to allow the injection of reactive power.

It is worth noting that, in practice, unsymmetrical faults, such as single-line-to-ground faults, double-line-to-ground faults, line-to-line faults, etc. are much more frequent than three-phase symmetrical faults. During unsymmetrical faults, the voltages in three-phase will be different and in three-wire systems, negative sequence components will present. This negative sequence will cause double-line-frequency oscillation on the DC-link of PV converters. Under severe conditions, this oscillation can trigger the DC voltage protection of the PV inverter, leading to FRT failures. It is necessary to regulate the negative sequence current and realize steady active and reactive power. A thorough discussion of the coordination control of the positive and negative sequence of IFC current under unbalanced AC voltage will be discussed in Chapter 9.

Ride-through Control Example for Voltage Controlled DG When grid forming IFC control is used in grid-connected DG, droop control and virtual synchronous generator

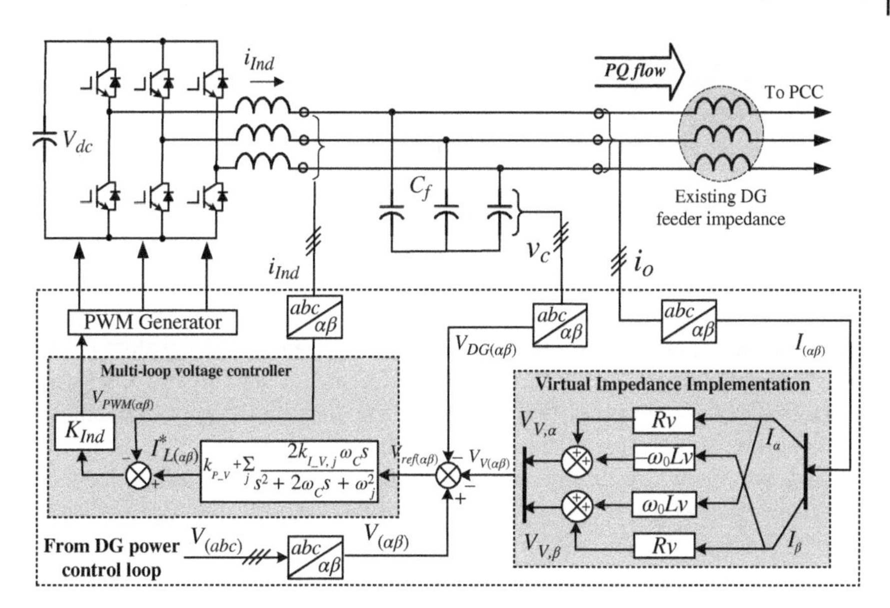

Figure 8.9 A typical example of virtual impedance-based voltage control for RT enhancement of DG.

control are usually adopted. For the voltage-controlled IFC, a grid voltage disturbance can cause a large transient output current. Different strategies have been proposed to improve the voltage-controlled IFC performance under grid disturbances in recent years. The virtual impedance-based voltage control method is practical and easy to implement.

An example from [9] of droop-controlled IFC with a virtual impedance loop for FRT is shown in Figure 8.9. The controller includes the cascaded outer voltage and inner current control loops and the virtual impedance to adapt the voltage control reference and fulfill different control requirements. The outer loop voltage reference comes from the droop control (see Chapter 4 for primary voltage–current control and droop control details). For the virtual impedance control, the voltage drop on the DG output virtual impedance can be expressed as:

$$V_V(s) = (V_{V,\alpha}(s) + jV_{V,\beta}(s)) = (R_V + j\omega_0 L_V)(I_\alpha(s) + jI_\beta(s)) \tag{8.1}$$

where I_α and I_β are the DG output current in the α-β frame, and R_V and L_V are the virtual resistor and inductor values. Under steady-state, the reference voltage comes from the power control loop, and the virtual impedance voltage drop is subtracted from the reference voltage to produce the virtual impedance effects.

During the grid voltage disturbance ride-through, the virtual impedance can also respond properly to adjust its value during the transient to limit the fault current. The voltage-controlled DG system is sensitive to PCC voltage disturbances.

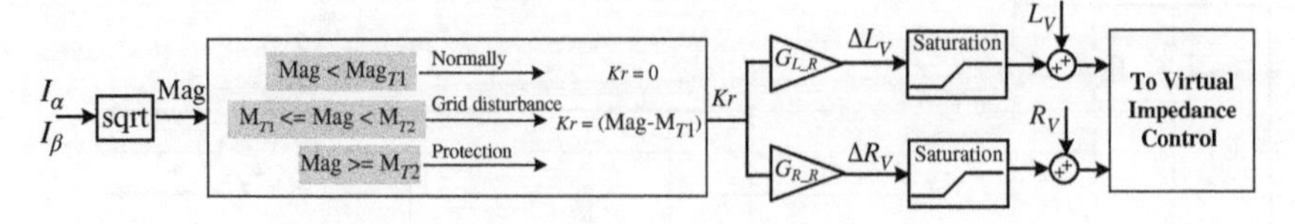

Figure 8.10 Grid disturbance ride-through using adaptive impedance.

Although the grid disturbance will be reflected as output power surges, the detection of PCC voltage disturbances through DG output power is too slow to protect the DG unit (due to the lower pass filtering and relatively slow power measurement). Therefore, the DG output current magnitude (measured without filtering as in the example virtual impedance implementation approach) should be used for the PCC voltage disturbance ride-through.

Figure 8.10 demonstrates the control strategy during grid voltage disturbances. The DG current magnitude (Mag) is first compared with the pre-set thresholds, Mag_{T1} and Mag_{T2}, which are determined by the DG ratings. When the current magnitude is lower than Mag_{T1}, the grid condition is considered normal, and only the designed impedance is implemented. When the current magnitude is higher than Mag_{T2}, a severe fault condition is detected and the DG system will trip for protection. When the current is between these two thresholds, the ride-through function is enabled, and the transient resistance (ΔR_V) and inductance (ΔL_V) are determined as $K_r G_\text{R\_R}$ and $K_r G_\text{L\_R}$ respectively. Here the two gains, $G_\text{R\_R}$ and $G_\text{L\_R}$, are related to the resistive and reactive parts of the virtual impedance. The coefficient K_r is determined by the difference between line current magnitude and lower threshold $\text{Mag}_{T1.}$ As a result, the virtual impedance will be tuned adaptively during the PCC voltage disturbances, where a current-magnitude-dependent series impedance is produced between the DG system and PCC. Considering a high current magnitude will produce large transient impedances, the current introduced by the PCC voltage disturbance will be limited to a lower value by the adaptive impedance control. Finally, as the power control loops will adjust the DG output voltage, the virtual impedance will slowly drop to the initially designed value. It is worth mentioning that enough damping during grid voltage disturbances should be provided by selecting the appropriate gain $G_\text{R\_R}$ in Figure 8.10. After riding through the grid voltage disturbances, the converter can turn to inject necessary reactive power to support the gird.

Figure 8.11 describes the DG system's performance during 9% grid voltage sags with/without adaptive impedance control. When the fixed virtual impedance is adopted in the simulation, the peak-to-peak current is around 19.5 A, as shown in Figure 8.11a. As expected, when the disturbance ride-through function is enabled in Figure 8.11b, the adaptive transient impedance ($\omega_0(L_\text{V}+\Delta L_\text{V})$, $(R_\text{V}+\Delta R_\text{V})$) can be tuned according to line current magnitude. Therefore, the current surge

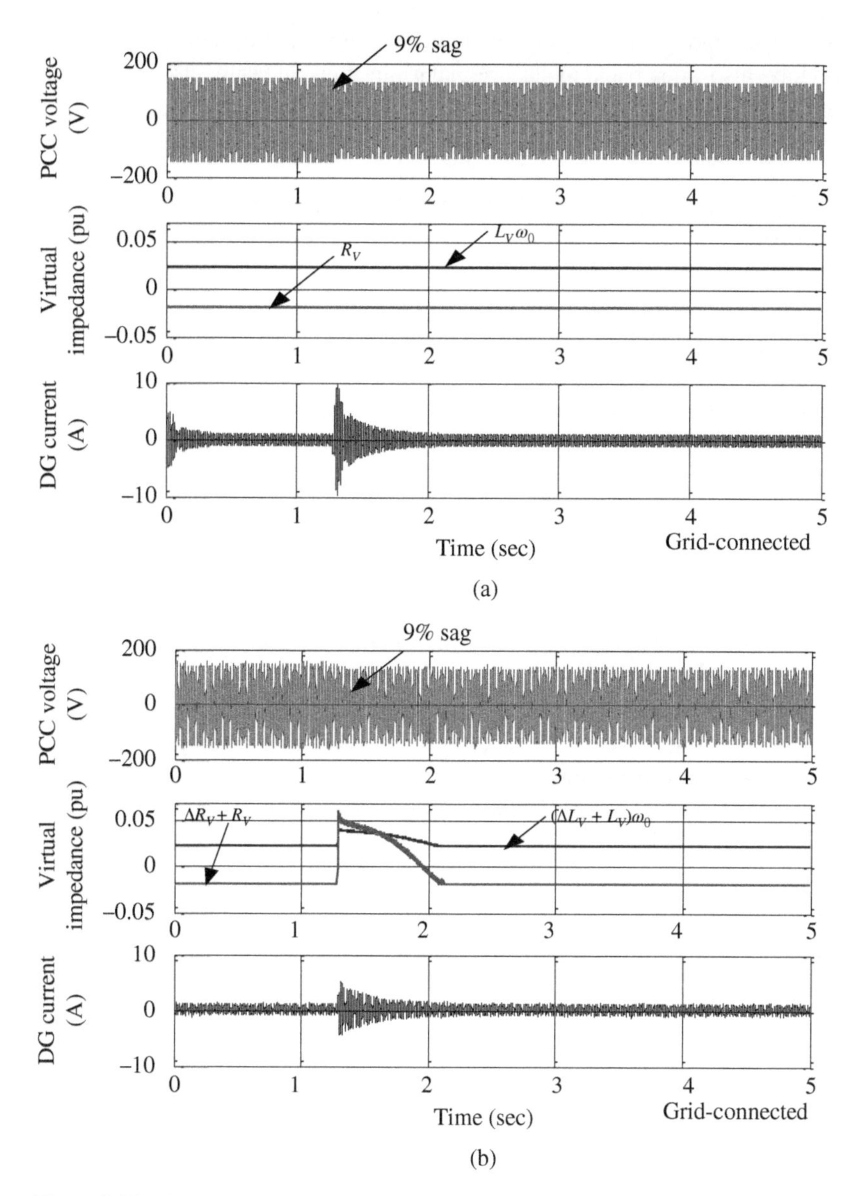

Figure 8.11 Experimental DG performance during PCC voltage sag: (a) with constant virtual impedance; (b) with adaptive virtual impedance.

during a grid voltage sag is obviously reduced. After the transient, the transient impedance also goes back to its original value automatically. The simulated system parameters are provided in Table C.2 of Appendix C.

8.4 Fault Current Contribution and Protection Coordination

DG implementation can negatively impact utilities and microgrids, especially their distribution protection system, during their fault-ride-through operation. For example, the DG fault ride-through operation may contribute to the fault current and cause sympathetic tripping of overcurrent protection relays. In this case, when a fault happens on a feeder adjacent to the DG feeder, the DG may feed the fault current. Since traditional protection systems have been designed based on the assumption of a unidirectional current, the DG feeder's protection will operate, disconnect the DG feeder from the substation, and consequently interrupt power delivery to the DG's feeder even though it has no fault on it.

The DG fault ride-through operation can also negatively impact the fuse-recloser coordination in the fuse-saving protection scheme. In this scheme, the feeder's recloser is designed to operate sooner than the fuse to prevent fuse damage during temporary faults. However, the DG contribution in a fault condition may increase the fuse current and make it operate faster than the recloser, leading to the failure of the fuse-recloser coordination.

To mitigate DG impacts on the traditional protection scheme, several methods have been proposed, which can be classified into four major categories:

- Limiting DG capacity: these methods try to find the DG's maximum capacity so that the DG does not impact the protection system.
- Modifying the protection system: these methods are based on using extra breakers or reclosers, reconfiguring the network, or using distance or directional relays, which are not commonly used in distribution system protection.
- Using adaptive protection: in contrast to traditional protection, in adaptive protection, relays can communicate with each other and have access to remote measurements. Also, their settings can be dynamically adjusted by using a fast processing unit.
- Utilizing FCLs: FCL devices are series elements that show negligible impedance during the network's normal operation. In contrast, during the fault condition, their impedances increase immediately to restrict the current flow through their branches.

While effective for mitigating the DG impact on the protection system, these solutions have some obvious disadvantages. For example, limiting the DG capacity is not a desirable solution since doing so also limits the DG penetration level. Modifying the protection system or add FCLs can be costly and make the protection procedure more complicated.

Compared to those methods, it is more desirable if the DG fault current contribution can be effectively controlled to avoid protection interference. While the rotating machine-based DG (mostly synchronous generator) does not have much current control capability, quick excitation control and adding damping to the excitation circuit can limit the fault current contrition [10]. On the other hand, inverter-based DG does not have a very high fault current due to the inverters' current limitation. However, the aggregate contributions of many small DG units or a few large units are sufficient to increase short circuit level, at least for the first few cycles, which may lead to protection problems [11]. The DG, however, has the flexibility to control its output current in a smart way during the fault ride-through operation and solve protection-related problems without restricting DG utilization during normal conditions.

In this section, the impacts of inverter-based DG systems on fuse-recloser coordination are introduced. Various fault conditions and the effects of different DG locations are discussed. The effects of DG reactive power injection, which are required during fault ride-through operation, on the protection scheme are also analyzed. An example control strategy that limits the DG output current according to the DG terminal voltage is presented to limit the DG impact.

8.4.1 Impact of DG on Fuse-recloser Coordination

A typical fuse-saving protection scheme has the objectives of (i) no interruption in power delivery occurs due to temporary faults, and (ii) fuse burning and replacement are needed only if the fault is quasi-permanent or permanent. When a fault occurs on a lateral like the one shown in Figure 8.12, the recloser first operates once or more times based on its fast time–current curve. Most of the faults in the distribution system are temporary and will be cleared during fast reclosing actions. If the fault is quasi-permanent, the fuse is supposed to clear the fault instead. The time-delayed operation of the recloser will occur if the fuse fails to interrupt the fault current. In the case of a permanent fault, the fuse is set to melt between an automatic recloser's fast and time-delayed operation.

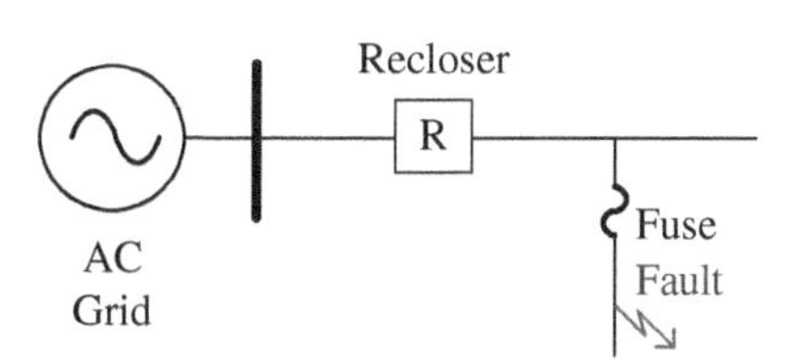

Figure 8.12 Fuse-recloser protection scheme in distribution feeders.

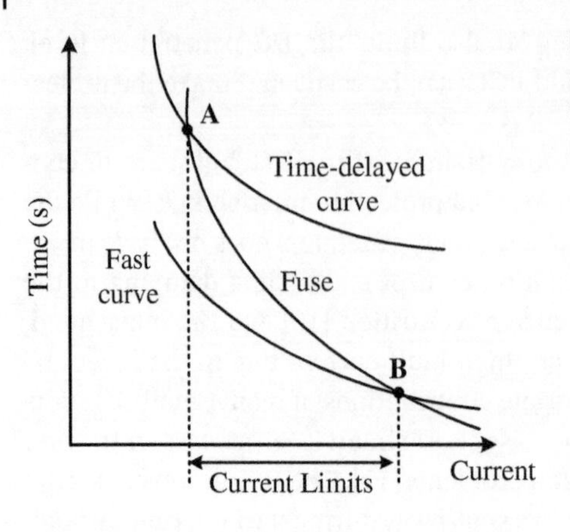

Figure 8.13 Time–current characteristic curves of recloser superimposed on a fuse curve.

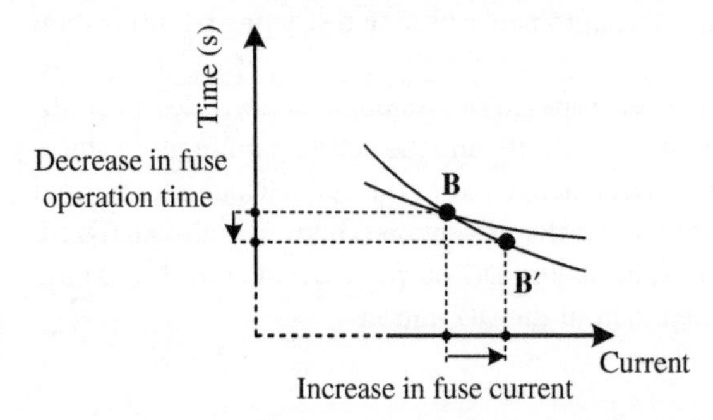

Figure 8.14 Impact of DG on operation of protection system under low impedance fault situation.

For proper coordination, the fuse and recloser curves are selected and set so that for all possible faults, fuse and recloser fault currents remain within the limits shown in Figure 8.13.

The DG fault ride through operation changes the fault current experienced by the fuse and recloser. For example, for low-impedance faults, adding DG to the system may increase the fuse's fault current. This will push the fuse current to the right side of point *B*, as shown in Figure 8.14. In this case, the fuse will melt simultaneously or faster than the recloser's operation, and an undesirable permanent interruption will occur on the lateral, even for temporary faults.

For further analysis, Figure 8.15 shows an equivalent circuit for a system with an inverter-based DG when a fault occurs downstream of the DG. In this figure, the network upstream of the distribution substation is modeled as an equivalent ideal voltage source V_g, and an equivalent impedance Z_{grid}. In this model, loads are neglected (considered as open circuits) during the fault. Besides, Z_1 and Z_2 are feeder impedances from the substation to the PCC and from the PCC to the fault location, respectively. R represents the recloser, and R_f is the fault resistance.

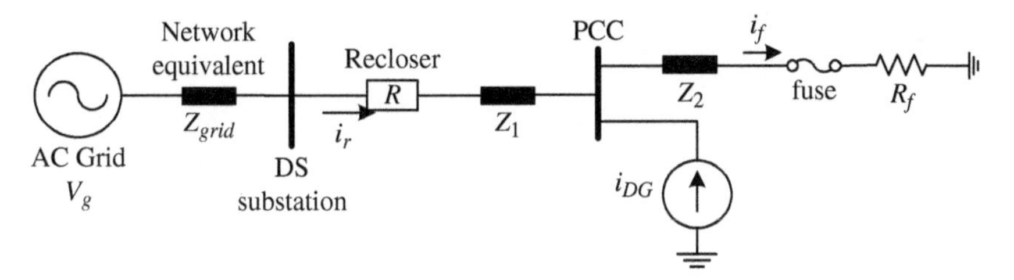

Figure 8.15 Equivalent circuit of a distribution system with inverter-based DG, during a downstream fault.

In this system, the three-phase short circuit per-unit current before implementing DG can be estimated as:

$$i_r = i_f = \frac{V_g}{Z_{grid} + Z_1 + Z_2 + R_f}. \tag{8.2}$$

After adding the DG, the currents through recloser and fuse can be obtained as:

$$i_r = \frac{V_g}{Z_{grid} + Z_1 + Z_2 + R_f} - i_{DG}\frac{R_f + Z_2}{Z_{grid} + Z_1 + Z_2 + R_f} \tag{8.3}$$

$$i_f = \frac{V_g}{Z_{grid} + Z_1 + Z_2 + R_f} + i_{DG}\frac{Z_{grid} + Z_1}{Z_{grid} + Z_1 + Z_2 + R_f}. \tag{8.4}$$

At point B of Figure 8.13, where a low-impedance fault occurs near the PCC ($R_f + Z_2 \ll Z_{net} + Z_1$), the recloser and fuse currents can be approximated as:

$$i_r \approx \frac{V_g}{Z_{grid} + Z_1 + Z_2 + R_f} \tag{8.5}$$

$$i_f \approx \frac{V_g}{Z_{grid} + Z_1 + Z_2 + R_f} + i_{DG}. \tag{8.6}$$

From (8.5) and (8.6), it can be concluded that adding DG increases the fuse current in low-impedance faults, while the fault current experienced by the recloser is almost constant. Thus, in this case, the fuse may operate faster than the recloser, leading to coordination failure. Suppose that before adding DG, point B of Figure 8.13 is reached for a fault with resistance R_{f1}. After adding DG, point B will be reached for a fault with resistance R_{f2}, where necessarily $R_{f2} > R_{f1}$. Consequently, fuse-recloser protection cannot be applied for faults with resistances between R_{f1} and R_{f2}.

On the other hand, for high-impedance faults ($R_f + Z_2 \gg Z_{net} + Z_1$) the recloser and fuse current can be estimated as:

$$i_r \approx \frac{V_g}{Z_{grid} + Z_1 + Z_2 + R_f} - i_{DG} \tag{8.7}$$

$$i_{\rm f} \approx \frac{V_{\rm g}}{Z_{\rm grid} + Z_1 + Z_2 + R_{\rm f}}. \tag{8.8}$$

Equations (8.7) and (8.8) show that for high-impedance faults (like point A of Figure 8.13), the current experienced by the recloser is reduced after adding DG, while the fuse current is almost constant. Under this condition, the fuse operation time remains constant, but the recloser time-delayed operation occurs with an additional delay. However, this delay will not cause miscoordination since the fuse operates sooner than the recloser's time-delayed operation.

Furthermore, analysis of the situation of a fault that occurs upstream to the DG can be conducted similarly. It has been found that the situation matches closely with that of a downstream fault. This means the location of faults does not play a significant role in the fuse-recloser coordination. Instead, the DG current and fault impedance will have more impact on the coordination. Compared to a high-impedance fault situation, where the coordination is usually maintained, a low-impedance fault tends to cause more problems.

8.4.2 Impact of Reactive Power Injection on Fuse-recloser Coordination

The provision of reactive power has been considered as an important ancillary service offered by inverter-based DG. However, if not controlled and appropriately coordinated, this reactive power injection may worsen the protection miscoordination.

Consider a low-impedance downstream fault as an example, where the fuse current has two parts, as shown in (8.6). In the active power injection from the DG, in (8.3), the first part of the fuse current $\frac{V_{\rm g}}{Z_{\rm net} + Z_1 + Z_2 + R_{\rm f}}$ is about 90° out of phase with the PCC voltage, while the second part (the DG current) $I_{\rm DG}$ is forced to be in phase with the PCC voltage. As a result, the two terms have a phase shift of around 90°, and the relationship in (8.9) can be obtained:

$$\left|\frac{V_{\rm g}}{Z_{\rm grid} + Z_1 + Z_2 + R_{\rm f}} + i_{\rm DG}\right| < \left|\frac{V_{\rm g}}{Z_{\rm grid} + Z_1 + Z_2 + R_{\rm f}}\right| + |i_{\rm DG}|. \tag{8.9}$$

On the other hand, in reactive power injection from DG, both terms in the fuse current ($\frac{V_{\rm g}}{Z_{\rm grid} + Z_1 + Z_2 + R_{\rm f}}$ and $I_{\rm DG}$) are about 90° out of phase with the PCC voltage. Consequently, the relationship in (8.10) can be obtained:

$$\left|\frac{V_{\rm g}}{Z_{\rm grid} + Z_1 + Z_2 + R_{\rm f}} + i_{\rm DG}\right| \approx \left|\frac{V_{\rm g}}{Z_{\rm grid} + Z_1 + Z_2 + R_{\rm f}}\right| + |i_{\rm DG}|. \tag{8.10}$$

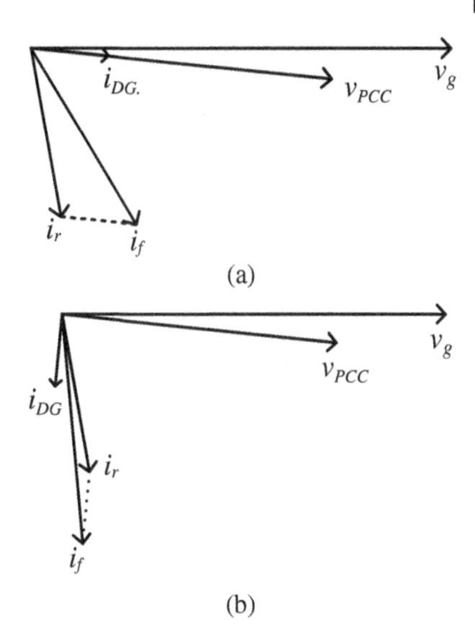

Figure 8.16 Voltage and current vector diagram of (a) DG provides active power, (b) DG provides reactive power.

For further illustration, Figure 8.16 shows the vector diagrams of the above voltages and currents (note that the direction of DG current is to the PCC). Figures 8.16a, b show that the fuse's fault current is larger when the DG provides reactive power. This means that in comparison to active power injection, reactive power injection worsens the fuse-recloser coordination situation.

Due to the LVRT requirement, the DG must stay connected and provide reactive power for the network during voltage sag. Since the fuse saving scheme is a commonly used protection method in the medium-voltage distribution feeders, satisfying these DG requirements could worsen the miscoordination problem. As a result, grid code compatibility for DG interconnection to the grid with over-current protection system should be carefully studied, and a suitable DG control strategy that can satisfy both requirements is necessary.

8.4.3 Example of Inverter Current Control Strategy under RT

To mitigate the impacts of DG on the fuse-recloser coordination, a simple and effective DG current control strategy is presented in this section [11].

One idea to simultaneously solve the miscoordination problem and ride through short-term disturbances is to reduce the converter current according to the severity of the abnormality. In this case, the converters near the fault location, which will produce the greatest impact on the protection system, experience the most voltage deviation from the normal boundaries and should significantly decrease their fault current contribution. On the other hand, the more distant DG units, which have no substantial effect on the protection system, can continue their power delivery.

According to the DG terminal voltage, the DG's reference current can be determined as in (8.11) to implement the above-mentioned current-control strategy.

$$\begin{cases} I_{\text{ref.}} = \dfrac{P_{\text{desired}}}{V_{\text{PCC}}} & \text{for } V_{\text{PCC}} \geq 0.88\,\text{p.u.} \\ I_{\text{ref.}} = kV_{\text{PCC}}^{n} I_{\max} & \text{for } V_{\text{PCC}} < 0.88\,\text{p.u.} \end{cases} \tag{8.11}$$

where I_{ref} is the converter reference current, $I_{\max}$ is the maximum output current that happens at $V_{\text{PCC}} = 0.88$ p.u. (the lower boundary for normal operation according to IEEE 1547 [2]), V_{PCC} is the voltage at the DG connection node, P_{desired} is the output desired power and k and n are constants to be determined.

Generally, the value of n determines the sensitivity of the control scheme to a voltage change. A larger value of n leads to a more obvious output current reduction with a voltage sag. However, too large n will cause the control scheme to be over sensitive to even a small voltage disturbance. With the above consideration, $n = 3$ is selected as a trade-off. Once the value of n is chosen, the coefficient k can be determined so that the reference current in (8.11) remains a continuous function around $V_{\text{PCC}} = 0.88$ p. u.; i.e. k can be obtained from (8.12), which gives $k = 1.4674$ when $n = 3$.

$$k(0.88)^{n} I_{\max} = \frac{P_{\text{desired}}}{0.88} \Rightarrow k = \frac{P_{\text{desired}}}{(0.88)^{n+1} I_{\max}}. \tag{8.12}$$

Figure 8.17 shows the flow chart illustrating the reference current determination procedure.

Test Cases Study

Performance during a Low-impedance Fault Condition To investigate the aforementioned strategy's ability to maintain fuse-saving coordination, a 13-node test

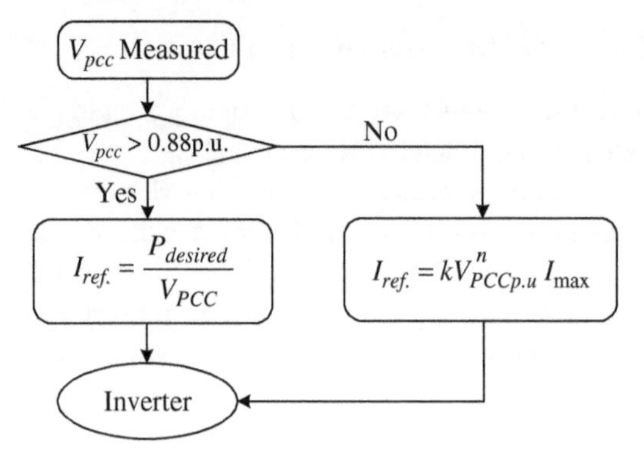

Figure 8.17 Control strategy to determine inverter reference current.

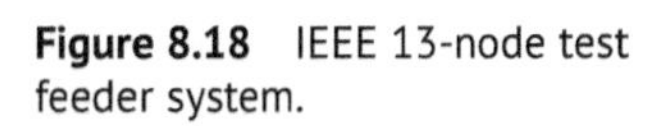

Figure 8.18 IEEE 13-node test feeder system.

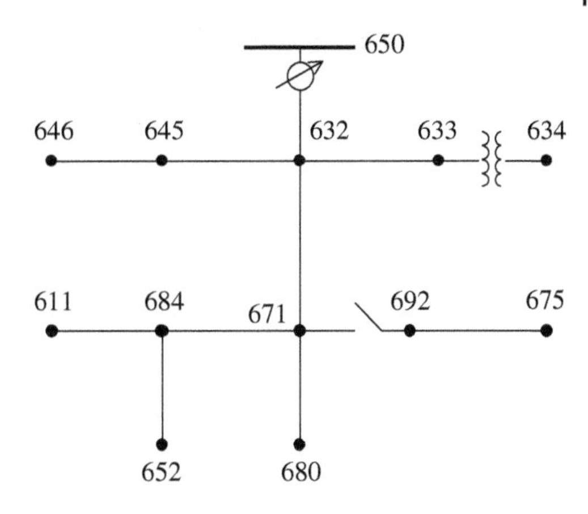

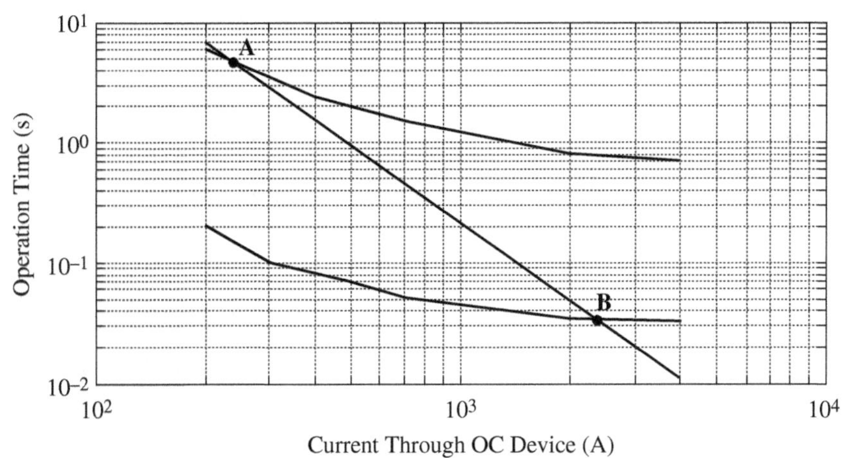

Figure 8.19 Fuse-recloser coordination in the simulated system.

feeder system (see Figure 8.18) [12] is constructed. A recloser was mounted at the substation, and an inverter-based DG was connected at node 645. The simple current-controlled voltage source inverter model is considered in this work. The DG's effect on the coordination between the recloser and the fuse on 645–646 was studied for faults in the middle of 645–646. Figure 8.19 shows the time–current characteristic curves of the over current (OC) devices of recloser and fuse used in simulations.

In the simulated system, point B is related to a fault resistance of 0.01 Ω, and point A corresponds to a fault resistance of 11.5 Ω. In other words, the fuse and recloser operate properly for fault resistances between 0.01 and 11.5 Ω. For faults with resistance lower than 0.01 Ω, the fuse will operate faster than the recloser, and the use of the fuse-saving scheme is not feasible.

Table 8.2 Fuse-recloser coordination study summary.

Case study scenarios	DG penetration level that miscoordination happens
0.01 Ω fault impedance	10%
0.1 Ω fault impedance	70%
0.01 Ω fault impedance, and reactive power compensation	10%
0.1 Ω fault impedance, reactive power injection	30%
With the control strategy in Figure 8.17	Always coordinated

Main study results related to different fault impedance, DG penetration level (penetration level= $\frac{P_{DG}}{P_{\text{load}}} \times 100$), and reactive power generation are summarized in Table 8.2, with the main conclusions as below:

- With a low impedance fault (0.01 Ω) occurs, low penetration levels of DG of 10% will make the fuse operates faster than the recloser, and the protection coordination is lost.
- With higher fault resistance of 0.1 Ω, the presence of DG at high penetration levels of 70% will cause miscoordination.
- When DG is controlled to inject a fully reactive current when it experienced a voltage sag ($V_{PCC} < 0.88$ p.u.), the coordination is more complex. For example, with fault impedance of 0.1 Ω, with reactive power injection, miscoordination occurred at low penetration levels of 30% (compared to 70% without reactive power injection).
- When the above-mentioned control scheme is implemented, it can successfully solve the miscoordination in low-resistance faults due to DG injection.

Performance During a High-impedance Fault Condition The converter's operation controlled by the control strategy during medium- and high-impedance faults are studied here. In the first case, three types of DG at a 30% total penetration level (10% each) were implemented on nodes 633, 645, and 675, and a 2 Ω fault was simulated in the middle of 645–646. Figure 8.20 shows the converters' output current. This figure reveals that the more distant converters experienced lower voltage sags and, consequently, provided more current during a fault.

In the second case, a DG has been installed at node 634 with a capacity equal to the spot load at this node, and an 8 Ω fault has been simulated from 0.2 to 0.3 s in the middle of nodes 645–646. Figure 8.21 shows the inverter output power in this situation. Figure 8.22 shows the PCC voltage, and Figure 8.23 shows the fuse current. According to these results, the converter with the control strategy can

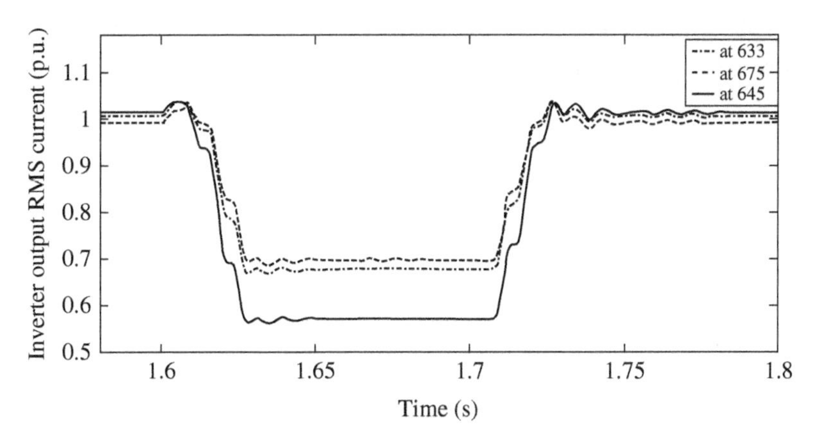

Figure 8.20 Output current of inverters for a 2 Ω fault in the middle of 645–646.

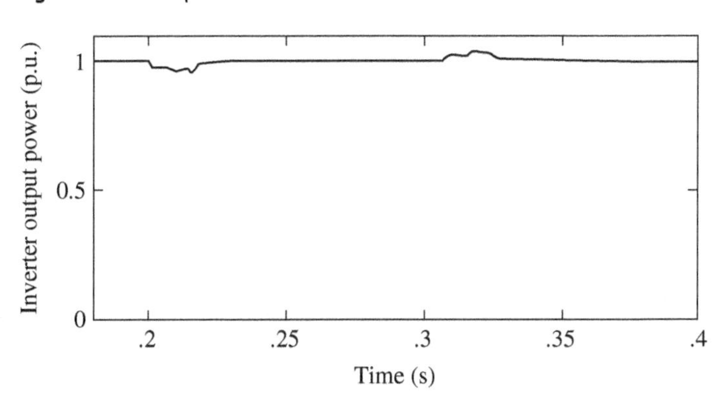

Figure 8.21 Inverter output power for a 8 Ω fault in the middle of 645–646.

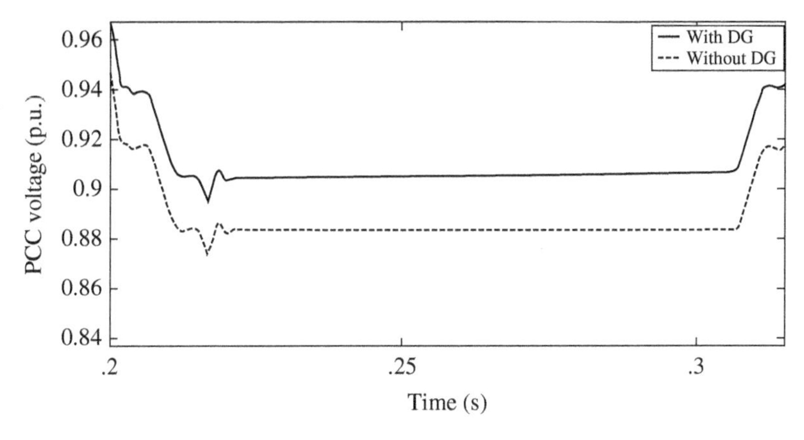

Figure 8.22 PCC voltage for a 8 Ω fault in the middle of 645–646.

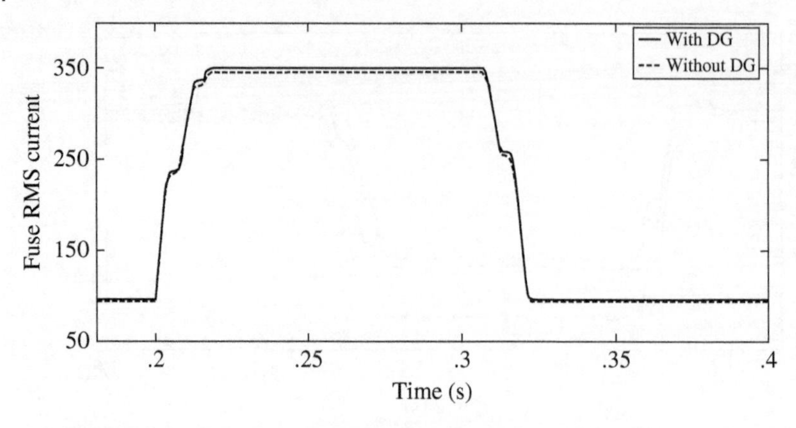

Figure 8.23 Fuse current for a 8 Ω fault in the middle of 645–646.

successfully ride through high impedance or distant faults without losing its power delivery capability and improves the voltages of local nodes without imposing any impact on the protection system.

8.5 Summary

The control of AC IFCs is critical to providing reliable and stable operation for smart microgrids during grid disturbances such as voltage sags and swells. First, different islanding detection methods are reviewed in this chapter, which is essential for IFCs to switch control strategies under a grid disturbance. Unlike the traditional anti-islanding operation, the FRT capability of IFCs, such as LVRT and HVRT, is increasingly essential for grid stability with the high penetration of DG. Details on the current FRT requirement as well as example FRT control strategies are provided. Finally, when the converter operates under ride-through mode, the fault current contribution to the grid may affect the traditional protection practice and, therefore, a study on this is also provided in this chapter.

References

1 (2003). IEEE standard for interconnecting distributed resources with electric power systems. In: *IEEE Std 1547-2003*, 1–28. https://doi.org/10.1109/IEEESTD.2003.94285.

2 (2018). IEEE standard for interconnection and interoperability of distributed energy resources with associated electric power systems interfaces. In: *IEEE*

Std 1547–2018 (Revision of IEEE Std 1547-2003), 1–138. https://doi.org/10.1109/IEEESTD.2018.8332112.

3 Kim, M.-S., Haider, R., Cho, G.-J. et al. (2019). Comprehensive review of islanding detection methods for distributed generation systems. *Energies* 12 (5): 837.

4 Ray, P.K., Kishor, N., and Mohanty, S.R. (2012). Islanding and power quality disturbance detection in grid-connected hybrid power system using wavelet and S-transform. *IEEE Transactions on Smart Grid* 3 (3): 1082–1094. https://doi.org/10.1109/TSG.2012.2197642.

5 Boemer, J.C. and Walling, R. (2018). DER ride-through performance categories and trip settings. *Proceedings of the PJM Ride-Through Workshop, Philadelphia*, PA, USA.

6 Reliability Guideline Bulk Power System Reliability Perspectives on the Adoption of IEEE 1547–2018. NERC Report, March (2020).

7 Tsuda, M., Mitani, Y., Tsuji, K., and Kakihana, K. (2001). Application of resistor based superconducting fault current limiter to enhancement of power system transient stability. *IEEE Transactions on Applied Superconductivity* 11 (1): 2122–2125. https://doi.org/10.1109/77.920276.

8 Lu, X., Wang, J., Guerrero, J.M., and Zhao, D. (2018). Virtual-impedance-based fault current limiters for inverter dominated AC microgrids. *IEEE Transactions on Smart Grid* 9 (3): 1599–1612. https://doi.org/10.1109/TSG.2016.2594811.

9 He, J. and Li, Y.W. (2011). Analysis, design, and implementation of virtual impedance for power electronics interfaced distributed generation. *IEEE Transactions on Industry Applications* 47 (6): 2525–2538. https://doi.org/10.1109/TIA.2011.2168592.

10 Yazdanpanahi, H., Xu, W., and Li, Y.W. (2014). A novel fault current control scheme to reduce synchronous DG's impact on protection coordination. *IEEE Transactions on Power Delivery* 29 (2): 542–551. https://doi.org/10.1109/TPWRD.2013.2276948.

11 Yazdanpanahi, H., Li, Y.W., and Xu, W. (2012). A new control strategy to mitigate the impact of inverter-based DGs on protection system. *IEEE Transactions on Smart Grid* 3 (3): 1427–1436. https://doi.org/10.1109/TSG.2012.2184309.

12 Kersting, W.H. (2001). Radial distribution test feeders. In: *2001 IEEE Power Engineering Society Winter Meeting. Conference Proceedings* (Cat. No.01CH37194), Columbus, OH, USA, vol. 2, 908–912. https://doi.org/10.1109/PESW.2001.916993.

9

Unbalanced Voltage Compensation in Smart Hybrid Microgrids

9.1 Introduction

In hybrid AC/DC microgrids, an unbalanced condition is caused by the ever-increasing unbalanced distribution of single-phase/unbalanced loads, single-phase/unbalanced distributed generation (DG), and remote grid faults. The unbalanced voltage has adverse effects on the power system and equipment, including electrical machine overheating, transformer overloading, capacity limitation of power electronics devices, more losses and less stability of the power system, and negative impacts on induction motors and adjustable speed drives. Also, the unbalanced voltage introduces adverse effects on the power electronic converters, such as increased peak current of the converter and double-frequency power oscillations at the output of three-phase converters. The power oscillations are also reflected as ripples in the DC link voltage. These adverse effects may lead to instability or system protection if the DC bus voltage exceeds the maximum limit, affecting DC converters and loads in the DC subgrid, and may result in over-current protection.

In general, the unbalanced voltage can be compensated for by installing additional devices in the power systems, such as a unified power quality conditioner (UPQC), static synchronous compensators (STATCOMs), active power filters, shunt capacitors, to name a few. For detailed information about power quality issues and compensation methods, please refer to Chapter 7. However, installing additional equipment increases the total investment cost. Also, centralized power quality compensation does not work well as the unbalance in a hybrid AC/DC microgrid is usually caused by distributed loads and generations. With the increasing penetration of power electronics converters equipment in hybrid AC/DC microgrids, they can be smartly controlled to improve the unbalanced condition in addition to their power management purposes. One critical piece of equipment is the interfacing converter (IFC), which can be in three-phase or single-phase configuration from DGs and energy storage systems (ESSs). It can

Smart Hybrid AC/DC Microgrids: Power Management, Energy Management, and Power Quality Control, First Edition. Yunwei Ryan Li, Farzam Nejabatkhah, and Hao Tian.

also link the AC and DC subgrids. In the case in which the output power/current of the DG/ESS connected to the AC subgrid is high, or high power/current is planned to be exchanged between AC and DC subgrids, parallel IFCs are used, which provide higher reliability.

9.2 Control of Individual Three-phase IFCs for Unbalanced Voltage Compensation

Three-phase IFCs are used to connect DGs/ESSs to the AC subgrid and interlink AC and DC subgrids. These converters provide an excellent opportunity for unbalanced voltage compensation of a three-phase AC subgrid in hybrid microgrids. On the other hand, the unbalanced voltage has adverse effects on IFC operations. In Figure 9.1, the performance of a three-phase IFC under unbalanced voltage is shown. The unbalanced voltage increases the IFC's output peak current with the same active and reactive power production. Also, power oscillations at twice the fundamental frequency are induced to the IFC's output active and reactive powers. This power oscillation is also reflected into the DC side of the IFC as ripples in the DC link voltage.

As shown in Figure 9.1, when three-phase IFCs operate under unbalanced voltage in hybrid microgrids, the unbalanced voltage causes adverse effects on their operation, such as increasing peak current, power oscillations, and DC link voltage ripples. However, actively compensating for the unbalanced or negative sequence voltage in the microgrid may aggravate the adverse effects on IFCs. As discussed in Chapter 4, for unbalanced compensation purposes, the IFC's virtual negative sequence impedance can be controlled to be much smaller than the grid side impedance and absorb more negative sequence current than the grid (since the grid negative sequence current is reduced, the unbalanced voltage of grid will reduce). Based on the desired compensation level, the converter's virtual impedance can be determined in the secondary microgrid or distribution system

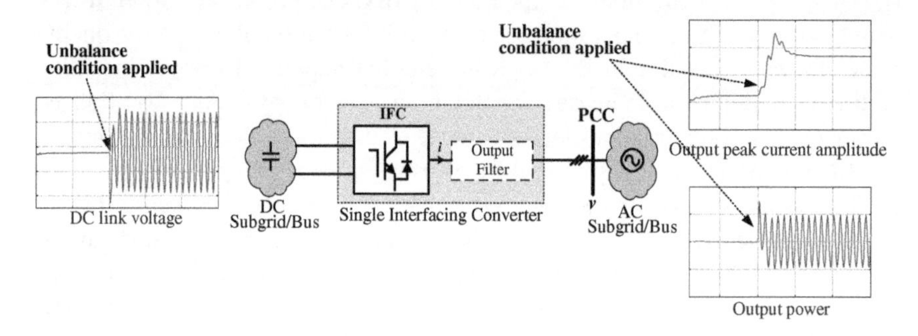

Figure 9.1 Typical three-phase IFC performance under unbalanced condition.

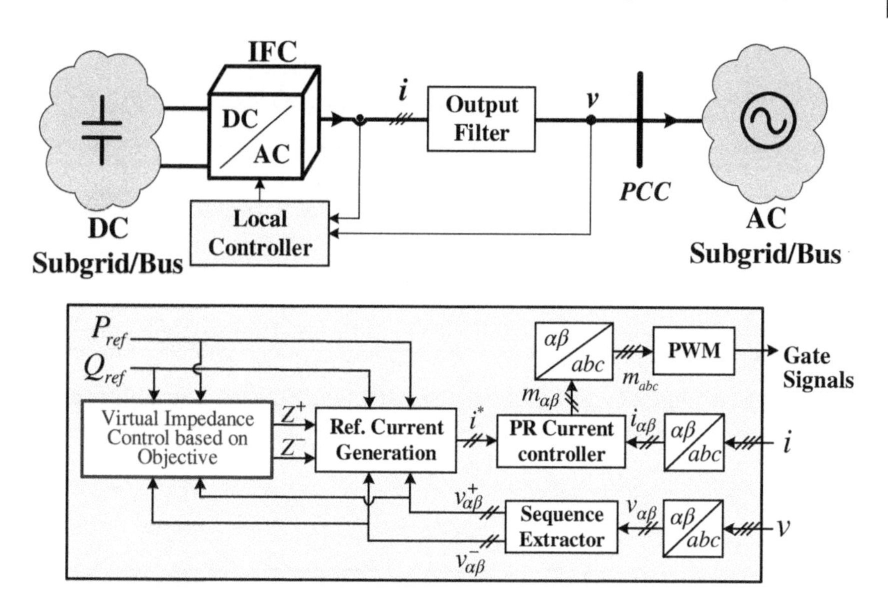

Figure 9.2 Typical individual three-phase IFC with control block diagram under unbalanced voltage.

controller. The impedances can then be used to update the IFC's reference current in the primary controller (for IFC current control) for unbalanced condition compensation. In this chapter, IFC primary control with current control mode (CCM) is considered, typically used in the grid-following converters. Please refer to Chapters 4 and 6 for more detail on the converter's primary and secondary control.

Figure 9.2 shows a typical individual three-phase IFC with a control block diagram under unbalanced voltage. In this example, the IFC local control system contains the outer loop (responsible for the IFC's reference current generation) and the inner current control loop (responsible for quick and accurate tracking of the reference current). The control is implemented in the stationary $\alpha\beta$ reference frame to avoid multiple frame transformations for different sequence component control. A sequence extractor can separate the point of common coupling (PCC) voltage into positive and negative sequences in the outer loop. This sequence extractor can be realized in many ways, such as advanced phase-locked loops, delay signal cancelation, to name a few [1]. In the inner control loop, the IFC is controlled in CCM using the proportional-resonant controller.

9.2.1 Three-phase IFC Model under Unbalanced Voltage

Figure 9.3 shows a three-phase IFC connected to the AC subgrid at the PCC with LCL filter, and an unbalanced load is connected to the PCC as a source of unbalanced voltage. Both positive and negative sequences virtual impedances of the IFC

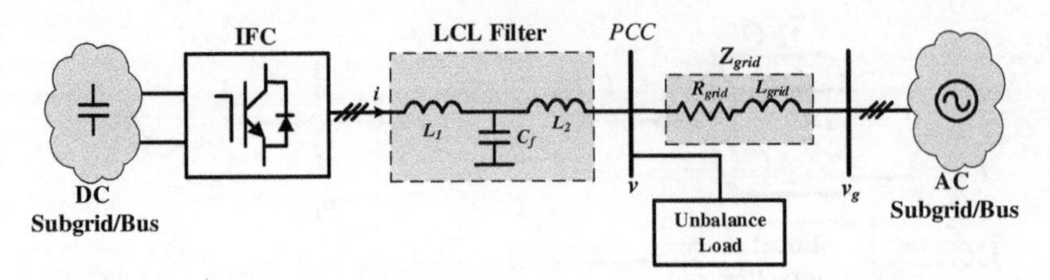

Figure 9.3 Single three-phase interfacing converter.

at fundamental frequency can be controlled for different compensation purposes, which can be achieved through the IFC reference current control.

In (9.1), the reference current of three-phase IFC is established to flexibly control both positive and negative sequence current injection to the grid [2, 3]. A more detailed derivation of this is provided in Appendix A.

$$i^* = \underbrace{\frac{Pk_1}{|v^+|^2}v^+}_{i_p^+} + \underbrace{\frac{P(1-k_1)}{|v^-|^2}v^-}_{i_p^-} + \underbrace{\frac{Qk_2}{|v^+|^2}v_\perp^+}_{i_q^+} + \underbrace{\frac{Q(1-k_2)}{|v^-|^2}v_\perp^-}_{i_q^-} \tag{9.1}$$

where P and Q are the IFC's output average active and reactive powers, v^+ and v^- are positive and negative sequence vectors of the PCC voltages, and $v_\perp^+$ lags v^+ by 90°. Also, i_p^+ and i_p^- are the positive and negative sequence vectors of active currents while i_q^+ and i_q^- are the positive and negative sequence vectors of reactive currents.

As is clear from (9.1), each positive and negative sequence current can be decomposed into active and reactive components. This concept can be extended into a virtual impedance, where each positive and negative sequence virtual impedance can be decomposed into active and reactive virtual impedances. From (9.1), the portions of positive and negative sequences active and reactive currents or virtual impedances can be adjusted by:

$$k_1 = \frac{P^+}{P} \tag{9.2}$$

$$k_2 = \frac{Q^+}{Q} \tag{9.3}$$

which can be appropriately designed to achieve different control objectives such as unbalanced voltage compensation and control of the IFC output power oscillations, peak current, and DC link voltage ripples.

From (9.1), the IFC's instantaneous active and reactive powers injected into the grid can be achieved as:

$$\begin{aligned} p &= \left(\underbrace{v^+.i^+}_{P^+} + \underbrace{v^-.i^-}_{P^-} \right) + \left(\underbrace{v^+.i^- + v^-.i^+}_{\Delta P} \right) \\ &= P + \left(\frac{Pk_1}{|v^+|^2} + \frac{P(1-k_1)}{|v^-|^2} \right) v^+ \cdot v^- + \left(\frac{Qk_2}{|v^+|^2} - \frac{Q(1-k_2)}{|v^-|^2} \right) v_\perp^+ \cdot v^- \end{aligned} \tag{9.4}$$

$$q = \left(\underbrace{v_{\perp}^{+}.i^{+}}_{Q^{+}} + \underbrace{v_{\perp}^{-}.i^{-}}_{Q^{-}} \right) + \left(\underbrace{v_{\perp}^{+}.i^{-} + v_{\perp}^{-}.i^{+}}_{\Delta Q} \right) = Q + \left(\frac{Pk_1}{|v^{+}|^2} - \frac{P(1-k_1)}{|v^{-}|^2} \right) v^{+} \cdot v_{\perp}^{-}$$
$$+ \left(\frac{Qk_2}{|v^{+}|^2} + \frac{Q(1-k_2)}{|v^{-}|^2} \right) v_{\perp}^{+} \cdot v_{\perp}^{-} \tag{9.5}$$

where P and ΔP are average and oscillatory terms of instantaneous active power, P^{+} and P^{-} are the positive and negative sequences of the IFC's average active power, Q and ΔQ are average and oscillatory terms of the instantaneous reactive power, and Q^{+} and Q^{-} are the positive and negative sequences of the IFC's average reactive power, respectively.

The positive and negative sequence virtual impedances of the IFC (at the fundamental frequency) can be derived using (9.6) and (9.7):

$$i^{+^*} = Y^{+}v^{+} = \left(\frac{P\,k_1}{|v^{+}|^2} - \frac{Q\,k_2}{|v^{+}|^2} e^{j\frac{\pi}{2}} \right) v^{+} \tag{9.6}$$

$$i^{-^*} = Y^{-}v^{-} = \left(\frac{P\,(1-k_1)}{|v^{-}|^2} + \frac{Q\,(1-k_2)}{|v^{-}|^2} e^{j\frac{\pi}{2}} \right) v^{-}. \tag{9.7}$$

From (9.6) and (9.7), the sequence network model of an individual three-phase IFC can be developed. Figure 9.4 shows the equivalent negative sequence model of an individual grid-connected three-phase IFC (no zero sequence in the system). In this figure, the positive sequence equivalent network together with unbalanced load are represented by an equivalent current source i_{load}^{-}. (Note that this value depends on the unbalanced load configuration as well as the positive sequence network parameters.) Moreover, in Figure 9.4, the IFC is represented as a virtual impedance Z^{-} in order to emulate its behavior in the negative sequence circuit (see (9.7)). The grid negative sequence impedance (Z_{grid}^{-}) is shown in parallel configuration, and the currents directions are assumed as shown in the figure for better explanation.

If the three-phase unbalanced load is connected to the PCC in a star connection (Z_{L1}, Z_{L2}, Z_{L3}), i_{load}^{-} can be represented as (9.8). (Note that using three-phase

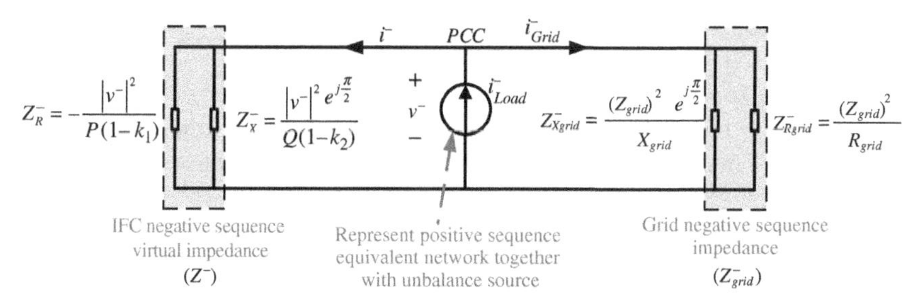

Figure 9.4 Sequential model of a three-phase IFC connected to the AC subgrid.

network symmetrical component analysis, i^-_{load} can be extended to other load types, e.g. single-phase load.)

$$i^-_{\text{load}} = \frac{v^+(a^2Z_{L1} + aZ_{L2} + Z_{L3})}{Z_{L1}Z_{L3} + Z_{L2}Z_{L3} + Z_{L1}Z_{L2} + Z^-_{\text{PCC}}Z_L + Z^+_{\text{PCC}}Z_L + 3Z^+_{\text{PCC}}Z^-_{\text{PCC}}} \tag{9.8}$$

where Z_L is the summation of load impedances ($Z_L = Z_{L1} + Z_{L2} + Z_{L3}$), and a is the rotation operator $\left(a = e^{j\frac{2\pi}{3}}\right)$. Moreover, v^+ represents the voltage of the PCC under balanced condition (without the presence of the unbalanced load), and Z^+_{PCC} and Z^-_{PCC} are positive and negative sequence impedances seen from the PCC. Considering Figures 9.3 and 9.4, these parameters can be achieved as follows:

$$v^+ = \frac{Z^+_{\text{PCC}}}{Z^+_{\text{PCC}} + Z^+_{\text{grid}}} \times v_g \tag{9.9}$$

$$Z^+_{\text{PCC}} = Z^+ \| Z^+_{\text{grid}} \tag{9.10}$$

$$Z^-_{\text{PCC}} = Z^- \| Z^-_{\text{grid}} \tag{9.11}$$

where Z^+_{grid} is the grid positive sequence impedance, v_g is the grid (AC subgrid) voltage, and Z^+ is the IFC positive sequence virtual impedance that can be expressed in (9.12) using (9.6).

$$Z^+ = Z^+_R \| Z^+_X = \left(-\frac{|v^+|^2}{Pk_1}\right) \| \left(\frac{|v^+|^2}{Qk_2}e^{-j\frac{\pi}{2}}\right) \tag{9.12}$$

In the case that weighting factors k_1 and k_2 are equal to one ($k_1 = 1$ and $k_2 = 1$), just positive sequence active and reactive powers are injected into the grid, and the IFC output AC current is balanced with only positive sequence components.

Considering the sequence network equivalent circuit in Figure 9.4, two scenarios are discussed.

***Scenario 1**: The IFC produces balanced AC current without compensation of unbalanced voltage.* In the case that weighting factors k_1 and k_2 are selected to be one ($k_1 = 1$ and $k_2 = 1$), just positive sequence active and reactive powers are injected into the grid. From Figure 9.4, both Z^-_R and Z^-_X are open circuit, and the negative sequence voltage of PCC can be expressed in (9.13)

$$|v^-| = |i^-_{\text{load}}| \times |Z^-_{\text{grid}}|. \tag{9.13}$$

***Scenario 2**: The IFC produces unbalanced current for compensation of unbalanced voltage.* In the case that weighting factors k_1 and k_2 are not equal to one ($k_1 \neq 1$ and $k_2 \neq 1$), both positive and negative sequences active and reactive powers are injected into the grid. From Figure 9.4, the negative sequence voltage of the PCC can be achieved:

$$|v^-| = |i^-_{\text{load}}| \times |Z^-_{\text{grid}}//Z^-|. \tag{9.14}$$

From (9.13), (9.14), and Figure 9.4, it can be seen that injecting the negative sequence active and reactive powers into the grid using a three-phase IFC could compensate for the PCC unbalanced voltage. In other words, the negative sequence virtual impedance of the IFC (Z^-) can be controlled to be much smaller than the grid impedance, directing i_{load}^- to flow to the IFC side and leading to an improved PCC voltage. On the other hand, such control of IFC negative sequence virtual impedance may also degrade the IFC performance. This influence is elaborated in the following sections.

9.2.2 Control of Unbalanced Voltage Adverse Effects on IFC Operation

As mentioned, the PCC unbalanced voltage can increase the IFC's output peak current and induce active and reactive power oscillations at the IFC output, which transfer to the DC side as ripples. The IFC's output current control typically considers the reference tracking of active and reactive power, unbalanced compensation, and minimization of the unbalanced adverse effects. The active and reactive power reference following is typical of the highest priority for the smart converters. The additional control objectives could be unbalanced compensation or minimizing adverse effects under unbalance, or both, depending on the operational objectives. As discussed in Section 9.2.1, different objectives can be realized by designing parameters k_1 and k_2 properly.

Here, a simple design example of the parameters k_1 and k_2 is provided, when the control objective of adverse effects, such as oscillating power and DC link ripple, is a priority.

Considering (9.1) and (9.4), if two constraints in (9.15) are satisfied, the active power oscillations at the output of IFC are canceled, which can therefore provide oscillation-free DC link voltage.

$$\begin{aligned}
&\left(\frac{Pk_1}{|v^+|^2} + \frac{P(1-k_1)}{|v^-|^2}\right) v^+ \cdot v^- + \left(\frac{Qk_2}{|v^+|^2} - \frac{Q(1-k_2)}{|v^-|^2}\right) v_\perp^+ \cdot v^- = 0 \\
&\Longrightarrow \frac{Pk_1}{|v^+|^2} + \frac{P(1-k_1)}{|v^-|^2} = 0 \to k_1 = \frac{|v^+|^2}{|v^+|^2 - |v^-|^2} \\
&\Longrightarrow \frac{Qk_2}{|v^+|^2} - \frac{Q(1-k_2)}{|v^-|^2} = 0 \to k_2 = \frac{|v^+|^2}{|v^+|^2 + |v^-|^2}.
\end{aligned} \tag{9.15}$$

This control strategy is tested in experiments, and the results are shown in Figure 9.5. Before applying the oscillation power control method, $k_1 = k_2 = 1$ is used meaning only positive current is injected from the IFC without any negative sequence current compensation (i.e. scenario 1 in Section 9.2.1). As is clear from the results, when the control in (9.15) is adopted, the active power oscillation is

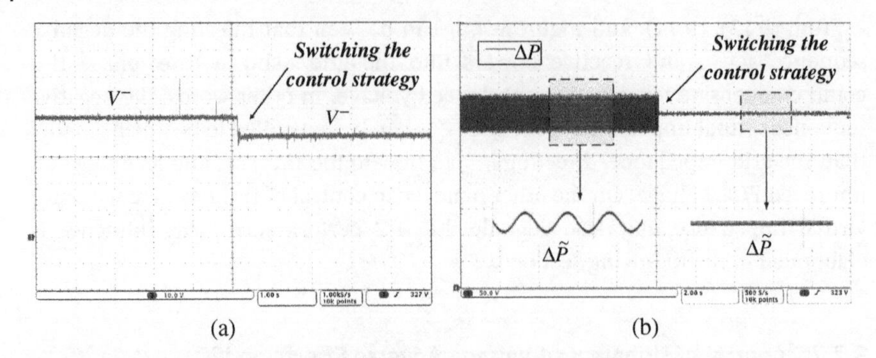

Figure 9.5 Performance of the IFC under unbalanced voltage when its output active power is canceled: (a) negative sequence voltage of the PCC, and (b) IFC active power oscillation.

zero. However, although the PCC negative sequence voltage is reduced with this method, the negative sequence voltage level cannot be adjustably controlled.

In addition to active power oscillation cancelation, IFC virtual impedance control can focus on peak current increase. For instance, from the IFC reference current in (9.1), if $k_1 = k_2 = 1$, the negative sequence current component is zero (virtual negative sequence impedance is infinite), and only balanced three-phase current is injected into the grid, even under the unbalanced PCC voltage condition. Please refer to Appendix B for more discussions on peak current control of single and parallel IFCs under unbalanced conditions.

The abovementioned control strategies for power oscillation cancelation and peak current minimization have one major limitation: the unbalanced voltage level cannot be adjustably controlled. This is very inconvenient when an adjustable compensation level is required, especially considering that the IFC available rating for unbalanced compensation may change due to the variation of active power flow. To overcome such difficulties, a control strategy is presented in the next section, where the unbalanced voltage can be controlled with an adjustable compensation level. Simultaneously, the adverse effects on the IFC operation, such as power oscillation, are minimized.

9.2.3 Adjustable Unbalanced Voltage Compensation with IFC Active Power Oscillation Minimization

In this section, the method for compensation of unbalanced voltage with an adjustable compensation level using a three-phase IFC is achieved. Then, at each compensation level, this method minimizes the IFC's output active power oscillation due to the presence of unbalanced voltage [2]. Similarly, in this control strategy, the peak current increase in the IFC can also be controlled.

To compensate for the unbalanced voltage (negative sequence PCC voltage control) with adjustable compensation level, the IFC's negative sequence current amplitude should be controlled with adjustable references, which can be achieved from (9.1) and Figure 9.4 as follows:

$$|i^{*-}| = \frac{|v^-|}{|Z^-|} = \sqrt{\left(\frac{P(1-k_1)}{|v^-|}\right)^2 + \left(\frac{Q(1-k_2)}{|v^-|}\right)^2}. \tag{9.16}$$

From (9.16), $|i^{-*}|$ can be set directly through the desired IFC negative sequence virtual impedance, considering the IFC's available rating for unbalanced compensation (which controls the IFC output peak current).

From (9.4), the active power oscillation can be expressed as (9.17).

$$\begin{aligned}\Delta P &= \left(\frac{Pk_1}{|v^+|^2} + \frac{P(1-k_1)}{|v^-|^2}\right) v^+ \cdot v^- + \left(\frac{Qk_2}{|v^+|^2} - \frac{Q(1-k_2)}{|v^-|^2}\right) v_\perp^+ \cdot v^- \\ &= \frac{3}{2}\left(\frac{Pk_1}{|v^+|^2} + \frac{P(1-k_1)}{|v^-|^2}\right) |v^+||v^-| \cos(2\omega t + \theta) \\ &\quad + \frac{3}{2}\left(\frac{Qk_2}{|v^+|^2} - \frac{Q(1-k_2)}{|v^-|^2}\right) |v^+||v^-| \sin(2\omega t + \theta)\end{aligned} \tag{9.17}$$

where $\theta^+ - \theta^- = \theta$ is the phase angle between positive and negative sequences. The following objective function is defined to minimize sinusoidal active power oscillation in (9.17):

$$\left[\left(\frac{Pk_1}{|v^+|^2} + \frac{P(1-k_1)}{|v^-|^2}\right) |v^+||v^-|\right]^2 + \left[\left(\frac{Qk_2}{|v^+|^2} - \frac{Q(1-k_2)}{|v^-|^2}\right) |v^+||v^-|\right]^2. \tag{9.18}$$

This objective function should be minimized subject to the constraint in (9.16), which provides controllable unbalanced voltage compensation. With the assumption of $k_1 - 1 = l_1$ and $k_2 - 1 = l_2$, the objective function in (9.18) and the constraint in (9.16) can be expressed as (9.19) and (9.20), respectively.

$$J(l_1, l_2) = A^2 l_1^2 + B^2 l_2^2 + Cl_1 + El_2 + F \tag{9.19}$$

$$P^2 l_1^2 + Q^2 l_2^2 = D^2 \tag{9.20}$$

where A, B, C, D, E, and F are defined as $A = \frac{P(|v^+|^2 - |v^-|^2)}{|v^+||v^-|}$, $B = \frac{Q(|v^+|^2 + |v^-|^2)}{|v^+||v^-|}$, $C = \frac{2P^2|v^-|^2(|v^+|^2 - |v^-|^2)}{|v^+|^2|v^-|^2}$, $D = |v^-|\,|i^{*-}|$, $E = \frac{-2Q^2|v^-|^2(|v^+|^2 + |v^-|^2)}{|v^+|^2|v^-|^2}$, $F = \frac{|v^-|^2(P^2+Q^2)}{|v^+|^2|v^-|^2}$.

The above optimization problem can be solved online in each time step in a digital controller using the Lagrangian method. The control strategy's performances under different IFC power factors in inductive and resistive grids are verified (the experiment setup parameters are provided in Table C.3 of Appendix C), and the results are shown in Figures 9.6 and 9.7. In the results, before applying the control method to the three-phase IFC, $k_1 = k_2 = 1$, where just positive sequence active and

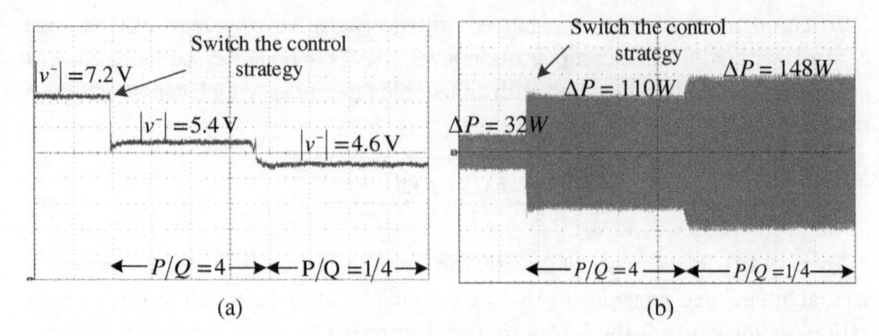

Figure 9.6 Performance of the PCC unbalanced voltage compensation and IFC active power oscillation minimization in an inductive grid: (a) negative sequence voltage of the PCC, and (b) IFC active power oscillation.

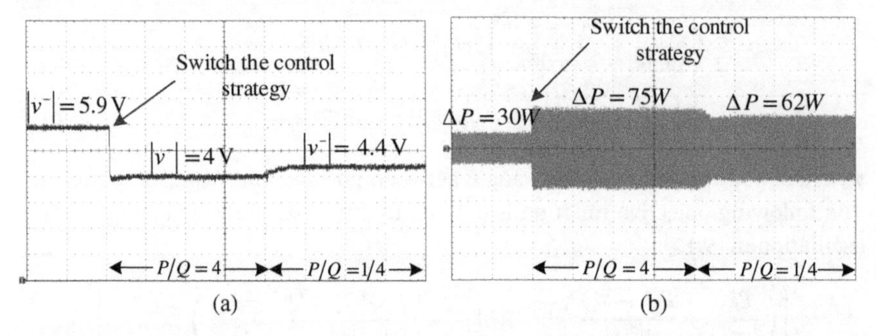

Figure 9.7 Performance of the PCC unbalanced voltage compensation and IFC active power oscillation minimization in a resistive grid: (a) negative sequence voltage of the PCC, and (b) IFC active power oscillation.

reactive currents are injected into the grid without any compensation. After applying the control method, the negative sequence voltage is reduced (unbalanced voltage is compensated), and the active power oscillation is minimized around that operating point. This control strategy is applicable for IFCs with different PFs under various grid conditions.

For more details about the optimal design of k_1 and k_2 for individual three-phase IFC control under different optimization objectives, please refer to [2].

9.3 Control of Parallel Three-phase IFCs for Unbalance Voltage Compensation

In hybrid AC/DC microgrids, if the output power/current of DG/ESS connected to the AC subgrid is high, parallel IFCs are typically used. Parallel IFCs are also used to link AC and DC subgrids where high power/current is planned to be exchanged or high reliability is required. In the parallel IFCs with a common DC link, the adverse effects of unbalanced voltage on the IFCs operation can be aggregated. For example, the IFC active power oscillation (ΔP), DC link ripples, and peak current increase can be potentially multiplied by the number of paralleled IFCs. However, if appropriately controlled, individual IFCs adverse effects might cancel out each other. This could help to improve both the AC and DC subgrids power quality. The active power oscillation cancelation in the parallel IFCs operation, which provides oscillation-free DC link voltage, is a desirable smart IFC control strategy.

In parallel IFCs, appropriate control signals are generated in the supervisory control center (SCC), a separate controller or a master IFC, based on control objectives. Then, the optimized parameters are sent to primary controllers of IFCs to update reference currents (assuming IFCs with CCM). For closely located IFCs, a local controller of one of the IFC can be used as an SCC (master–slave structure of SCC).

In Figure 9.8, the parallel IFCs with a control block diagram under unbalanced voltage are shown. In this figure, it is assumed that the SCC has a centralized structure. The closed-loop current control with a practical proportional-resonant controller in the stationary $\alpha\beta$ reference frame is adopted. A sequence extractor can decompose PCC voltage into positive and negative sequences. In next section, more detail of this control strategy is illustrated.

9.3.1 Parallel Three-phase IFCs Model under Unbalanced Voltage

Figure 9.8 shows n-parallel three-phase IFCs with a common DC link connected to the PCC with output filters. Since the control of adverse effects of unbalanced voltage on IFCs operation is desired, the reference current in (9.21) can be used to easily control/cancel out the three-phase IFC power oscillation under unbalanced conditions. It should be clarified that although the power oscillations can be canceled out by controlling k_1 and k_2 (as indicated in the reference current (9.1)), (9.21) with newly introduced variables, i.e. k_{pi} and k_{qi}, provides an easier way for power oscillation cancelation, especially when parallel IFCs are used (please refer to Appendix A for detailed information).

$$i_i^* = i_{pi}^* + i_{qi}^* = \left(\frac{P_i}{|v^+|^2 + k_{pi}|v^-|^2} v^+ + \frac{P_i k_{pi}}{|v^+|^2 + k_{pi}|v^-|^2} v^- \right)$$

$$+ \left(\frac{Q_i}{|v^+|^2 + k_{qi}|v^-|^2} v_\perp^+ + \frac{Q_i k_{qi}}{|v^+|^2 + k_{qi}|v^-|^2} v_\perp^- \right). \tag{9.21}$$

In this section, the PCC voltage is represented by v, the ith IFC output current and the average active and reactive powers are represented by i_i, P_i, and Q_i. Moreover, parallel IFC collective current vector is represented by i_{IFCs}. These parameters

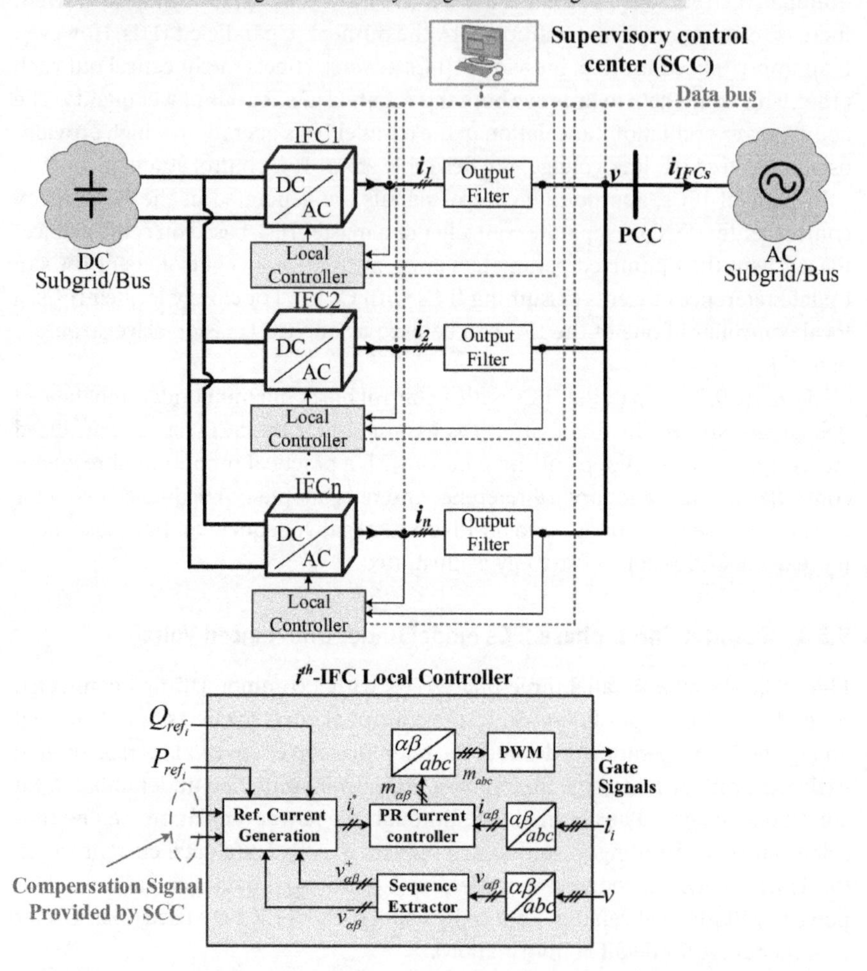

Figure 9.8 Typical parallel three-phase IFCs with control block diagram under unbalanced voltage and centralized structure of the supervisory control center.

are shown in Figure 9.8. From (9.21), the ith IFC instantaneous output active and reactive powers can be modeled as:

$$p_i = P_i + \underbrace{\frac{P_i(1 + k_{pi})(v^+.v^-)}{|v^+|^2 + k_{pi}|v^-|^2}}_{\Delta P_{pi}} + \underbrace{\frac{Q_i(1 - k_{qi})\left(v_\perp^+.v^-\right)}{|v^+|^2 + k_{qi}|v^-|^2}}_{\Delta P_{qi}} \tag{9.22}$$

$$q_i = Q_i + \underbrace{\frac{P_i(1 - k_{pi})\left(v^+.v_\perp^-\right)}{|v^+|^2 + k_{pi}|v^-|^2}}_{\Delta Q_{pi}} + \underbrace{\frac{Q_i(1 + k_{qi})\left(v_\perp^+.v_\perp^-\right)}{|v^+|^2 + k_{qi}|v^-|^2}}_{\Delta Q_{qi}}. \tag{9.23}$$

From (9.21)–(9.23), it can be concluded that:

- Under unbalanced voltage, the ith IFC current will increase, and the output active and reactive powers will oscillate.
- The IFC output current will be balanced under $k_p = k_q = 0$.
- The two components of active power oscillations (ΔP_{pi} and ΔP_{qi}) are orthogonal, and $k_{pi} = -1$ and $k_{qi} = 1$ result in zero active power oscillations.
- The aforementioned operating points are independent of average active and reactive power flow directions.

As an example, in Figure 9.9, a typical individual IFC's output active power and three-phase currents under different k_{pi} and k_{qi} in unity PF and zero PF operations are shown, which verify the aforementioned discussions. From Figure 9.9, it is clear that k_{pi} and k_{qi} can control the active power oscillation and IFC output peak current.

Using individual IFC relations in (9.21)–(9.23), the current reference vector and instantaneous active and reactive powers of n-parallel IFCs in Figure 9.8 can be achieved as follows:

$$\begin{aligned} i^*_{\text{IFCs}} = \sum_{i=1}^{n} i^*_i = \sum_{i=1}^{n} \frac{P_i}{|v^+|^2 + k_{p_i}|v^-|^2} v^+ + \sum_{i=1}^{n} \frac{P_i k_{p_i}}{|v^+|^2 + k_{p_i}|v^-|^2} v^- \\ + \sum_{i=1}^{n} \frac{Q_i}{|v^+|^2 + k_{q_i}|v^-|^2} v_\perp^+ + \sum_{i=1}^{n} \frac{Q_i k_{q_i}}{|v^+|^2 + k_{q_i}|v^-|^2} v_\perp^- \end{aligned} \tag{9.24}$$

$$p = \sum_{i=1}^{n} P_i + \underbrace{\sum_{i=1}^{n} \frac{P_i(1 + k_{p_i})(v^+.v^-)}{|v^+|^2 + k_{p_i}|v^-|^2}}_{\Delta P_p} + \underbrace{\sum_{i=1}^{n} \frac{Q_i(1 - k_{q_i})\left(v_\perp^+.v^-\right)}{|v^+|^2 + k_{q_i}|v^-|^2}}_{\Delta P_q} \tag{9.25}$$

$$q = \sum_{i=1}^{n} Q_i + \underbrace{\sum_{i=1}^{n} \frac{P_i(1 - k_{p_i})\left(v^+.v_\perp^-\right)}{|v^+|^2 + k_{p_i}|v^-|^2}}_{\Delta Q_p} + \underbrace{\sum_{i=1}^{n} \frac{Q_i(1 + k_{q_i})\left(v_\perp^+.v_\perp^-\right)}{|v^+|^2 + k_{q_i}|v^-|^2}}_{\Delta Q_q}. \tag{9.26}$$

To cancel out the collective active power oscillations of n-parallel IFCs in the control strategy, the following constraint should be satisfied based on (9.25):

$$\begin{aligned} \sum_{i=1}^{n} \Delta P_{p_i} = \sum_{i=1}^{n} \frac{P_i(1 + k_{p_i})(v^+.v^-)}{|v^+|^2 + k_{p_i}|v^-|^2} = 0 \\ \Rightarrow \sum_{i=1}^{n} \frac{P_i}{|v^+|^2 + k_{p_i}|v^-|^2} = \frac{\sum_{i=1}^{n} P_i}{|v^+|^2 - |v^-|^2} \end{aligned} \tag{9.27}$$

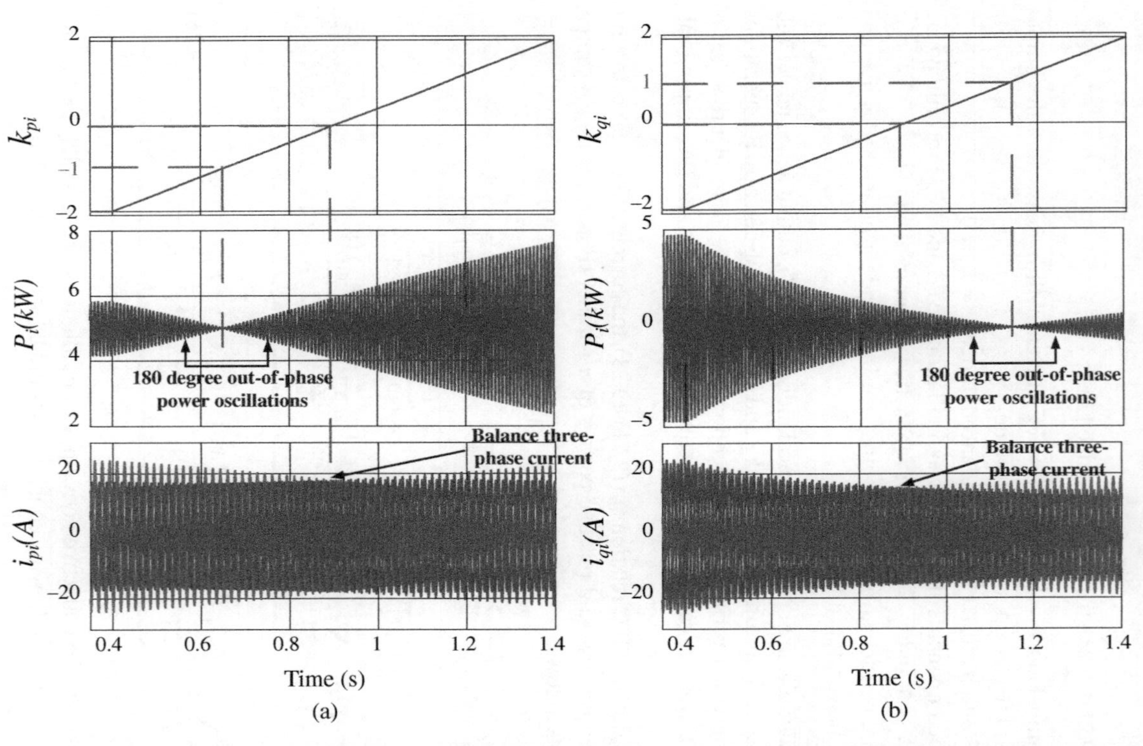

Figure 9.9 Individual IFC output active power and three-phase currents under unbalanced voltage condition: (a) different k_{pi} in unity PF operation mode and (b) different k_{qi} in zero PF operation mode.

$$\sum_{i=1}^{n} \Delta P_{q_i} = \sum_{i=1}^{n} \frac{Q_i(1-k_{q_i})\left(v_{\perp}^{+}.v^{-}\right)}{|v^{+}|^2 + k_{q_i}|v^{-}|^2} = 0$$
$$\Rightarrow \sum_{i=1}^{n} \frac{Q_i}{|v^{+}|^2 + k_{q_i}|v^{-}|^2} = \frac{\sum_{i=1}^{n} Q_i}{|v^{+}|^2 + |v^{-}|^2}. \tag{9.28}$$

Considering (9.24)–(9.28), the current reference vector and instantaneous active-reactive powers of n-parallel IFCs under zero active power oscillations are obtained as:

$$i^{*}_{\text{IFCs}}\big|_{\Delta P=0} = \frac{\sum_{i=1}^{n} P_i}{|v^{+}|^2 - |v^{-}|^2} v^{+} + \frac{-\sum_{i=1}^{n} P_i}{|v^{+}|^2 - |v^{-}|^2} v^{-} + \frac{\sum_{i=1}^{n} Q_i}{|v^{+}|^2 + |v^{-}|^2} v_{\perp}^{+} + \frac{\sum_{i=1}^{n} Q_i}{|v^{+}|^2 + |v^{-}|^2} v_{\perp}^{-} \tag{9.29}$$

$$p = \sum_{i=1}^{n} P_i \tag{9.30}$$

$$q = \sum_{i=1}^{n} Q_i + \frac{2\left(v^{+}.v_{\perp}^{-}\right)}{|v^{+}|^2 - |v^{-}|^2} \sum_{i=1}^{n} P_i + \frac{2\left(v_{\perp}^{+}.v_{\perp}^{-}\right)}{|v^{+}|^2 + |v^{-}|^2} \sum_{i=1}^{n} Q_i \tag{9.31}$$

with main observations as:

- From (9.29), under zero active power oscillations, the current is independent of k_{pi} and k_{qi}, and is affected by $\sum_{i=1}^{n} P_i$ and $\sum_{i=1}^{n} Q_i$ variations.
- From (9.31), the reactive power oscillations are independent of k_{pi} and k_{qi} under $\Delta P = 0$.

From the above discussions, the control strategies of parallel IFCs under unbalanced voltage are classified into two groups:

- In the first group, one IFC, named the redundant IFC, is dedicated to cancel out active power oscillations of the other IFCs (the redundant IFC produces 180° out-of-phase active power oscillations compared to the other IFCs), while all IFCs peak currents are also controlled not to exceed their rating limits [4, 5].
- In the second group, all IFCs participate in ΔP cancelation according to their power ratings. Like the first control option, peak currents of all IFCs are controlled not to exceed their rating limits [6].

In the following, two control strategies, one for each group, are provided. In both control strategies, it is assumed that the IFCs operate under unity power factors.

9.3.2 Parallel Three-phase IFCs Control under Unbalanced Voltage: Redundant IFC for ΔP Cancelation

In this control strategy, one IFC among parallel IFCs, named the redundant IFC, cancels out active power oscillations produced by all other parallel IFCs, resulting

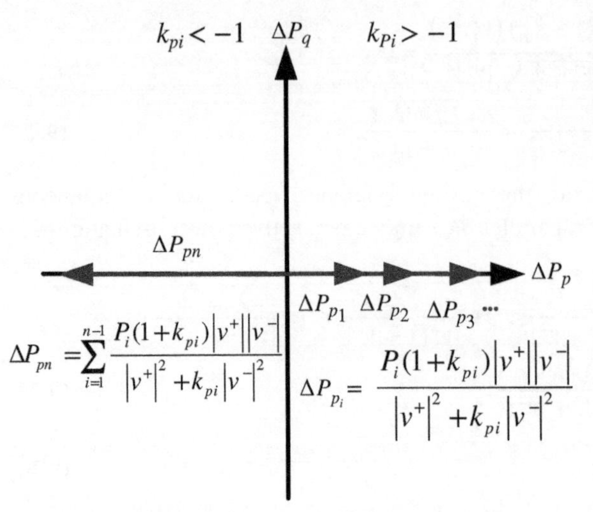

Figure 9.10 Vector representation of ΔP cancelation using the redundant IFC (IFC under unity PF operation).

in the DC subgrid/link voltage oscillation cancelation. In Figure 9.10, a vector representation of active power oscillations cancelation using the redundant IFC is shown. In this figure, it is assumed that $P_i > 0$ $i = 1, \ldots, n$, $\theta = \theta^+ - \theta^- = 0$, and $k_{pi} \geq -1$ (except the redundant IFC). As a result, redundant IFC (converter n) works under $k_{pn} \leq -1$ to produce 180° out-of-phase power oscillations to cancel out active power oscillations.

Based on this control strategy, all IFCs, except the redundant IFC, operate independently under zero active power oscillations ($k_{pi} = -1$). When the peak current of a converter exceeds its current rating limit, its k_{pi} moves toward zero to reduce the output peak current (please see Figure 9.9). As a result, power oscillations will be generated by this converter. In this case, the redundant IFC is controlled considering (9.27) for active power oscillation cancelation.

One main challenge of this control strategy is that the redundant IFC's power rating should be large enough to cancel other IFCs power oscillations (the redundant IFC's peak current may need to be much higher than the other IFCs). If all IFCs peak currents are kept in the same phase with a collective peak current of n-parallel IFCs (for example, all in phase b), the redundant IFC's peak current will reduce (since IFCs peak current amplitudes summation reduce). It should be considered that the collective peak current of parallel IFCs is a constant value under fixed values of active powers (independent of k_{pi} [see (9.29)]).

Based on a detailed analysis of peak currents of individual and parallel IFCs under unbalanced voltage (please refer to Appendix B for more information), it is concluded that for all IFCs, if constraint (9.32) is satisfied, the peak current of the

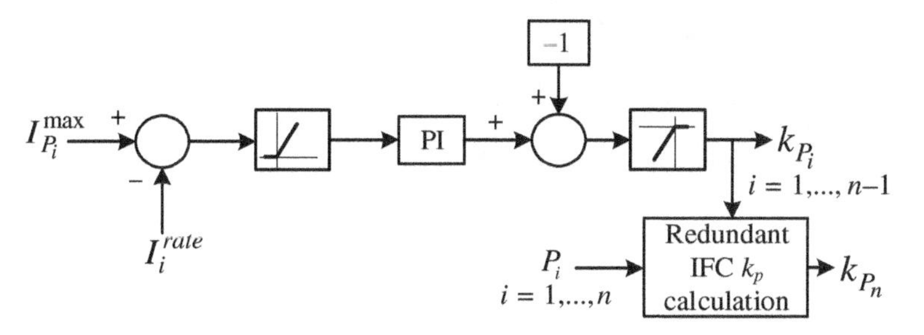

Figure 9.11 Parallel IFCs control strategy under unity PF operation when one redundant IFC is focusing on ΔP cancelation.

individual ith IFC and collective peak current of parallel IFCs will be in the same phase.

$$k_{pi} < 0 \qquad i = 1, \dots, n. \tag{9.32}$$

A block diagram of this control strategy is shown in Figure 9.11. In this control strategy, I_{Pi}^{max} is the peak current of ith IFC, and it can be measured or calculated using (9.21) (see Appendix B for a detailed analysis). For all n IFCs, the control is started under $k_{pi} = -1; i = 1, 2, \dots, n$ where $\Delta P_{pi} = 0$. If the peak current of each of the first ordinary $n-1$ IFC (except the redundant IFC) exceeds its rating current limit (I_i^{rate}), its k_{pi} will can be controlled toward zero to limit its peak current at the rating value. As a result, the peak currents of ordinary IFCs (except the redundant one) are constant values under fixed average active powers in different operation conditions (under $k_{pi} = -1$ or under their rating values). All information of those ordinary $n-1$ IFCs (P_i and k_{pi}) is sent to the redundant IFC's controller. Based on this information, k_{pn} of the redundant converter is determined using (9.27) to cancel active power oscillations.

It should noted that the variation of the IFC average power flow direction and values will not affect the control strategy performance, although the operating point will be changed.

Three parallel IFCs have been simulated using the above control strategy in MATLAB/Simulink, and the results are provided in Figure 9.12. The simulated system parameters are provided in Table C.4 of Appendix C. In the simulations, the third IFC with the largest power/current rating is considered the redundant converter. A two-phase unbalanced fault is applied to the system at $t = 0.15$ s as a source of unbalanced voltage. During $0.15\text{ s} < t < 0.3$ s, the k_{pi} of all IFCs is set to zero. At $t = 0.3$ s, the above control strategy is applied where the k_{pi} of all converters is set to -1. Although the initial set point provides zero active power oscillations for IFCs, the first and second IFC peak currents exceed their rating limits. Therefore,

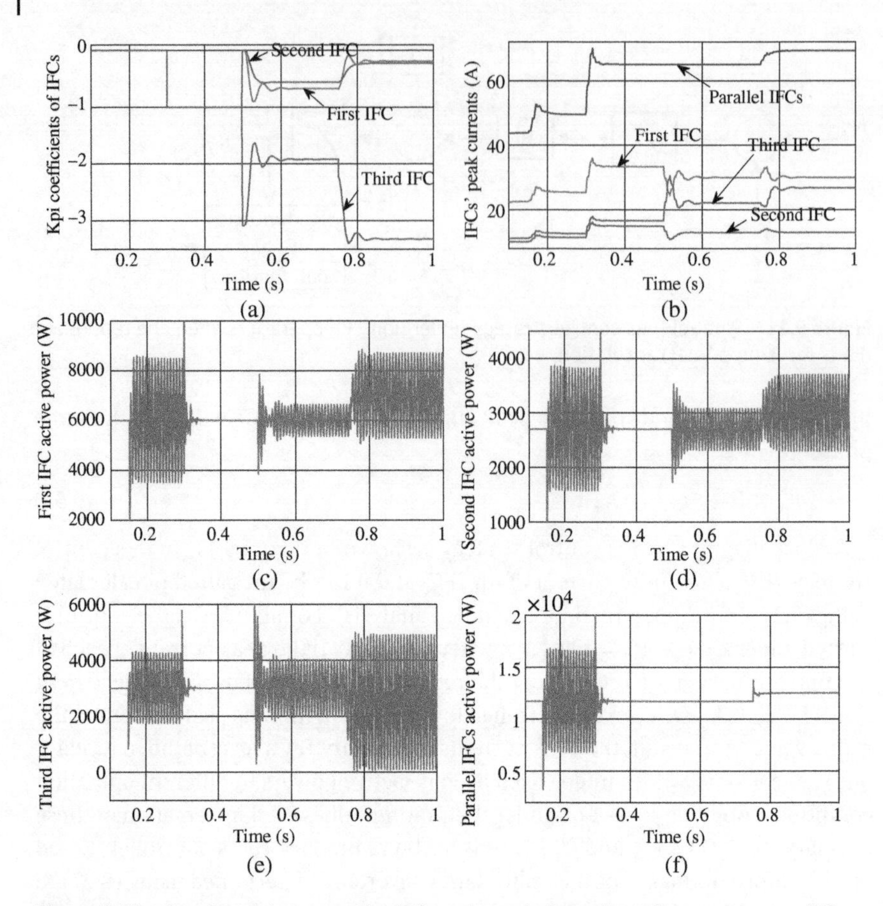

Figure 9.12 Results of three parallel IFCs under unbalanced condition when one IFC cancel out ΔP oscillations: (a) k_{pi} coefficient factors, (b) IFC output peak currents, (c) first IFC output active power, (d) second IFC output active power, (e) third IFC output active power, and (f) parallel IFC output active power.

at $t = 0.5$ s, the control strategy sets the first and second IFC peak currents on their rating limits by moving k_{p1} and k_{p2} toward zero. The third IFC (redundant IFC) cancels out active power oscillations produced by the IFCs peak current control. After average power variations at $t = 0.75$ s, the control again sets the first and second IFCs peak currents on their rating limits since they hit limits, and provides zero collective active power and DC link voltage oscillations using the redundant IFC. From the results, it is clear that the collective peak current of parallel IFCs is a constant value under fixed values of active powers (independent of k_{pi} [see (9.29)]).

A more detailed design and study of this control strategy can be found in [4] and [5].

9.3.3 Parallel Three-phase IFCs Control under Unbalanced Voltage: All Parallel IFCs Participate in ΔP Cancelation

As discussed in Section 9.3.2, since only the redundant IFC is used to cancel out the power oscillations of the parallel IFCs under unbalanced voltage, this IFC's power rating should be large enough. On the other hand, the collective peak current of the parallel IFCs (which is a constant under the fixed average active and reactive powers) is not shared among the IFCs based on their power ratings. Therefore, some IFCs worked at their rating limits while others operated far from their rating limits. Thus, a control strategy presented in this section aims to share the active power oscillations cancelation and collective peak current among all parallel IFCs according to their respective ratings. Here, it is assumed that all IFCs operate under unity PF. In detail, the objectives of the control strategy presented in this section are as follows:

- All parallel IFCs participate in active power oscillation cancelation.
- Provide oscillation-free DC link voltage.
- Share collective peak current among n-parallel IFCs according to their ratings.
- Maximize power/current transfer capability of IFCs by minimizing the peak current of each IFC.

To achieve the first three objectives, the following relationships should be satisfied:

$$\sum_{i=1}^{n} \frac{P_i}{|v^+|^2 + k_{p_i}|v^-|^2} = \frac{\sum_{i=1}^{n} P_i}{|v^+|^2 - |v^-|^2} \tag{9.33}$$

$$U_1 I_{p1}^{\max} = U_2 I_{p2}^{\max} = \ldots = U_n I_{pn}^{\max} \tag{9.34}$$

$$U_i = \frac{S_1}{S_i} \qquad i = 1, 2, \ldots, n \tag{9.35}$$

where (9.33) ensures zero active power oscillation, and (9.34) and (9.35) are for peak current sharing according to each converters ratings.

Based on a thorough study of IFCs peak current under unbalanced voltage in Appendix B, to maximize the parallel IFCs power/current transfer capability and to achieve the last objective above, the following constraints should be satisfied:

- Peak currents of all individual IFCs and collective peak current of n-parallel IFCs should be in the same phase (e.g. all of them in phase b).
- Peak currents of all individual IFCs and collective peak current of n-parallel IFCs should be in-phase (phase angle difference between their phasors is zero).

With these conditions, the summation of n-parallel IFCs peak currents is minimized to maximize parallel IFCs power/current transfer capability. In the presented control strategy, individual IFCs peak currents are adjusted based on their power ratings (see (9.34) and (9.35)). Thus, parallel IFCs power/current transfer capability is maximized, and individual IFCs operation is optimized.

Additional analysis of the above two conditions is provided below:

1. *Peak currents of all individual IFCs and collective peak current are in the same phase.* As mentioned in Section 9.3.2 (it is proven in Appendix B), if $k_{pi} < 0$; $i = 1, \ldots, n$ for all parallel IFCs under unity PF operation mode, all individual IFCs peak currents would be in the same phase with the collective peak current of n-parallel IFCs.
2. *Peak currents of all individual IFCs and collective peak current are in-phase.* Based on IFC peak current study under unbalanced voltage, the individual ith IFC peak current phase angle under unity PF can be achieved as:

$$\delta_i\big|_{I_{pi}^{\max}} = \tan^{-1}\left(\cot\gamma\left(-1 - \frac{2k_{pi}|v^-|}{|v^+| - k_{pi}|v^-|}\right)\right) \tag{9.36}$$

where γ is the rotation angle which is equal to $\rho, \rho + \pi/3$ and $\rho - \pi/3$ for the abc axes, respectively. Also, the n-parallel IFCs collective peak current phase angle under unity PF can be achieved as:

$$\delta\big|_{I_{p-\mathrm{IFCs}}^{\max}}\Big|_{\Delta P=0} = \tan^{-1}\left(\cot\gamma\left(-1 + \frac{2|v^-|}{|v^+| + |v^-|}\right)\right). \tag{9.37}$$

From (9.36) and (9.37), the phase angles of individual IFCs maximum currents in each phase depend on k_{pi} while in parallel IFCs under $\Delta P = 0$, the phase angles of collective maximum currents in each phase are independent of k_{pi}. However, since in parallel IFCs with common DC and AC links, $|v^+|$, $|v^-|$ and γ are common values in individual and parallel IFCs, and the variations of $\delta_{xi}\big|_{I_{xpi}^{\max}}$ (when k_{pi} deviates from $k_{pi} = -1$) is small enough (under $k_{pi} = -1$, $\delta_{xi}\big|_{I_{xpi}^{\max}} = \delta_x\big|_{I_{xp-\mathrm{IFCs}}^{\max}}\Big|_{\Delta P=0}$; $x = a, b, c$), the $\delta_{xi}\big|_{I_{xpi}^{\max}}$ in each phase can be assumed constant under k_{pi} variations and equal to $\delta_x\big|_{I_{xp-\mathrm{IFCs}}^{\max}}\Big|_{\delta P=0}$ with a good approximation.

With the discussions above, under unity PF, when

$$k_{pi} \leq 0 \qquad i = 1, 2, \ldots, n \tag{9.38}$$

the peak currents of all individual IFCs and the collective peak current are in the same phase and in-phase, which maximizes the power/current transfer capability of parallel IFCs and minimizes the summation of their peak currents.

Variable k_{pi} of all IFCs can be determined by solving the set of non-linear equations in (9.33)–(9.35) considering (9.38). The determined k_{pi} of the IFCs provides the zero collective active power oscillation of parallel IFCs using (9.33), shares collective peak current of parallel IFCs among them based on their power ratings using (9.34) and (9.35), and assures the maximum power/current transferring capability of parallel IFCs by (9.38). If the determined k_{pi} from (9.33)–(9.35) is greater than zero, which leads the IFCs peak currents to different phases, they will be set to zero, and the calculation will be repeated for the rest of IFCs for k_{pi} determinations. It is worth mentioning that the sharing factors in (9.35) are independent of the converters' operating points, but are related to their ratings for peak current sharing.

In this control strategy, the operating point information of all IFCs and the PCC voltage are sent to the central controller for all IFCs k_{Pi} determination (see Figure 9.8). Due to computational complexity in solving non-linear equations, the control strategy may have challenges for online operation.

To reduce the computation burden, the IFC peak current analysis can be introduced. As discussed, when $k_{pi} < 0; i = 1, \ldots, n$, peak currents of all individual IFCs and collective peak current of parallel IFCs are in the same phase and in-phase. Thus, the following relation among peak currents of individual IFCs and the collective peak current of parallel IFCs with unity PF operation can be derived:

$$I_{p-\text{IFCs}}^{\max} \cong \sum_{i=1}^{n} I_{pi}^{\max}. \tag{9.39}$$

From (9.39), it is concluded that the collective peak current of n-parallel IFCs, which is a constant value under zero active power oscillations and fixed output active powers, can be shared linearly among individual IFCs. Also, the active power oscillation cancelation constraint of (9.33) is embedded in (9.39). Thus, the non-linear relations can be replaced by (9.40)–(9.43).

$$I_{p1}^{\max} + I_{p2}^{\max} + \ldots + I_{pn}^{\max} \cong I_{p-\text{IFCs}}^{\max}\Big|_{\Delta P=0} \tag{9.40}$$

$$U_1 I_{p1}^{\max} = U_2 I_{p2}^{\max} = \ldots = U_n I_{pn}^{\max} \tag{9.41}$$

$$U_i = \frac{S_1}{S_i} \qquad i = 1, 2, \ldots, n \tag{9.42}$$

$$k_{pi} \leq 0 \qquad i = 1, 2, \ldots, n. \tag{9.43}$$

From (9.40)–(9.43), the $I_{pi}^{\max}$ for the individual ith IFC under $\Delta P = 0$ can be calculated as:

$$I_{pi}^{\max} \cong \left(\frac{S_i}{S_1 + S_2 + \ldots + S_n} \right) \times I_{p-\text{IFCs}}^{\max}\Big|_{\Delta P=0}. \tag{9.44}$$

After each IFC reference peak current calculation, k_{pi} can be determined using the following equation (only k_{pi} is unknown in (9.45)):

$$I_{xpi}^{\max} = \sqrt{\left(\frac{P_i}{|v^+|^2 + k_{pi}|v^-|^2}\right)^2 \times \left(|v^+|^2 + |v^-|^2 k_{pi}^2 + 2|v^+||v^-|k_{pi}\cos(2\gamma)\right)}$$

$$x = a, b, c. \tag{9.45}$$

It should be considered that k_{pi} of all IFCs should be less than zero to keep all IFCs peak currents in the same phase with a collective peak current of parallel IFCs. Also, from (9.45), the minimum possible value of $I_{pi}^{\max}$ is $P_i/|v^+|$ ($k_{pi} = 0$ in (9.45); see Appendix B). And if any calculated $I_{pi}^{\max}$ is less than $P_i/|v^+|$, it will be restricted to that value, and the calculation will be repeated for the rest of the IFCs.

Considering (9.45), it can be concluded that if $P_i/|v^+| \leq I_{pi}^{\max} \leq P_i/|v^-|$, the $I_{pi}^{\max} - k_{pi}$ quadratic equation will have two different real roots for k_{pi} under each determined $I_{pi}^{\max}$ and different values of γ. These two roots have different signs. As mentioned in this control method, all IFCs peak currents are controlled to be $I_{pi}^{\max} \geq P_i/|v^+|$. In addition, $I_{pi}^{\max} \leq P_i/|v^-|$ is always satisfied since $|v^-|$ is a small percentage of $|v^+|$, and $P_i/|v^-|$ exceeds the current rating limit of the IFC. As a result, $P_i/|v^+| \leq I_{pi}^{\max} \leq P_i/|v^-|$ is always satisfied, which leads to two districts real roots for k_{pi} with different signs. For more detailed discussions and design of this control strategy, please refer to [6].

In general, this control strategy reduces the computational burden of active power oscillation cancelation and collective peak current sharing of parallel IFCs and can be run online easily in the distributed control structure of parallel IFCs (the k_{Pi} of IFCs is generated individually in the outer control layer of the IFCs local controllers). Due to approximate calculations, this control strategy may have small errors in the active power oscillation cancelation. Similarly, individual IFCs peak currents should be recalculated under variations of average active power flow directions and values.

The abovementioned control strategy performance is verified, and the results are provided in Figure 9.13. In this study, three IFCs are paralleled and tested under unbalanced voltage conditions. The three IFCs have the rating of $S_1 = 9$ kVA, $S_2 = 4$ kVA, and $S_3 = 10$ kVA. More simulation system parameters can be found in Table C.5 of Appendix C. After applying the unbalanced condition at $t = 0.15$ s, first, zero active power oscillations of IFCs are provided at $t = 0.3$ s by setting $k_{p1} = k_{p2} = k_{p3} = -1$. As can be seen, the parallel IFCs collective peak current is not shared between them based on their power ratings. The control method discussed above ((9.40)–(9.43)) is applied at $t = 0.5$ s. At $t = 0.75$ s, the output

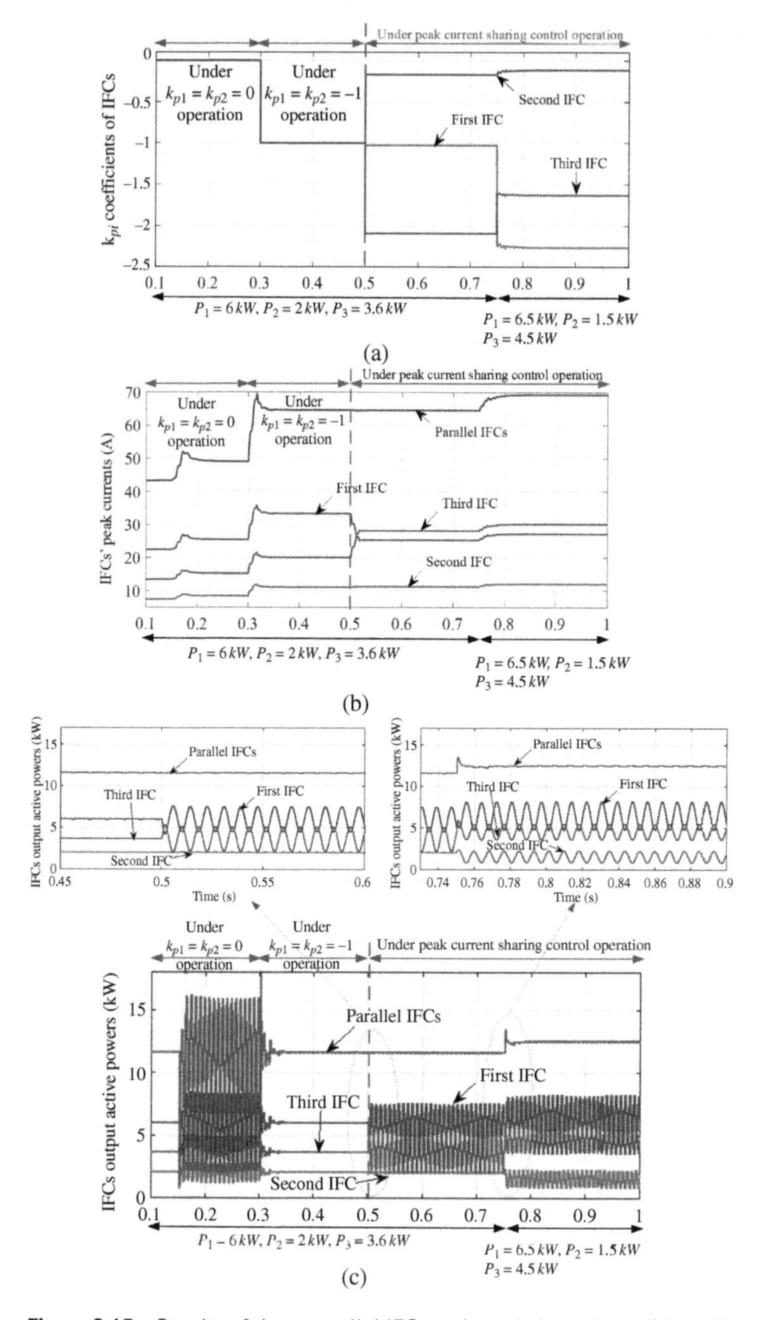

Figure 9.13 Results of three parallel IFCs under unbalanced condition when all IFCs participate in ΔP cancelation: (a) k_{pi} coefficient factors, (b) IFCs output peak currents, and (c) IFCs output active powers.

average active powers of IFCs are modified. It is clear that all IFCs peak currents are kept in the same phase with a collective peak current of parallel IFCs to maximize power transferring capability ($k_{pi} \leq 0$), the collective peak current is shared among parallel IFCs based on their power ratings, and the collective active power oscillation is zero.

9.4 Control of Single-phase IFCs for Three-phase System Unbalanced Voltage Compensation

The high penetration level of single-phase IFCs in hybrid AC/DC microgrids provides an excellent opportunity for three-phase system unbalanced voltage compensation. Such single-phase IFCs can be either DG/ESS IFCs connected to the AC subgrid or DC/AC subgrid linking converters.

In general, an effective solution for the unbalanced condition compensation is balancing the active powers in different phases, e.g. using the active power curtailment of PV systems or using single-phase IFCs from ESSs. Moreover, the reactive power control can further help the reduction/mitigation of unbalanced voltages [7, 8]. In both strategies, the SCC provides the references of average active/reactive powers to the primary controllers of the IFCs. The primary controller updates the reference currents of IFCs for compensation purposes. It should be mentioned that the SCC usually has a centralized structure, and the primary controller of the IFC is under CCM.

In Figure 9.14, a typical example of single-phase DGs connected to the AC subgrid is shown. The three-phase currents and voltages at the PCC and operating powers of single-phase IFCs are measured and sent to the SCC. The optimization problem is solved online to determine each single-phase IFC's reference powers in the SCC. The reference powers are transferred into local controllers of the IFCs, which track reference active and reactive powers. It is worth mentioning that communication is needed in this control strategy.

9.4.1 System Model with Embedded Single-phase IFCs under Three-phase Unbalanced Voltage

Here, the AC-coupled hybrid microgrid is connected to the main grid at the PCC (see Figure 9.14). It is assumed that the PCC three-phase voltages and currents are unbalanced due to single-phase loads and generators, shown in Figure 9.15. In this section, v_{PCC} and i_{PCC} are considered as the PCC current and voltage vectors, respectively.

Using instantaneous power theory, the total reference current of phase-x at the PCC from a single-phase perspective could be defined as in (9.46). (Please refer to

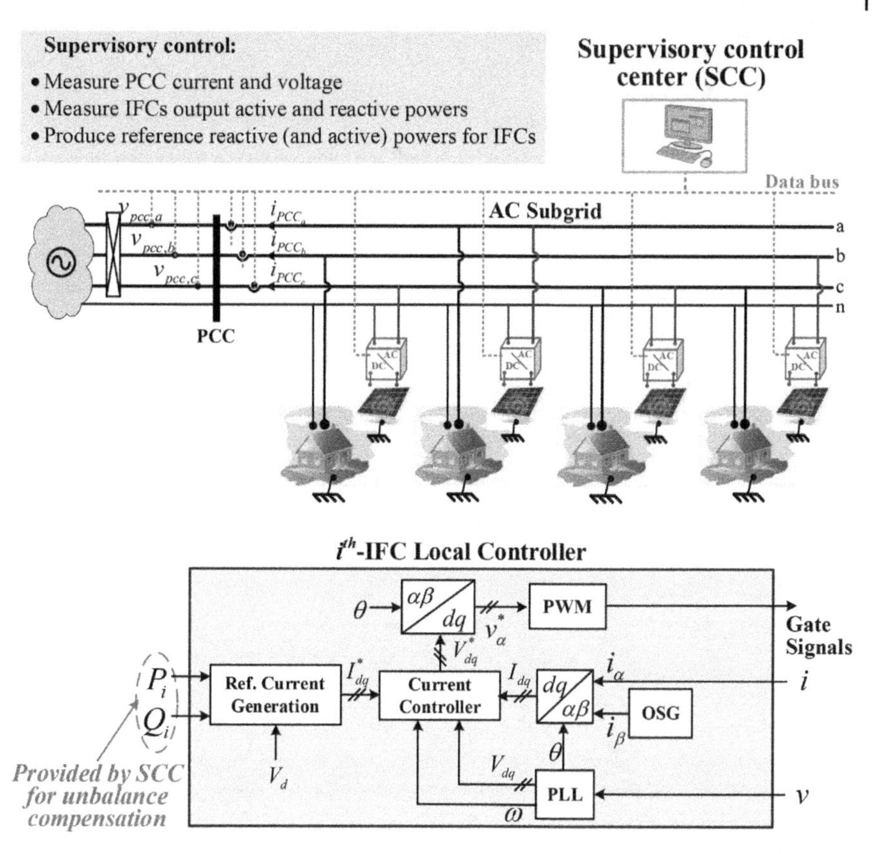

Figure 9.14 Hybrid AC/DC microgrids with single-phase DGs' IFCs connected to the AC subgrid.

Appendix A for detailed discussions of the instantaneous power theory.)

$$i_{\mathrm{PCC}_x} = \underbrace{\frac{P_{\mathrm{PCC},x}}{|v_{\mathrm{PCC}_x}|^2} v_{\mathrm{PCC}_x}}_{i_{\mathrm{PCC}_{xp}}} + \underbrace{\frac{Q_{\mathrm{PCC},x}}{|v_{\mathrm{PCC}_x}|^2} v_{\mathrm{PCC}_{x\perp}}}_{i_{\mathrm{PCC}_{xq}}} \quad x = a, b, c \tag{9.46}$$

where $i_{\mathrm{PCC}_{xp}}$ is the active current of PCC in phase-x, and $i_{\mathrm{PCC}_{xq}}$ is the reactive current of PCC in phase-x. The active and reactive currents generate the active and reactive powers in three phases, respectively. As concluded from (9.46), the active and reactive powers of single-phase IFCs in each phase can be controlled to adjust the current at the PCC.

From (9.46), the negative sequence component of the PCC current vector can be derived as follows:

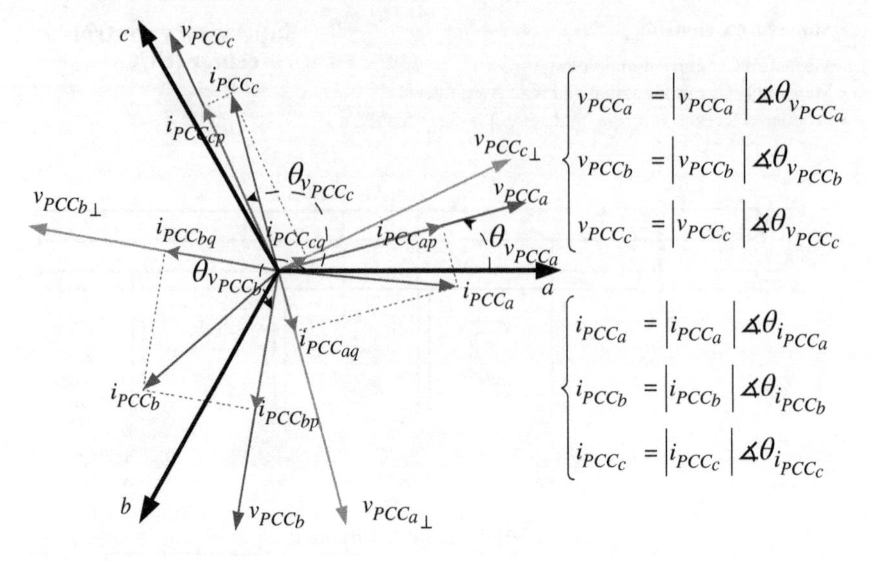

Figure 9.15 PCC voltage and current.

$$\begin{aligned}
i_{\text{PCC}}^{-} &= i_{\text{PCC}_p}^{-} + i_{\text{PCC}_q}^{-} \\
&= \frac{1}{3}\left(\frac{P_{\text{PCC},a}}{|v_{\text{PCC}_a}|^2} v_{\text{PCC}_a} + \frac{P_{\text{PCC},b}}{|v_{\text{PCC}_b}|^2} e^{j\frac{4\pi}{3}} v_{\text{PCC}_b} + \frac{P_{\text{PCC},c}}{|v_{\text{PCC}_c}|^2} e^{j\frac{2\pi}{3}} v_{\text{PCC}_c}\right) \\
&+ \frac{1}{3}\left(\frac{Q_{\text{PCC},a}}{|v_{\text{PCC}_a}|^2} v_{\text{PCC}_{a\perp}} + \frac{Q_{\text{PCC},b}}{|v_{\text{PCC}_b}|^2} e^{j\frac{4\pi}{3}} v_{\text{PCC}_{b\perp}} + \frac{Q_{\text{PCC},c}}{|v_{\text{PCC}_c}|^2} e^{j\frac{2\pi}{3}} v_{\text{PCC}_{c\perp}}\right).
\end{aligned} \tag{9.47}$$

In (9.47), the negative sequence of the PCC current vector $\left(i_{\text{PCC}}^{-}\right)$ encompasses negative sequence components of active and reactive current vectors ($i_{\text{PCC}_p}^{-}$ and $i_{\text{PCC}_q}^{-}$). Also, the unequal values of not only three-phase average active powers but also reactive powers contribute to the negative sequence current. Considering (9.47), equalizing average active powers in three phases could reduce the negative sequence current drastically. For example, single-phase energy storage or active power curtailment of single-phase PV systems can be used to equalize three-phase average active powers $P_{\text{PCC},a} \approx P_{\text{PCC},b} \approx P_{\text{PCC},c}$; thus the active current negative sequence value $\left(i_{\text{PCC}_p}^{-}\right)$ goes down obviously. The value of $i_{\text{PCC}_q}^{-}$ can also be considered for minimizing i_{PCC}^{-}.

For better illustration, the negative sequence model of the hybrid system seen from the PCC is shown in Figure 9.16. As clear from the figure, when the three phases average active powers are equalized, $i_{\text{PCC}_p}^{-}$ will be very small, close to zero.

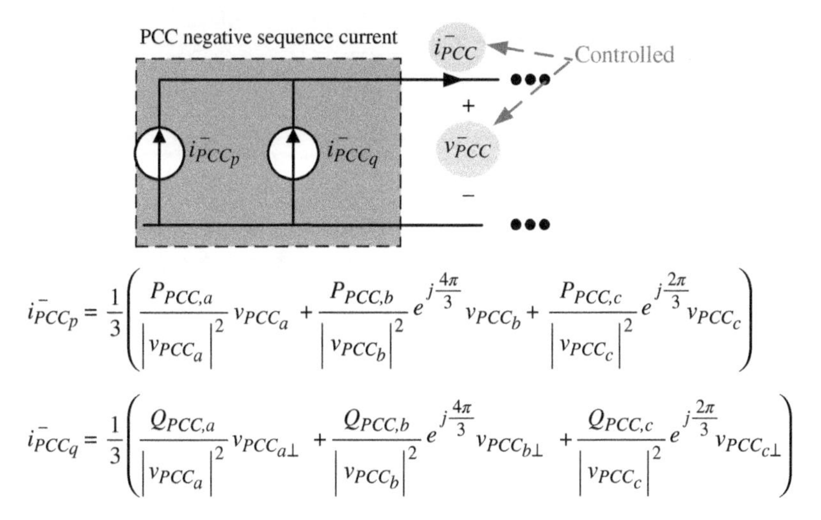

Figure 9.16 Negative sequence model of the system seen from the PCC.

Moreover, $i_{\text{PCC}_q}^-$ can be controlled to absorb $i_{\text{PCC}_p}^-$ in order to reduce/minimize the value of i_{PCC}^-. When $i_{\text{PCC}_q}^- = -i_{\text{PCC}_p}^-$, the negative sequence current is zero.

In Figure 9.17, an example of the PCC negative sequence current minimization, with and without three-phase average active power equalization, is shown. From the figure, when three-phase average active powers are not equal, large $i_{\text{PCC}_q}^-$ is needed to minimize i_{PCC}^-. While a very small $i_{\text{PCC}_q}^-$ is required for i_{PCC}^- minimization when the three-phase average active power have been equalized.

The zero sequence component of the PCC current vector can be achieved from the reference current in (9.46) as follows:

$$i_{\text{PCC}}^0 = i_{\text{PCC}_p}^0 + i_{\text{PCC}_q}^0 =$$
$$\frac{1}{3}\left(\frac{P_{\text{PCC},a}}{|v_{\text{PCC}_a}|^2} v_{\text{PCC}_a} + \frac{P_{\text{PCC},b}}{|v_{\text{PCC}_b}|^2} v_{\text{PCC}_b} + \frac{P_{\text{PCC},c}}{|v_{\text{PCC}_c}|^2} v_{\text{PCC}_c}\right) +$$
$$\frac{1}{3}\left(\frac{Q_{\text{PCC},a}}{|v_{\text{PCC}_a}|^2} v_{\text{PCC}_{a\perp}} + \frac{Q_{\text{PCC},b}}{|v_{\text{PCC}_b}|^2} v_{\text{PCC}_{b\perp}} + \frac{Q_{\text{PCC},c}}{|v_{\text{PCC}_c}|^2} v_{\text{PCC}_{c\perp}}\right). \quad (9.48)$$

From (9.48), unequal values of three-phase average active and/or reactive powers produce zero sequence current. Similarly, equalizing average active powers in three phases can reduce the zero sequence current $\left(i_{\text{PCC}}^0\right)$ drastically. Also, reactive powers in three phases can be controlled to reduce/minimize the value of i_{PCC}^0.

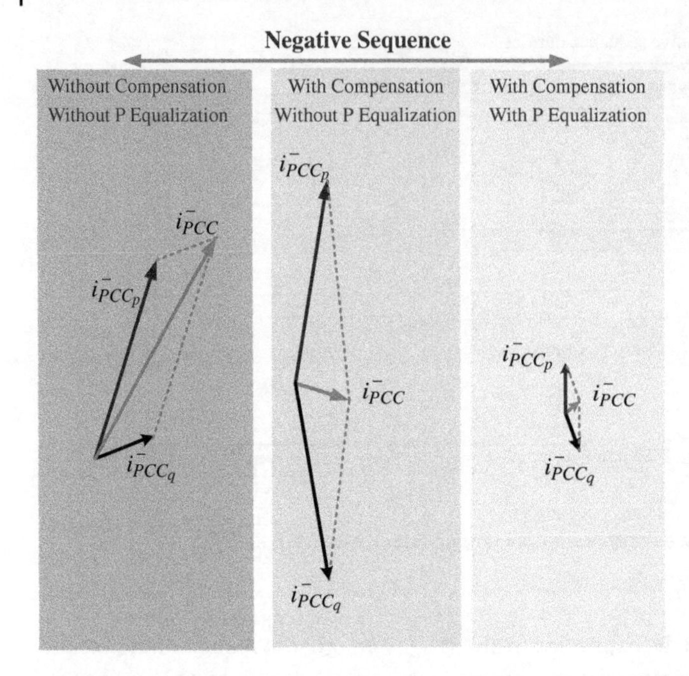

Figure 9.17 An example of the PCC zero-sequence current minimization, with and without three-phase average active power equalization.

In Figure 9.18, the zero sequence model of the hybrid system seen from the PCC is shown. As seen from this figure, when the three-phase average powers are equalized ($P_{\text{PCC},a} \approx P_{\text{PCC},b} \approx P_{\text{PCC},c}$), e.g. using single-phase ESSs, $i^0_{\text{PCC}_p}$ will be very close to zero. Moreover, $i^0_{\text{PCC}_q}$ can be controlled to absorb the zero sequence of active current to minimize i^0_{PCC}. The desired value of $i^0_{\text{PCC}_q}$ is $-i^0_{\text{PCC}_p}$ where i^0_{PCC} is zero.

9.4.2 Reactive Power Control of Single-phase IFCs for Three-phase AC Subgrid Unbalanced Voltage Compensation

As discussed, the first option for the three-phase AC system unbalanced condition compensation is to equalize the average active powers in three phases, e.g. using single-phase IFCs from ESSs or using the active power curtailment of PV systems. Also, reactive powers can be controlled in three phases for further compensation purposes. This section assumes that the three phases' average active powers are constant values (equalized or not after active power control effort has been exhausted), and the IFC reactive powers are further controlled for adjustable compensation of the unbalanced condition. In the following, the minimization of negative and zero sequence currents at the PCC is discussed.

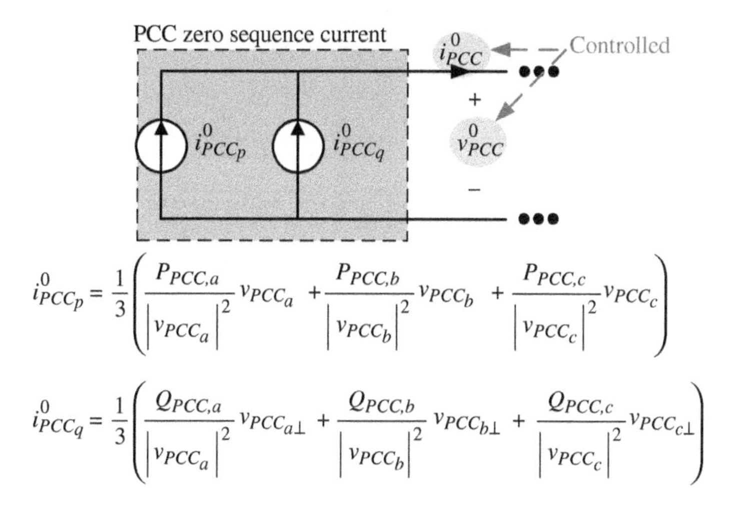

Figure 9.18 Zero sequence model of the system seen from the PCC.

Objective Function

In order to minimize the negative sequence component of the PCC current vector (i^{-}_{PCC}) in (9.47), the square of its amplitude is defined as an objective function for minimization as follows:

$$F_1(Q_{\text{PCC},a}, Q_{\text{PCC},b}, Q_{\text{PCC},c}) = C_1 Q^2_{\text{PCC},a} + C_2 Q^2_{\text{PCC},b} + C_3 Q^2_{\text{PCC},c} + C_4 Q_{\text{PCC},a}\, Q_{\text{PCC},b} + C_5 Q_{\text{PCC},a}\, Q_{\text{PCC},c} + C_6 Q_{\text{PCC},b}\, Q_{\text{PCC},c} + C_7 Q_{\text{PCC},a} + C_8 Q_{\text{PCC},b} + C_9 Q_{\text{PCC},c} + C_{10} \tag{9.49}$$

where C_1 to C_{10} are constant values at each operating point, which are provided in Table C.6 of Appendix C and $Q_{\text{PCC},\,a}$, $Q_{\text{PCC},\,b}$, and $Q_{\text{PCC},\,c}$ are three control variables.

To minimize the zero sequence component of the PCC current (i^{0}_{PCC}) in (9.48), the square of zero sequence current amplitude in (9.50) is minimized. Similar to (9.49), three control variables $Q_{\text{PCC},\,a}$, $Q_{\text{PCC},\,b}$, and $Q_{\text{PCC},\,c}$ are adjusted for minimization purposes. Also, D_1 to D_{10} are constant values at each operating point, provided in Table C.7 of Appendix C.

$$F_2(Q_{\text{PCC},a}, Q_{\text{PCC},b}, Q_{\text{PCC},c}) = D_1 Q^2_{\text{PCC},a} + D_2 Q^2_{\text{PCC},b} + D_3 Q^2_{\text{PCC},c} + D_4 Q_{\text{PCC},a}\, Q_{\text{PCC},b} + D_5 Q_{\text{PCC},a}\, Q_{\text{PCC},c} + D_6 Q_{\text{PCC},b}\, Q_{\text{PCC},c} + D_7 Q_{\text{PCC},a} + D_8 Q_{\text{PCC},b} + D_9 Q_{\text{PCC},c} + D_{10}. \tag{9.50}$$

From (9.49) and (9.50), to flexibly compensate the negative and zero sequence components of the PCC current, the objective function in (9.51) is defined to be minimized [7]. In (9.51), k^- and k^0 are two controllable weighting factors that are related as (9.52).

$$F(Q_{PCC,a}, Q_{PCC,b}, Q_{PCC,c}) = k^{-}F_1(Q_{PCC,a}, Q_{PCC,b}, Q_{PCC,c}) + k^{0}F_2(Q_{PCC,a}, Q_{PCC,b}, Q_{PCC,c}) \tag{9.51}$$

$$k^{-} + k^{0} = 1. \tag{9.52}$$

From (9.51) and (9.52), k^{-} and k^{0} can be controlled to flexibly compensate for the negative and zero sequence currents to keep them at their desired level. For example, when $k^{-} = 1$ ($k^{0} = 0$), three variables $Q_{PCC,\,a}$, $Q_{PCC,\,b}$, and $Q_{PCC,\,c}$ are controlled to minimize the negative sequence current; otherwise $k^{-} = 0$ ($k^{0} = 1$) leads to minimization of the zero sequence current. In the case that $k^{-} \neq 1$ and $k^{0} \neq 1$, both negative and zero sequences currents are compensated for in which their compensation levels are determined by k^{-} and k^{0}.

In general, different criteria can be considered for weighting factor (k^{-} and k^{0}) determination in the distribution power system. These include distribution power system topology, the values of negative and zero sequences current, existing compensation strategies of unbalanced condition, to name a few. For instance, in a European medium voltage distribution power system, which is a three-phase three-wire system, the integrated single-phase DGs are considered for the negative sequence current compensation ($k^{-} = 1$ and $k^{0} = 0$). As another example, in a three-phase three-wire oilfield electric power distribution system, single-phase DGs can be used for negative sequence current compensation. In a three-phase four-wire distribution system, if a negative sequence current is already compensated for (for example, with three-phase DG), the single-phase DGs will be focused on zero sequence current compensation. As another criterion, the ratio of negative sequence over positive sequence current as well as the zero sequence over positive sequence current can be considered for k^{-} and k^{0} determination.

In Figure 9.19, the phasor diagrams of negative and zero sequence current are shown when (a) the negative sequence current is minimized and (b) the zero sequence current is minimized. From the figure, it is concluded that (i) under negative sequence minimization, the controlled $i^{-}_{PCC_q}$ cancels out $i^{-}_{PCC_p}$, which leads to i^{-}_{PCC} minimization, but the resultant $i^{0}_{PCC_q}$ increases the i^{0}_{PCC}, (ii) under zero sequence minimization, although controlled $i^{0}_{PCC_q}$ could compensate for $i^{0}_{PCC_p}$ and minimize the i^{0}_{PCC}, the resultant $i^{-}_{PCC_q}$ increases i^{-}_{PCC}.

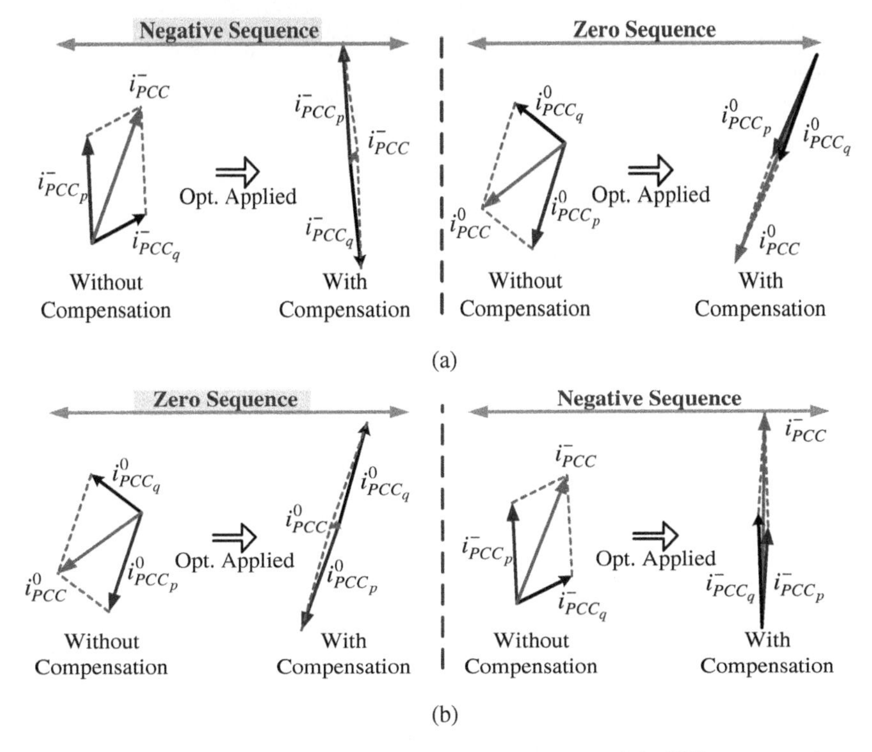

Figure 9.19 Phasor diagram of negative and zero sequences of the PCC current under (a) negative sequence current minimization $k^- = 1 (k^0 = 0)$ and (b) zero-sequence current minimization $k^- = 0\ (k^0 = 1)$.

Constraints

In this control strategy, three control variables ($Q_{\mathrm{PCC},a}$, $Q_{\mathrm{PCC},b}$, and $Q_{\mathrm{PCC},c}$) are restricted considering allowable phase voltages range and IFCs' available reactive power ratings.

Each phase's reactive power at the PCC is restricted between upper and lower limits to keep its voltage amplitude in the acceptable range. Considering Figure 9.14 and assuming that the PCC voltage phasor of phase-x is $|v_{\mathrm{PCC}_x}|\angle\theta_{v_{\mathrm{PCC}_x}}$, the grid voltage phasor of phase-x is $|v_{\mathrm{Grid}_x}|\angle\theta_{v_{\mathrm{Grid}_x}}$, and the grid coupling impedance of phase-x is $|Z_x| \angle \delta_x$, the upper and lower limits of phase-x

reactive power for regulating its voltage between $\left|v_{\mathrm{PCC}_x}^{\min}\right| \leq |v_{\mathrm{PCC}_x}| \leq \left|v_{\mathrm{PCC}_x}^{\max}\right|$ is achieved as:

$$Q_{\mathrm{PCC},x_{V-reg}}^{\max} = \frac{\left(\left|v_{\mathrm{PCC}_x}^{\max}\right|\right)^2}{|Z_x|}\sin(\delta_x) - \frac{\left|v_{\mathrm{PCC}_x}^{\max}\right| |v_{\mathrm{Grid}_x}|}{|Z_x|}\sin(\delta_x + \theta_{v_{\mathrm{PCC}_x}} - \theta_{v_{\mathrm{Grid}_x}}) \quad x = a, b, c \tag{9.53}$$

$$Q_{\mathrm{PCC},x_{V-reg}}^{\min} = \frac{\left(\left|v_{\mathrm{PCC}_x}^{\min}\right|\right)^2}{|Z_x|}\sin(\delta_x) - \frac{\left|v_{\mathrm{PCC}_x}^{\min}\right| |v_{\mathrm{Grid}_x}|}{|Z_x|}\sin(\delta_x + \theta_{v_{\mathrm{PCC}_x}} - \theta_{v_{\mathrm{Grid}_x}}) \quad x = a, b, c \tag{9.54}$$

where $\left|v_{\mathrm{PCC}_x}^{\max}\right|$ and $\left|v_{PCC_x}^{min}\right|$ are maximum and minimum acceptable rms voltages of the PCC in phase-x. Therefore, constraints in (9.55) should be considered for the PCC three-phase voltage regulation.

$$Q_{\mathrm{PCC},x_{V-\mathrm{reg}}}^{\min} \leq Q_{\mathrm{PCC},x} \leq Q_{\mathrm{PCC},x_{V-\mathrm{reg}}}^{\max} \quad x = a, b, c. \tag{9.55}$$

In (9.53) and (9.54), just the grid impedance in fundamental frequency is needed. In addition to the above constraints, the maximum available reactive powers of the single-phase IFCs should be considered and can be calculated using the nominal power ratings of IFCs and their operating active powers. This is important since the smart IFCs are normally already installed in the system and could reach their limited power handling capability when providing an unbalanced compensation function. The calculated powers are then used to restrict the PCC reactive power injection/absorption capability in that phase:

$$Q_{\mathrm{PCC,X_{IFC}}}^{\min} \leq Q_{\mathrm{PCC},x} \leq Q_{\mathrm{PCC,X_{IFC}}}^{\max} \quad x = a, b, c. \tag{9.56}$$

Finally, the interactions of the limitations of the reactive powers in (9.55) and (9.56) are used to determine each phase reactive power constraint at the PCC:

$$\max\left(Q_{\mathrm{PCC},x_{V-\mathrm{reg}}}^{\min}, Q_{\mathrm{PCC},x_{\mathrm{IFCs}}}^{\min}\right) \leq Q_{\mathrm{PCC},x} \leq \min\left(Q_{\mathrm{PCC},x_{V-\mathrm{reg}}}^{\max}, Q_{\mathrm{PCC},x_{\mathrm{IFCs}}}^{\max}\right)$$
$$\Rightarrow \quad Q_{\mathrm{PCC},x}^{\min} \leq Q_{\mathrm{PCC},x} \leq Q_{\mathrm{PCC},x}^{\max} \quad x = a, b, c. \tag{9.57}$$

The developed objective function in (9.51) is minimized subject to constraints in (9.57) using the Karush–Kuhn–Tucker (KKT) method [7]. Using this method, systems of three linear equations with three variables are obtained, which can be solved online easily due to their simplicity. Then, the determined reference reactive powers in each phase from optimization is shared among single-phase

IFCs of that phase considering their maximum available reactive powers, as follows:

$$Q_x = Q_{1_x} + Q_{2_x} + \dots + Q_{r_x} \qquad x = a, b, c \tag{9.58}$$

$$T_{1_x} Q_{1_x} = T_{2_x} Q_{2_x} = \dots = T_{r_x} Q_{r_x} \qquad x = a, b, c \tag{9.59}$$

$$T_{i_x} = \frac{Q_{1_x}^{\max}}{Q_{i_x}^{\max}} \qquad x = a, b, c \text{ and } i = 1, \dots, r \tag{9.60}$$

where Q_{IFCs_x} is the total reference reactive powers of single-phase IFCs in phase-x, calculated from the optimization problem results. Moreover, r_x is the number of single-phase IFCs in phase-x (they can be different in each phase), and Q_{r_x} is the reactive power of the rth IFC in phase-x. T_{i_x} is the sharing factor of the ith IFC in phase-x, and $Q_{i_x}^{\max}$ is the maximum available reactive power of the ith IFC in phase-x. Since T_{i_x} is based on maximum available reactive powers of IFCs, it provides the same available room for IFCs' operation.

This control strategy is applied to the IEEE 13-node test system, in which seven single-phase DGs are integrated into different buses, as is shown in Figure 9.20. The simulated system parameters can be found in Tables C.8 and C.9 of Appendix C. In the simulation, the single-phase DGs are used to compensate the negative sequence current at the PCC ($k^- = 1$ and $k^0 = 0$). The results are shown in Figure 9.21. As seen from the results, after applying the control strategy at $t = 0.2$ s, the negative sequence current is reduced from 185 A to 12 A, and the negative sequence voltage at the PCC is reduced from 72 V to 5 V.

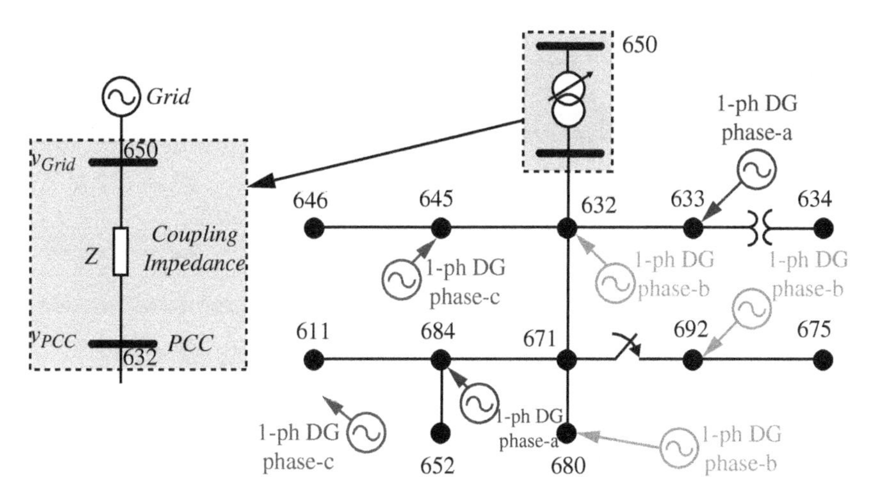

Figure 9.20 IEEE 13-node test system with embedded single-phase IFCs from DGs.

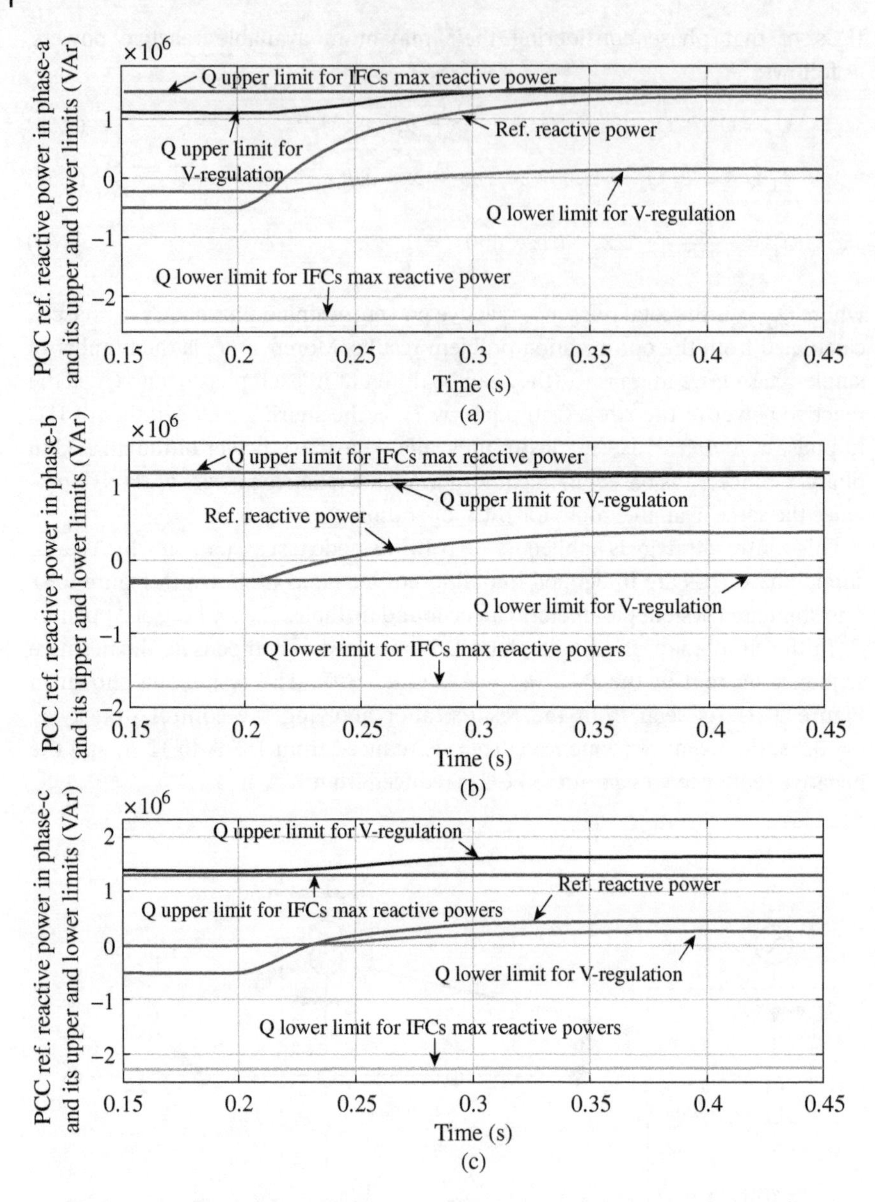

Figure 9.21 Performance of negative sequence current compensation ($k^- = 1$; $k^0 = 0$) using single-phase IFCs: (a) reference reactive power in phase-a at the PCC and its boundary limits, (b) reference reactive power in phase-b at the PCC and its boundary limits, (c) reference reactive power in phase-c at the PCC and its boundary limits, (d) total reference reactive power of phase-a DGs' IFCs, and IFC1 and IFC2 share, (e) total reference reactive power of phase-b DGs' IFCs, and IFC1, IFC2, and IFC3 share, (f) total reference reactive power of phase-c DGs' IFCs, and IFC1 and IFC2 share, (g) negative sequence current at the PCC, and (h) negative sequence voltage at the PCC.

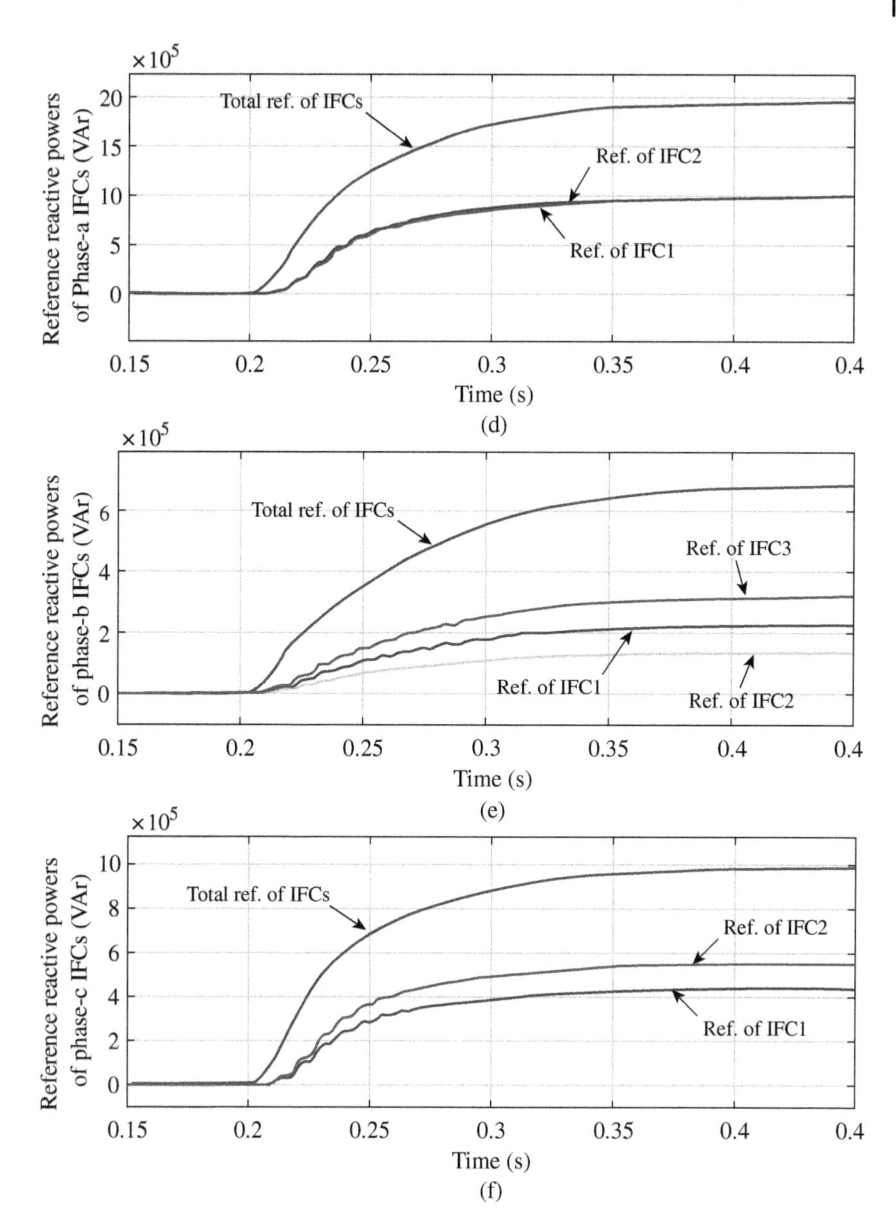

Figure 9.21 (*Continued*)

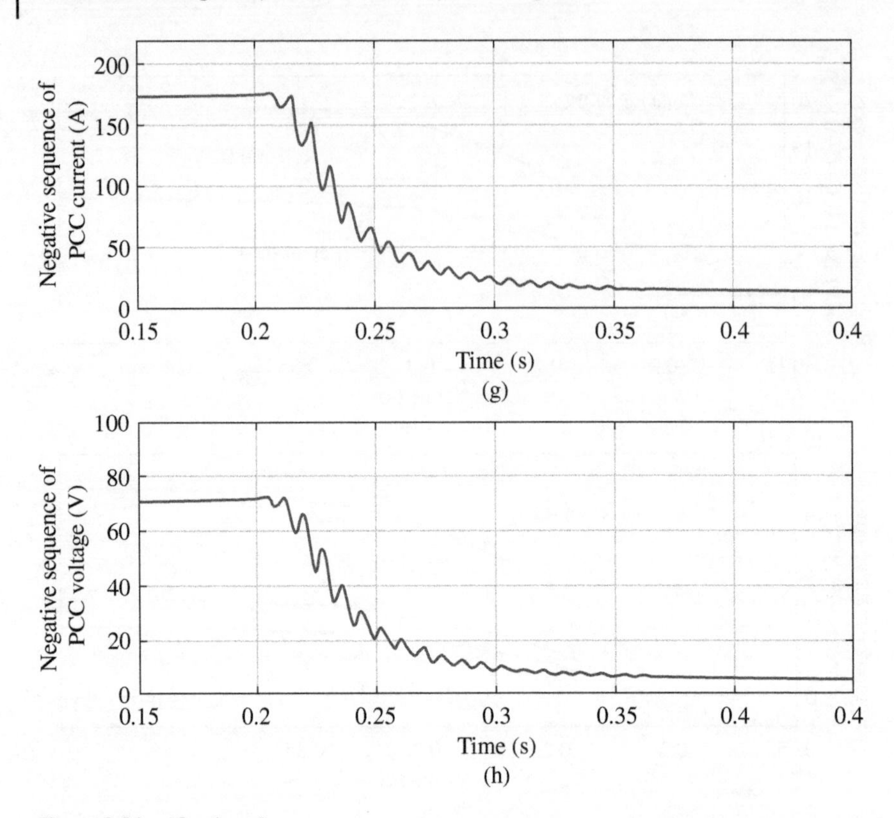

Figure 9.21 (*Continued*)

9.5 Summary

In this chapter, unbalanced voltage compensation in hybrid AC/DC microgrids has been studied. The smart IFCs from DGs and ESSs, and AC/DC subgrids' IFCs are introduced for unbalanced voltage compensation due to: (i) the sources of unbalanced conditions are distributed (e.g. single-phase/unbalanced loads, single-phase/unbalanced DG), (ii) installing additional compensation devices (e.g. STATCOM, UPQC) increases total investment costs. In this chapter, different control strategies of individual and parallel three-phase IFCs and single-phase IFCs for unbalanced voltage compensation have been provided. The presented control strategies have been supported by simulation and experimental results.

References

1 Wang, Y. and Li, Y.W. (2013). An overview of grid fundamental and harmonic components detection methods. In: *Conf. Rec. IEEE Energy Conversion Congress and Exposition (ECCE'13)*, Denver, USA, 5185–5192.
2 Nejabatkhah, F., Li, Y.W., and Wu, B. (2016). Control strategies of three-phase distributed generation inverters for grid unbalanced voltage compensation. *IEEE Transactions on Power Electronics* 31 (7): 5228–5241.
3 Teodorescu, R., Liserre, M., and Rodriguez, P. (2011). *Grid Converters for Photovoltaic and Wind Power Systems*, vol. 29. Wiley.
4 Nejabatkhah, F., Li, Y.W., and Sun, K. (2018). Parallel three-phase interfacing converters operation under unbalanced voltage in hybrid AC/DC microgrid. *IEEE Transactions on Smart Grid* 9 (2): 1310–1322.
5 Sun, K., Wang, X., Li, Y.W. et al. (2017). Parallel operation of bi-directional interfacing converters in a hybrid AC/DC microgrid under unbalanced grid voltage conditions. *IEEE Transactions on Power Electronics* 32 (3): 1872–1884.
6 Nejabatkhah, F., Li, Y.W., Sun, K., and Zhang, R. (2018). Active power oscillation cancellation with peak current sharing in parallel interfacing converters under unbalanced voltage. *IEEE Transactions on Power Electronics* 33 (12): 10200–10214.
7 Nejabatkhah, F. and Li, Y.W. (2019). Flexible unbalanced compensation of three-phase distribution system using single-phase distributed generation inverters. *IEEE Transactions on Smart Grid* 10 (2): 1845–1857.
8 Nejabatkhah, F., Li, Y.W., and Tian, H. (2019). Power quality control of smart hybrid AC/DC microgrids: an overview. *IEEE Access* 7: 52295–52318.

10

Harmonic Compensation Control in Smart Hybrid Microgrids

10.1 Introduction

In both DC and AC subgrids of a hybrid microgrid system, the presence of non-linear loads and switching mode power converters make harmonic pollution a significant concern. These harmonics, particularly the low-order harmonics, can cause extra loss, interference in the grid-connected apparatus, and even resonance in the microgrid. To mitigate the harmful harmonics, as discussed in Chapter 7, both passive power filters (PPFs) and active power filters (APFs) can be used to provide low-impedance paths for harmonics, improving the power quality of the other nodes in the microgrid. However, installing dedicated filters (passive or active) requires extra cost, and these centralized compensation solutions may not be that effective since the sources of harmonics are widely distributed in today's distribution systems.

Thanks to the development of smart interfacing converter (IFC) control strategies, harmonic compensation has been added as an ancillary function to both DC/AC and DC/DC IFCs, including distributed generations, energy storage, and loads, as well as interlinking converters of microgrids. Thus, the harmonics can be mitigated in both AC and DC subgrids in a distributed manner without extra cost.

In this chapter, the harmonics compensations based on the external virtual impedance concept in power systems are studied first. Then, control strategies of low-switching-frequency interfacing power electronics converters for harmonic compensation in AC subgrids are studied. After that, control of interfacing power converters for harmonics compensation in DC subgrids is addressed. Finally, coordinated control of multiple interfacing power converters for harmonic compensation in hybrid microgrids is discussed.

Smart Hybrid AC/DC Microgrids: Power Management, Energy Management, and Power Quality Control, First Edition. Yunwei Ryan Li, Farzam Nejabatkhah, and Hao Tian.

10.2 Control of Interfacing Power Converters for Harmonic Compensation in AC Subgrids

For IFCs participating in power quality improvement control, there are three harmonics compensation objectives in general [1]:

1. PCC voltage harmonics compensation
2. Local load harmonics compensation
3. IFC line current harmonics rejection.

To demonstrate the meaning of the above three objectives, a simplified example system with an IFC is shown in Figure 10.1, where the IFC is connected to the PCC with an LCL filter. There are two types of loads in the system: a PCC load directly connected at the PCC and local loads installed at the IFC unit LCL filter output terminals. The main AC grid is linked to the PCC through a grid side impedance. Together with the local load, the IFC unit can be considered a small microgrid in this case. Many other microgrids can be connected to the PCC in similar ways.

The IFC line current (i_{IFC}) is generally required to be free of distortion based on most grid codes for DG interconnections. In this case, the local and PCC non-linear loads can cause non-trivial harmonics in the PCC voltage (v_{PCC}) and microgrid current (i_{MG}).

When the IFC is used to compensate the harmonics, the three abovementioned compensation objectives can be explained as follows.

For PCC harmonic voltage compensation, the IFC absorbs the harmonic current of the PCC's non-linear load. Therefore, the grid will only supply linear current to

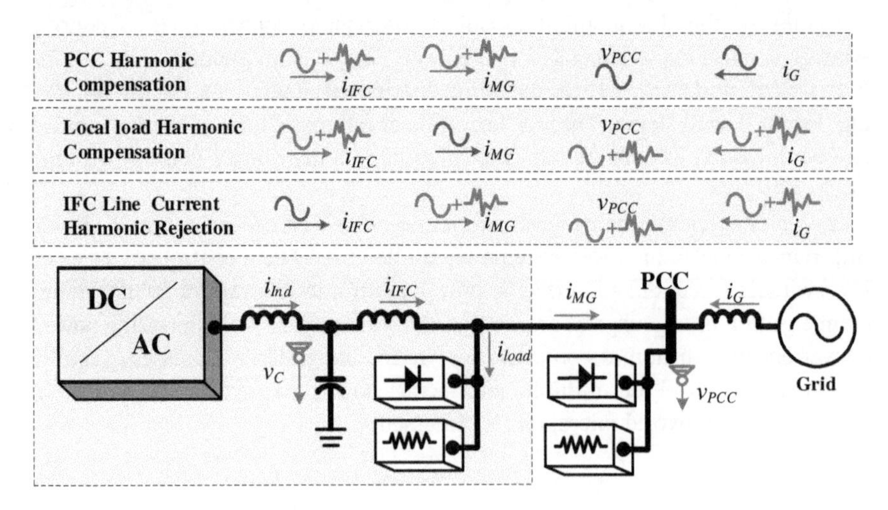

Figure 10.1 AC subgrid harmonic compensation using interfacing converter (IFC) units.

the PCC, resulting in good voltage quality at the PCC. In other words, if the IFC is modeled as a virtual impedance, the output impedance at the harmonic orders would be small enough to absorb non-linear load current for the PCC harmonic voltage compensation. This compensation is significant for a weak grid with a lower short-circuit capacity, where the harmonic current flowing through the high grid impedance can bring significant voltage distortions at the PCC.

For local harmonics compensation, the IFC will compensate the harmonic current drawn by the local non-linear loads. This compensation would be important when the IFC unit and local load form a microgrid, where the IFC locally compensates the non-linear load within the microgrid. As a result, the microgrid will behave like a linear load or source to the AC subgrid.

Finally, in the IFC line current rejection control, the IFC produces the sinusoidal current at the fundamental frequency according to the power control reference regardless of the distorted PCC voltage. In this control mode, the effects from the local and PCC non-linear loads are not compensated. This requires the IFC's output virtual impedance to be high enough at harmonic frequencies to block any harmonic current flow to the IFC. Many existing DG interconnection grid codes require the IFC to operate in line current rejection control due to the output current harmonics requirement.

The key to realizing the different harmonic control targets is to properly control the impedance of IFCs at harmonic frequencies. As discussed in Chapter 4, this can be achieved by applying the virtual impedance concept, which enables the IFCs to flexibly change their equivalent output impedance. For example, for harmonics compensation, the IFC's external virtual impedance at harmonic frequencies can be controlled to absorb the non-linear load harmonics current and improve the PCC voltage quality. In Figure 10.2, equivalent impedance models of IFCs, loads, and the grid are shown. The IFC is modeled as an impedance Z_{IFC}. The non-linear load at the PCC is modeled as a harmonic current source (i_{load}), and an impedance (Z_{load}) accounts for the fundamental frequency power. The grid is equivalent to a voltage source with impedances (Z_{grid}) in series. In such a system, the non-linear load injects harmonic current into the system, and the IFC and the grid share the harmonic currents according to their respective impedances. The grid side harmonic current will distort the PCC voltage due to the grid impedance voltage drop. If the non-linear load harmonic currents can be absorbed by the IFC (when IFC's equivalent impedance is controlled to be small enough at harmonics frequency), the voltage distortion at the PCC can be mitigated and the grid will see a grid-connected system with high power quality.

Generally, the virtual impedance control of the IFC for harmonics compensation can be realized through closed-loop control reference modifications in the current control method (CCM) for use in the grid-following converters, the voltage control method (VCM) for the grid-forming converters, and the hybrid control

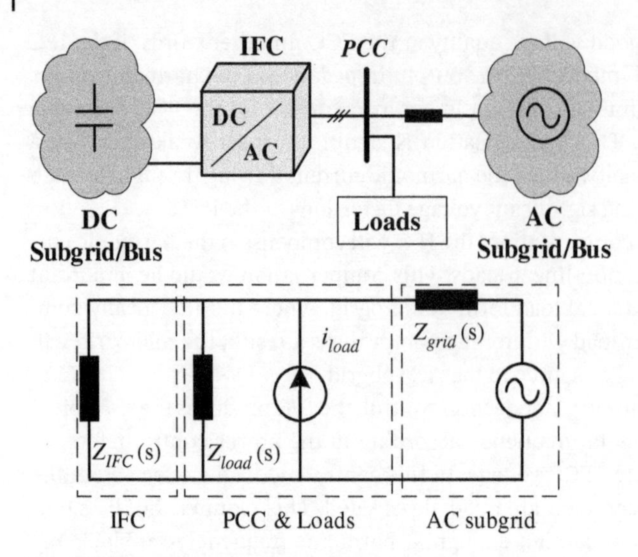

Figure 10.2 The equivalent circuit of the IFC that is connected to the AC subgrid under harmonics control.

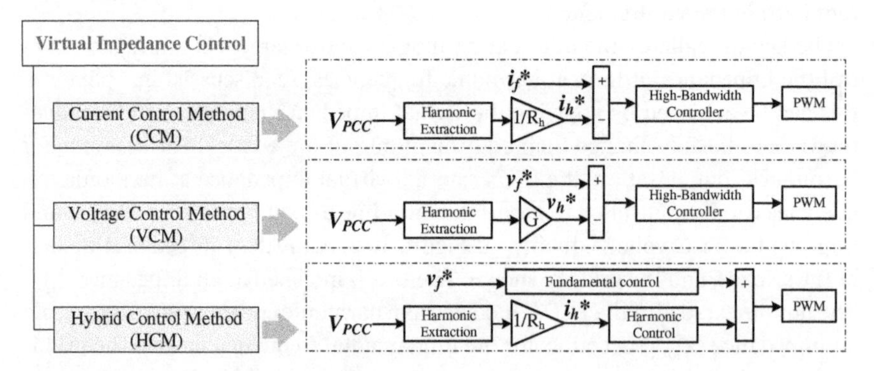

Figure 10.3 External virtual impedance control of IFCs through: the current control method (CCM), the voltage control method (VCM), and the hybrid control method (HCM).

method (HCM). Due to the consideration of low switching frequency and limited closed-loop control bandwidth, the feed-forward control method (FCM) can also be considered. In Figure 10.3, these control methods with virtual impedance shaping are shown. In this figure, R_h is the desired virtual impedance, which can be adjusted dynamically through online digital control. More explanations on these control strategies will be provided in the following sections.

Figure 10.1 shows the diagram with a single IFC unit. In practice, multiple IFC units can work together and share the harmonic compensation current.

For example, when multiple IFC units are working in PCC harmonic voltage compensation mode, it is desired that these IFC units share the PCC non-linear load current according to their available ratings and priorities. When there is a supervisory controller in the system, the distorted PCC voltage can be measured by the supervisory controller, and the harmonic compensation command is sent to the IFC local controller. At the IFC unit local controller, the voltage or current tracking schemes are used to ensure satisfactory harmonic control (for example, resonant controller (RSC), repetitive controller (RC), or deadbeat (DB) controller, discussed in Chapter 4). It should be mentioned that communication links increase the cost of the microgrid. To overcome this limitation, coordinated harmonic compensation without using communications can be considered. In this case, the IFC unit harmonic compensation command is determined according to the measured local signals, such as harmonic voltage at the IFC installation point, the IFC line harmonic current, and the IFC's available power rating. More explanation on IFC coordination for harmonics compensation will be provided later in this chapter.

Regarding the IFC local controller, traditionally there are two categories of control schemes: current-controlled scheme and voltage-controlled scheme. They are called the CCM and the VCM, respectively, in this book. The CCM is suitable for grid-connected operations in which the IFC line current is controlled according to the output power reference (such as in a grid-connected PV inverter system); therefore it is also referred to as grid-following control. However, for more direct voltage support, especially in autonomous microgrid operation mode, the VCM, or grid-forming control will be more suitable. Also, the VCM in microgrid operation enables droop control for load demand sharing without communication among the IFC systems. Compared to power control through current regulation, the voltage regulation method is more sensitive to IFC or grid voltage disturbances. However, there are ways to reduce this sensitivity through current limiting or virtual impedance control [2], where the fundamental and harmonic currents from the IFC are controlled indirectly. More discussions about current control and VCMs of power converters can be found in Chapter 4.

Other than the traditional CCM and VCM schemes, the HCM can control both the IFC voltage and line current but at different frequencies. Specifically, for the IFC with the HCM, the fundamental frequency voltage is controlled similarly to the VCM, while the harmonic currents are directly controlled to easily realize different harmonics compensation objectives. It will be shown that the HCM has very well decoupled voltage and current control branches. To realize the harmonics compensation functions in an IFC with the CCM, the VCM, and the HCM, the current or voltage IFC reference should be modified respectively according to the harmonic compensation objectives.

10.2.1 Harmonics Compensation with the Current Control Method (CCM)

The IFC with CCM has a high impedance at the fundamental frequency as the converter behaves like a current source, and it is convenient to add harmonic controllers to mitigate harmonics. The CCM-based IFC systems compensation strategy is shown in Figure 10.4. The current control scheme can be an RSC, RC, or deadbeat controller, which is selected according to the harmonic control requirement (see Chapter 4). In this example, the current-control scheme is based on the stationary reference frame with the parallel proportional and RSC at fundamental and harmonic frequencies. This current control can be equivalently implemented in the synchronous rotating reference frame. In the CCM-based IFC system, the line current harmonics rejection is inherently enabled as the default compensation mode when the reference current does not contain harmonics. Therefore, two other harmonics compensation modes to be realized are PCC voltage harmonics compensation and local harmonics compensation.

Compensate PCC Voltage Harmonics with the CCM

For the PCC voltage quality enhancement, the basic idea is to direct the non-linear load current to the IFC side, so the AC subgrid will supply only the sinusoidal fundamental current. As a result, the PCC voltage harmonics can be reduced as the harmonic voltage drops on the grid impedance will be mitigated. As direct compensation of the non-linear load currents at the PCC is difficult due to the

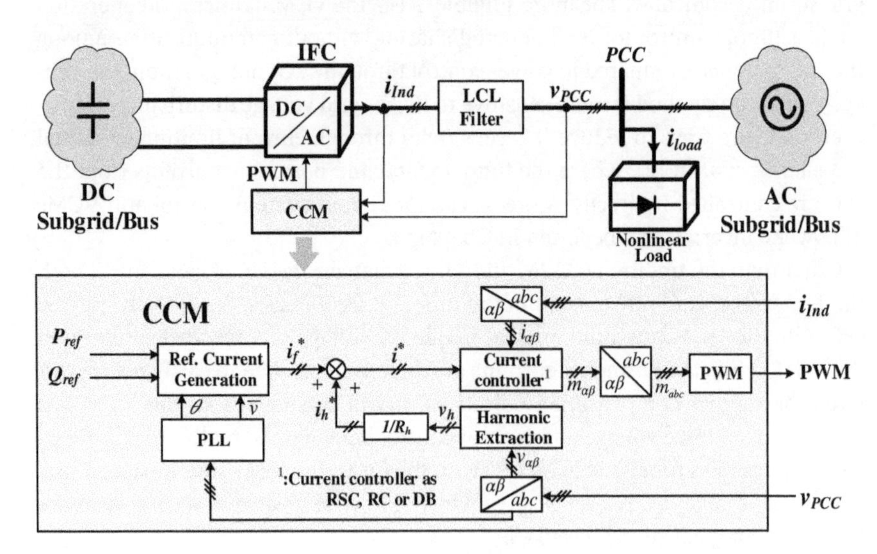

Figure 10.4 IFC control block diagram with CCM-based harmonic compensation.

distributed nature of the loads, a more popular method based only on local measurement is to use the resistive active power filter (R-APF) concept, where the IFC unit works as a small damping resistor at the selected harmonic frequencies.

In this virtual harmonic resistance control, the IFC unit current reference (i^*) will be modified by adding a harmonic current term as follows:

$$i^* = i_f^* + i_h^* = i_f^* + \underbrace{H_D(s)\cdot v}_{v_h}/R_h \tag{10.1}$$

where i_f^* is the fundamental current reference for IFC real and reactive power control, i_h^* is the harmonic current reference, and R_h is the equivalent IFC resistance at harmonic frequencies. $H_D(s)$ in (10.1) is the harmonic detector to extract PCC harmonic voltages. In this control scheme, the PCC voltage could be approximated by the IFC installation point voltage, and therefore remote measurement or communication can be avoided. However, for a weak grid with higher feeder impedance, the information on PCC voltage and feeder impedance is desired for more accurate harmonics compensation, especially when sharing harmonics among multiple IFC units is necessary. Note that when PCC voltage quality is improved using (10.1), the microgrid current i_{MG} (including IFC and local loads) may have a non-trivial harmonic current. In (10.1), the virtual damping resistance R_h will be determined according to the PCC harmonics compensation requirement, the IFC available rating, and with the consideration of the stability range of virtual impedance. In practice, R_h can also be adjusted to be an online "THD-R_h" loop [3]. With this method, the PCC harmonics voltage can be automatically tuned to ensure better performance.

Compensate Local Load Harmonics with the CCM

When local load harmonics compensation is desired, the IFC unit needs to absorb the harmonic current from local loads. As discussed earlier, with the local harmonics compensation, the microgrid with IFC and local loads will behave like a linear load/source seen from the rest of the AC subgrid as the local load non-linear currents being compensated for within the microgrid. The local load currents shall be measured to realize this task, and the harmonic components are detected and then added to the CCM current reference. The modified current reference i^*, in this case, is presented as follows:

$$i^* = i_f^* + i_h^* = i_f^* + H_D(s)\cdot i_{load} \tag{10.2}$$

Again, $H_D(s)$ is the harmonic detector to extract local load harmonic currents. Once the local load harmonic currents are properly absorbed by the IFC unit, the microgrid current (i_{MG}) between the microgrid and AC subgrid will be sinusoidal with little distortion. However, in this case, due to the impact of PCC non-linear loads, the PCC voltage distortion will not be compensated for entirely.

Reject IFC Line Current Harmonics with the CCM

As the default compensation mode, a pure sinusoidal current reference will be applied to the harmonic controllers to eliminate the IFC output current harmonics. In other words, in this control strategy $i_h^* = 0$, and the current reference is equal to the sinusoidal fundamental reference current.

Case Study Results with the CCM

In this case study, the IFC connected to the AC subgrid is used to compensate for the PCC voltage harmonics with the CCM. The study system parameters can be found in Table C.10 of Appendix C. The results are shown in Figure 10.5. At first, the system performance without harmonics compensation is tested. As seen in Figure 10.5a, since the IFC does not compensate the harmonics (the output current of the IFC is sinusoidal; default operation mode), the main AC grid/subgrid provides the non-linear load current, which leads to PCC voltage harmonics distortion. When the PCC voltage harmonics compensation is activated (see Figure 10.5b), the IFC absorbs the non-linear load current, effectively improving the grid current and significantly reducing the PCC voltage.

10.2.2 Harmonics Compensation with the Voltage Control Method (VCM)

Although most grid-connected converters use the CCM scheme, the VCM is increasingly used in IFCs due to grid-forming features (see Chapter 4).

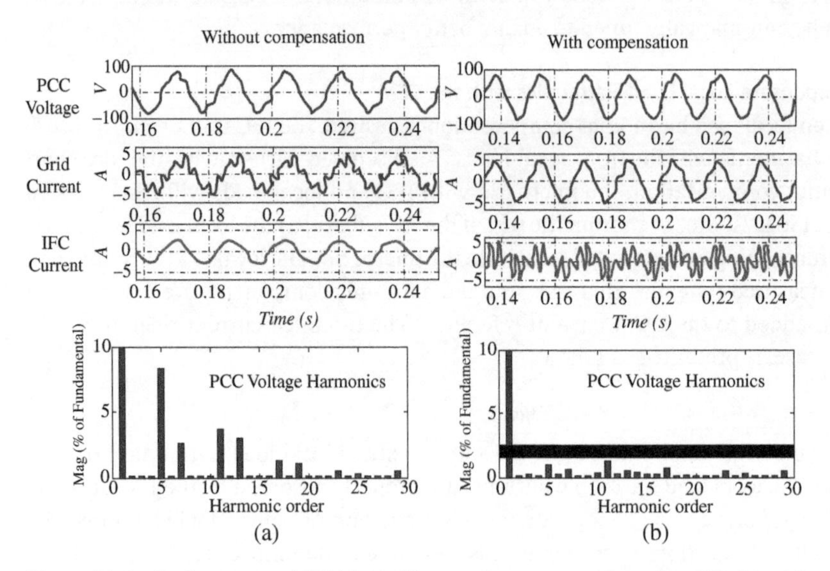

Figure 10.5 Performance of CCM-based harmonic compensation of an IFC: (a) without compensation activation and (b) with PCC voltage harmonics compensation.

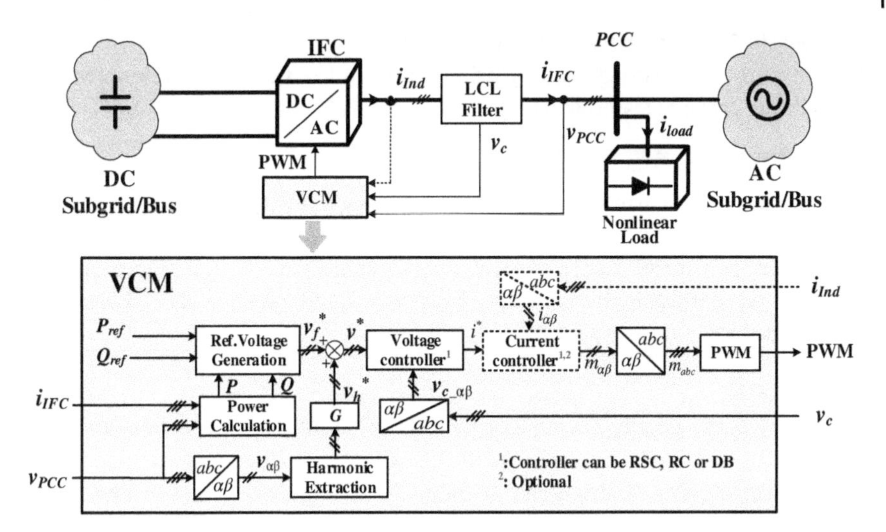

Figure 10.6 IFC control block diagram with VCM-based harmonic compensation.

In particular, the autonomous microgrid operation will require a VCM for microgrid voltage and frequency control. In this case, the voltage reference derived from the well-understood droop control further enables multiple IFC units to achieve decentralized power-sharing without communication between them. Nevertheless, the VCM cannot directly regulate the IFC unit line current harmonics as there is no closed-loop line current regulation in the control scheme. Therefore, the VCM-based IFC line current quality is sensitive to PCC voltage disturbances and the local harmonic loads. As a result, the VCM-based IFC is rarely used to address system harmonic problems. However, it is possible to realize the power quality compensation using the VCM with virtual harmonic impedance control. The diagram of a voltage-controlled IFC is presented in Figure 10.6.

For this VCM-based IFC, the droop controllers (real power–frequency droop and reactive power–voltage magnitude droop) can be used to derive instantaneous voltage reference (v_{f}^{*}). Note that to ensure zero steady-state reactive power tracking in the grid-connected operation, a reactive power integral control term will also be added to the voltage magnitude reference [4]. Afterward, a voltage-control scheme (generally, RSC, RC, or DB) will be employed to ensure satisfied voltage tracking. In the voltage tracking stage, a double-loop voltage controller is usually considered to obtain a better dynamic (outer voltage, inner current control). Here, the outer loop has a proportional and a resonant controller. To reduce the capacitor voltage distortions, parallel harmonic resonant compensators will also be used.

Compensate PCC Voltage Harmonics with the VCM

To control the PCC voltage quality using the VCM [5], the voltage reference can be modified through a simple feed-forward term as:

$$v^* = v_f^* + v_h^* = v_f^* + G.H_D(s) \cdot v_{PCC} \tag{10.3}$$

where $H_D(s)$ is the harmonic detector to extract PCC harmonic voltages, G is the gain of the feed-forward term, and v^* is the modified voltage reference.

With the control of (10.3), the IFC works as a small virtual impedance at the selected harmonic frequencies. The corresponding equivalent harmonic impedance of $Z_{IFC,\,eq}$ is expressed as:

$$Z_{IFC\_eq} = Z_{IFC}/(1 + G) \tag{10.4}$$

where Z_{IFC} is the original impedance without harmonics compensation, and it is equal to $Z_{IFC} = sL_2 + R_2$ (L_2 and R_2 are the LCL filter grid side inductor impedance).

Similar to the R-APF discussed in the CCM scheme, a small equivalent IFC impedance ($Z_{IFC,eq}$) using a high gain (G) can improve the PCC voltage quality by absorbing more harmonic currents. However, for the control in (10.3) and (10.4), the realized equivalent harmonic impedance ($Z_{IFC,eq}$) (or it can be called the virtual harmonic impedance) will be inductive at harmonic frequencies if a real number is used as the feed-forward gains (G). If a virtual resistance like in an R-APF is desired in this case, a complex number should be selected for G. Additionally, the final equivalent IFC impedance is also related to the existing grid side impedance and accurate control of the virtual impedance value will need the grid side impedance information.

Compensate Local Load Harmonic with the VCM

If local harmonic load compensation is desired, the VCM in (10.3) is incapable of realizing this control objective. Alternatively, the load harmonic current can be absorbed by adding a harmonic current feed-forward term to the inner current loop reference. With this method, the VCM is also able to compensate local load harmonic currents.

However, it should be pointed out that local harmonics compensation using the VCM requires high bandwidth inverter output current (i) tracking such as a hysteresis controller, predictive controller, multiple harmonic RSCs, etc. As a result, the inner loop proportional controller $G_{inner}(s)$ used for the conventional VCM in Figure 10.4 must be replaced. Indeed, these inner loop controllers increase the control complexity of the IFC unit. Also, the un-damped IFC output current response may adversely introduce some LCL filter resonances.

Reject IFC Line Current Harmonics with the VCM

For the conventional VCM, the IFC line current cannot be controlled directly. However, this current regulation can be achieved indirectly by using $G = -1$.

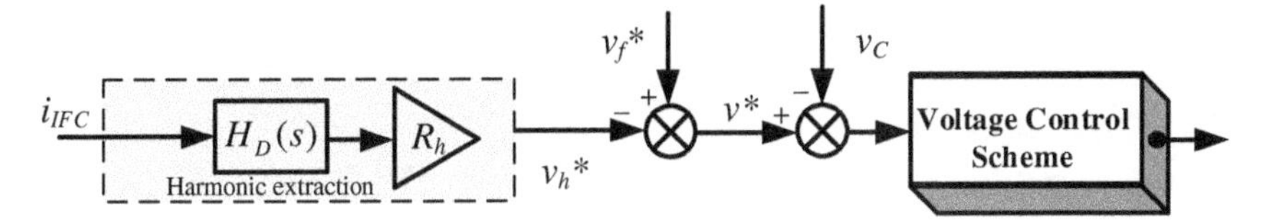

Figure 10.7 The IFC line current harmonic rejection with the VCM using IFC current feedback.

In this case, a large closed-loop harmonic impedance will be present at the IFC terminal, forcing the non-linear load current to flow to the grid side. As a result, the IFC line current can be controlled to be sinusoidal. This makes the IFC behave like a normal CCM-controlled IFC without any system harmonics compensation. Note that, in practice, a value of G slightly higher than -1 (e.g. -0.9) is needed to realize a large positive impedance in the digital controller.

Alternatively, with the help of virtual series impedance control using the IFC output current (i_{IFC}), the IFC can have a large equivalent impedance to reject the harmonic current. The realization of such a series harmonic impedance for line current harmonics rejection is shown in Figure 10.7. With this scheme, an external virtual impedance is added to the system, which adds a voltage drop v_{h}^* due to the harmonic current. As a result, the converter will behave like an impedance at harmonic frequencies present at the output terminal. The harmonic current cannot flow into the IFC due to the high equivalent impedance, avoiding distortions in IFC currents.

Case Study Results with the VCM

The VCM-based harmonic compensation performance has been evaluated in the simulation, and the results are shown in Figure 10.8 [6]. The simulated system parameters are provided in Table C.10 of Appendix C. In this study, the system's performance without harmonic compensation is shown in Figure 10.8a, where the feed-forward gain G is zero. It can be seen that the IFC and grid share the non-linear load current according to the converter side and grid impedance ratio. When the feed-forward gain G is positive, the harmonic can be absorbed by the IFC (non-linear load currents will flow to the converter), and the PCC voltage harmonics are reduced (see Figure 10.6b). While the negative gain ($G = -1$) will lead the non-linear load current to the grid side due to large impedance of IFC, enforcing sinusoidal output current.

10.2.3 Harmonics Compensation with the Hybrid Control Method (HCM)

The hybrid voltage and current control method (HCM) is realized through simultaneous control of the LCL filter capacitor voltage at the fundamental frequency

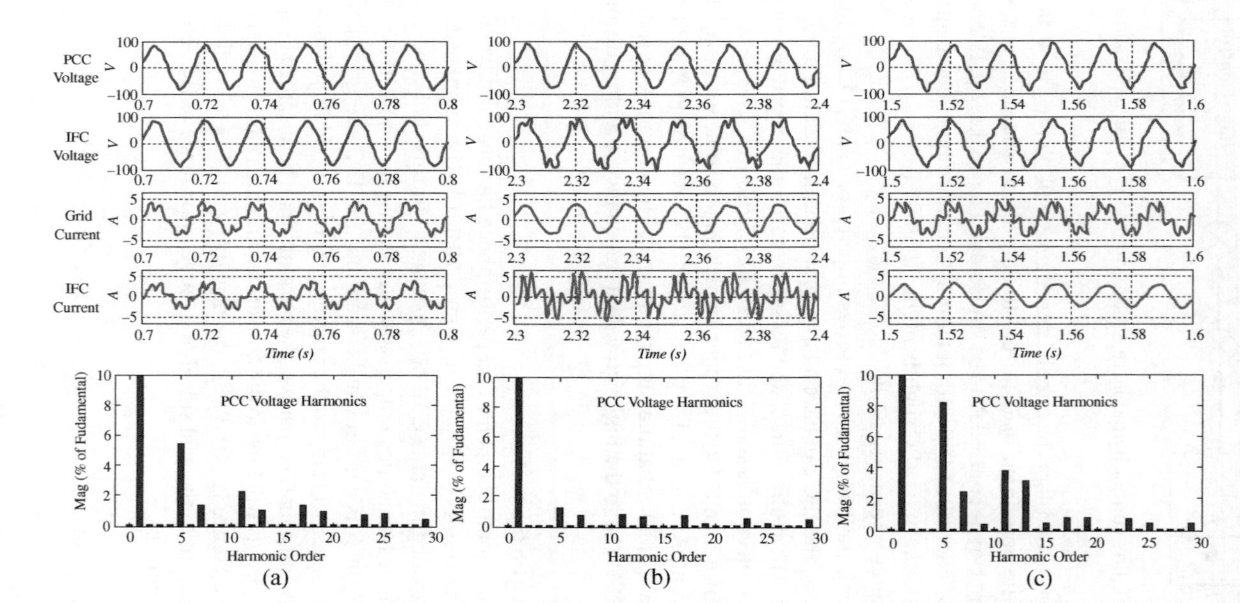

Figure 10.8 Performance of VCM-based harmonic compensation of an IFC: (a) without compensation activation ($G = 0$), (b) PCC voltage harmonics compensation ($G > 0$), and (c) IFC line current harmonics rejection ($G = -1$). Adapted from [6].

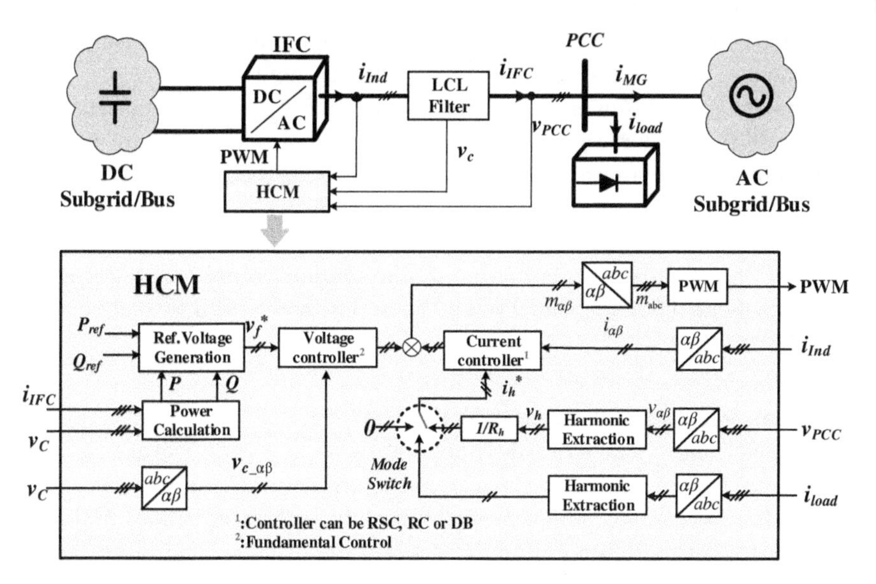

Figure 10.9 A typical individual IFC control block diagram with HCM harmonic compensation.

and the line current at the harmonic frequency. Like the VCM, the output power of the IFC is controlled by the regulation of the fundamental filter capacitor voltage and, therefore, the grid-forming function (e.g. with droop control) can be achieved. At the same time, a closed-loop harmonic current compensator regulates the line current harmonics.

By further transforming the cascaded structure of double-loop controller into a more straightforward single-loop parallel structure with multiple control branches, the HCM as shown in Figure 10.9 can be expressed as:

$$v_{\mathrm{out}}^{*} = G_{\mathrm{power}}(s) \cdot \left(v_{\mathrm{f}}^{*} - v_{\mathrm{c}}\right) + G_{\mathrm{harmonic}}(s) \cdot \left(i_{\mathrm{h}}^{*} - i_{\mathrm{Ind}}\right) + G_{\mathrm{damping}}(s) \cdot i_{\mathrm{IFC}} \quad (10.5)$$

where $v_{\mathrm{out}}{}^{*}$ is the expected output voltage, the first term is a closed-loop control term for the fundamental capacitor voltage, the second term is a closed-loop control term for the IFC line harmonic current, and the third term is an active damping term. Each term will be further explained as follows:

As shown in Figure 10.7, the voltage tracking controller is an RSC at the fundamental frequencies. Here, only a fundamental frequency RSC ($G_{\mathrm{power}}(s)$) is used to ensure good tracking of the fundamental capacitor voltage for power flow control. To minimize the harmonic current control term's interference, the proportional gain in the voltage controller is removed.

The second term in (10.5) is the harmonic current controller. It is mainly used to track the reference harmonic current at selected frequencies. Therefore, multiple RSCs at different harmonic frequencies ($G_{\mathrm{harmonic}}(s)$) are adopted. Considering that line currents may have some non-characteristic harmonic currents, a small proportional gain k_p is also used to achieve better harmonic tracking. Since there is no fundamental frequency RSC here and the gain k_p is normally small, the tracking of line harmonic current will not introduce any obvious disturbances to fundamental voltage control. Therefore, the fundamental IFC voltage and line harmonic current can be regulated separately without much cross-coupling using the first two control terms in (10.5).

Finally, the active damping term has only a proportional controller ($G_{damping}(s)$), which can damp the fundamental voltage control path and the harmonic current control path.

Compensate PCC Voltage Harmonics with the HCM

HCM-based control can be easily employed to improve PCC voltage quality. In this case, the harmonic current reference will be obtained similarly as shown in (10.1) ($i_h^* = -H_D(s).v_{PCC}/R_h$) where R_h is the virtual resistance and $H_D(s)$ is the harmonics extractor. In this case, the IFC unit works like a small virtual resistive impedance at the selected harmonics frequencies to improve the PCC voltage similarly to the CCM-controlled IFC.

Compensate Local Load Harmonics with the HCM

When the reference current i_h^* in (10.5) is selected as the harmonic content of local load currents, this term controls the IFC unit to compensate most of the harmonics produced by local loads, leaving an improved microgrid current (i_{MG}) to the PCC. Further considering that the harmonic current control loop has a very small gain at the fundamental frequency, it is possible that the measured total local load current (i_{load}) can be directly used as the reference current i_h^* for harmonic current control without the harmonics extraction block $H_D(s)$. Note that this contrasts with the PCC voltage compensation, as typically the current THD of a non-linear load is much higher than the PCC voltage THD. This harmonic compensation without using harmonic extraction is an obvious advantage of the HCM, making the method especially attractive for many low-power cost-effective IFC units with limited computing capability.

Reject IFC Line Current Harmonics with the HCM

When i_h^* in (10.5) is set to zero, the HCM can also reduce the IFC line current harmonics. In this control mode, the steady-state performance is similar to the case for an IFC unit using the conventional CCM without any compensation, where the line current is sinusoidal, and the main grid supplies all harmonic currents.

Finally, it is worth mentioning that the concept of the HCM can also be used for CCM-based IFC units, where a fundamental current control term can replace the first term in (10.5). As a result, the fundamental current and harmonic current are controlled separately in a decoupled manner. Compared to the traditional CCM base control, separating fundamental and harmonic current control brings the advantages of the HCM, such as no harmonics extraction block requirements for the local harmonics compensation, as discussed earlier.

Case Study Results with the HCM

In Figure 10.10, the performance of the HCM-based harmonic compensation is evaluated under three different operating modes: PCC voltage harmonics compensation, IFC line current harmonics rejection, and local load harmonics current compensation. The simulated system parameters are provided in Table C.10 of Appendix C. As seen in Figure 10.10a, the PCC voltage THD is reduced under the PCC voltage harmonics compensation. Since the control objective, in this case, is not focused on compensating the local harmonic load, the i_{MG} still has non-trivial ripples. In the IFC line current harmonics rejection, as shown in Figure 10.10b, the IFC unit line current (i_{IFC}) is almost distortion-free. Meanwhile, the harmonic currents of both the local and PCC loads are supplied by the main AC grid. Thus, the PCC voltage and the main grid current distortions are further aggravated. The local load harmonics current compensation results are shown in Figure 10.10c. Since the harmonic current of the local load is compensated by the IFC unit, the microgrid output current (i_{MG}) is distortion-free. Meanwhile, the IFC line current and main grid current are distorted, as expected.

10.2.4 Comparison of Harmonics Compensation with the CCM, the VCM, and the HCM

For application situations, the selection of harmonic compensation method mainly depends on the IFC operation mode and existing control schemes. Specifically, the CCM-based compensation method is recommended for current-controlled IFCs for grid-following operation and does not work well with a voltage-controlled IFC for grid-forming control, such as in a stand-alone system. The VCM- and HCM-based compensation methods are good fits for voltage-controlled IFCs. Therefore, they can be easily implemented in both grid-connected operations (working as a virtual synchronous generator) or in stand-alone operations, where voltage magnitude and frequency control are required. For example, a grid-connected, current-controlled photovoltaic converter or voltage-controlled energy storage system in stand-alone operation will need the CCM- and the VCM/HCM-based compensation methods, respectively. Additionally, the HCM-based method can also replace fundamental frequency voltage control with fundamental frequency current control and works as the current control mode similar to the CCM-based IFCs.

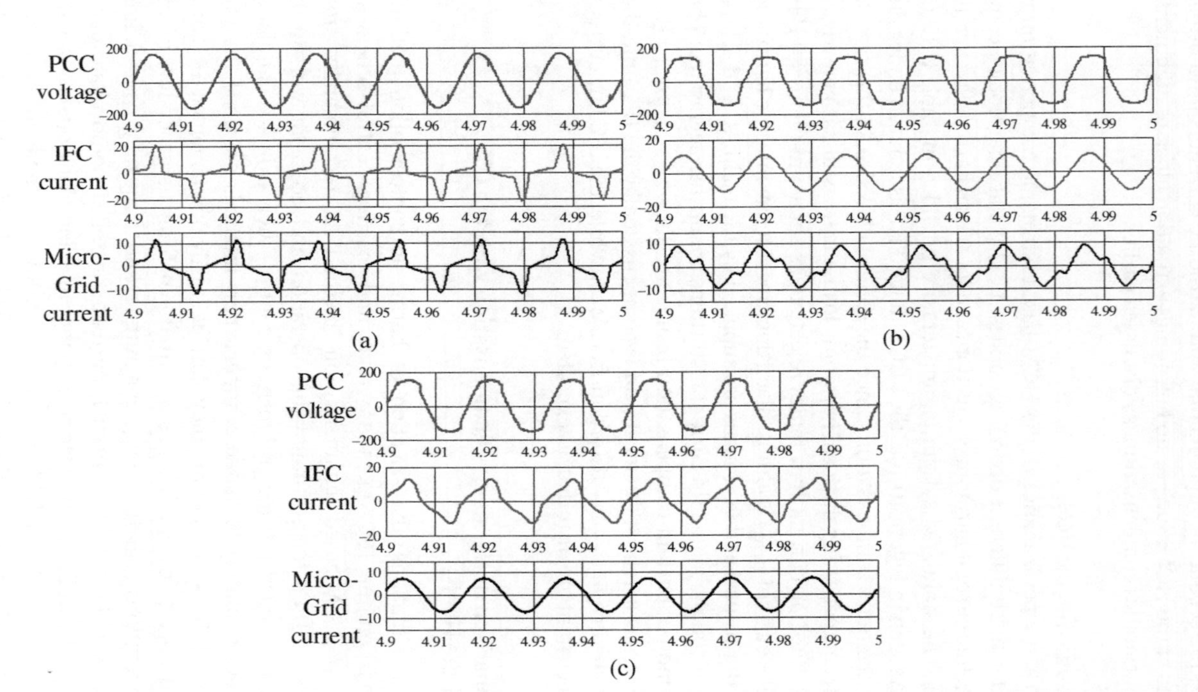

Figure 10.10 Performance of HCM-based harmonic compensation of an IFC: (a) PCC voltage harmonic compensation; (b) IFC line current harmonic rejection; (c) local load current harmonics compensation.

In terms of PCC voltage compensation, the virtual harmonics impedance in CCM- and HCM-based methods is realized by measuring the harmonics voltage and direct closed-loop current tracking. As a result, the performance for the CCM- and HCM-based methods are less sensitive to feeder impedance variations, which is different from the case of VCM-based harmonics compensation. This is because the virtual impedance realized by the VCM-based method is related to the existing IFC feeder impedance, as shown in (10.4). When multiple IFCs are adopted to compensate for the PCC harmonic voltage or provide damping to the system harmonics, the CCM- and HCM-based methods can easily emulate the resistor at harmonic frequencies. While the VCM-based method will again need the current impedance information to produce the desired harmonic impedance angle. Therefore, an accurate estimation or pre-knowledge of impedance is needed for harmonic control using the VCM.

For local harmonics compensation, the local load harmonics current can be directly used as the reference in the CCM- and HCM-based methods as long as the harmonic current tracking loop is designed properly. As discussed earlier, local harmonic compensation using the VCM requires high bandwidth inverter output current tracking, which means the inner current loop controller should have a high bandwidth, such as a hysteresis controller, a predictive controller, multiple harmonic RSCs, etc. Indeed, these inner loop controllers increase the control complexity and are without a damping term compared to the CCM- and HCM-based methods.

For line harmonic rejection, the traditional CCM-based IFC can realize this naturally by regulating the IFC output current with very low harmonics. The HCM-based method can also easily achieve this by simply setting the harmonics current control loop reference to zero. For the VCM-based method, this is realized indirectly by producing high harmonic impedance at the IFC unit output to minimize the harmonics current.

All three methods can be implemented with a very simple modification of the existing IFC control scheme. The computation demand for these methods does not vary very much. As mentioned earlier, the HCM-based method (with decoupled fundamental and harmonics compensation control) can realize harmonics compensation without a harmonic detection block, making it more computationally efficient than the other two methods.

However, the closed-loop-based method for harmonics compensation may face control bandwidth challenges, especially when the converter switching frequency is low. As those converters' main function is still for power generation and control, they usually have low switching frequency to reduce losses. Control techniques for dealing with this low switching IFC are discussed in the next section.

10.3 Control of Low-switching Interfacing Power Converters for Harmonics Compensation in an AC Subgrid

Typically, IFCs with CCM-, VCM-, and HCM-based harmonic control usually require a switching frequency higher than 5 kHz. However, in high-power IFCs (≥500 kW), low-switching frequency (<3 kHz) is generally employed to reduce power loss. In such IFCs, the harmonic control performance is significantly affected by the control delay and low sampling rate, and the harmonic compensation is limited to low-order harmonics, such as the fifth and seventh order harmonics. Therefore, it is preferable to perform harmonic control with a simple control structure instead of multiple-loop control. The bandwidth is thus not limited by inner control loops. Also, the sampling rate is expected to be increased while keeping low-switching frequency, requiring new sampling approaches to replace the well-accepted regular sampling method.

In the control system of low-switching IFCs, the abovementioned VCM and CCM can only be applied for harmonics compensation when the inner current/voltage control loop employs controllers with fast transient and wide bandwidth, such as DB control. Otherwise, the HCM is more effective as the harmonics can be directly regulated instead of feeding to cascaded inner control loops, which may degrade the transient performance and stability.

In addition to the CCM, VCM, and HCM harmonic control strategies, which are also called feedback harmonic control strategies, FCM paths aimed at harmonic mitigation can also be employed in low-switching IFCs. It is well known that adding a full feed-forward path of PCC voltage can help the IFCs resist PCC voltage interference [7]. If distortions exist in the PCC voltage, the converter can perform harmonic rejection due to the feed-forward control. In this case, any PCC voltage changes can directly impact the control signals without any controller regulation. As a result, the system can compensate for transients or long-term power quality disturbances. For example, the distorted PCC voltage will be directly added to the control signals; a corresponding voltage will be produced, and the distortion can be attenuated. However, system delay and high-frequency noises can also impact the control performance in such methods. So, improvements should be made by adding filters or lead compensators, as shown in Figure 10.11b.

Also, as shown in Figure 10.12, more flexible harmonic control can be obtained by feed-forwarding PCC voltage or grid-side current to reshape the IFC output impedance, enabling flexible harmonic current rejection and voltage harmonic compensation [8]. To achieve accurate impedance reshaping at desired harmonic orders for low-switching-frequency converters, it is necessary to take system delay

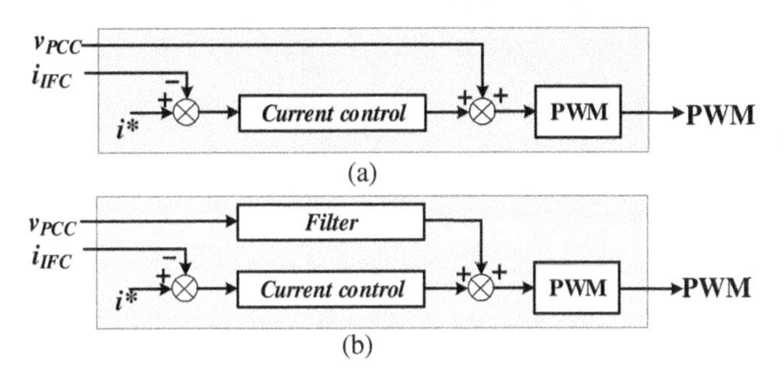

Figure 10.11 Feed-forward control methods: (a) full feed-forward of PCC voltage, and (b) feed-forward of filtered PCC voltage.

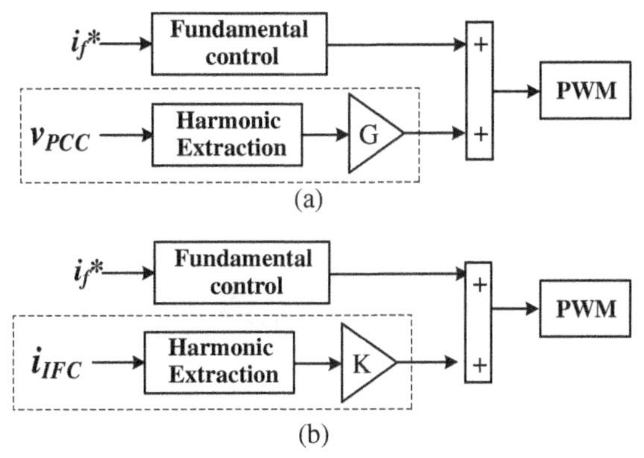

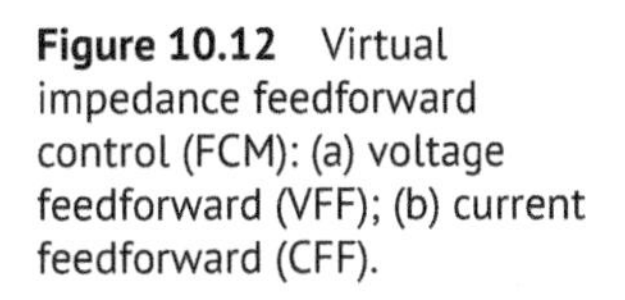
Figure 10.12 Virtual impedance feedforward control (FCM): (a) voltage feedforward (VFF); (b) current feedforward (CFF).

and feedback loop gains into consideration. The detailed principle and design approach will be discussed in Section 10.3.1.0.

Besides the control methods, the sampling method can also affect the harmonic control performance in low-switching IFCs. The widely applied regular sampling method limits the control sampling rate to be the same (symmetrical regular sampling) or twice (asymmetrical regular sampling) that of the switching frequency. When the switching frequency is low (<3 kHz), e.g. in high-power IFCs, the control sampling rate will be insufficient to extract the harmonics accurately. Also, the computation delay brought about by the low sampling rate will cause stability concerns. Therefore, new sampling methods should be used. In the following, three different sampling methods are introduced, which are compared with the asymmetrical regular sampling method.

10.3.1 Low-switching Interfacing Converters Sampling Methods

Traditionally, the regular sampling method is widely used in power electronics to avoid interferences of switching ripples. To perform the regular sampling, control

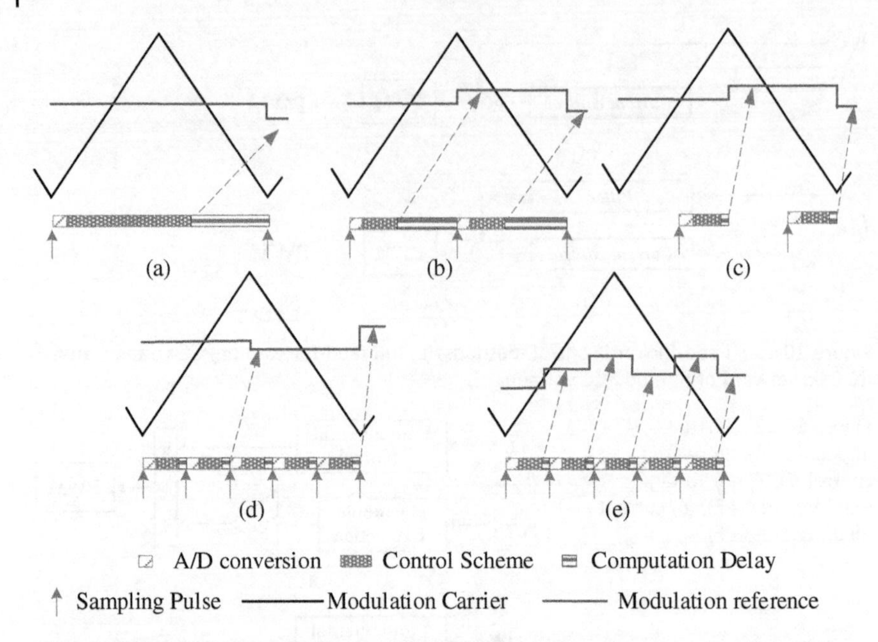

Figure 10.13 Different sampling and PWM methods: (a) synchronous sampling with single-update PWM; (b) synchronous sampling with double-update PWM; (c) real-time computation with double-update PWM; (d) oversampling with double-update PWM; (e) oversampled control and PWM.

variables will be sampled at the beginning or both the beginning and the middle of one switching cycle, as shown in Figures 10.13a, b. Then it will naturally sample the average value from a signal that contains switching-frequency ripples. This method is simple and effective for converters with relatively high switching frequency. However, in low-switching-frequency converters, these methods will not be employed for harmonic control as the insufficient sampling rate and large control delay will significantly degrade the harmonic control performance. For low-switching-frequency VSCs, three sampling schemes can be used to improve the harmonic control performance.

Real-time Computation Method

The real-time computation method is shown in Figure 10.13c. Compared to the regular sampling methods shown in Figures 10.13a,b, the sampling pulse is shifted forward for control computation. As a result, the modulation reference is generated by the newest available sampling results. The modulation reference is thus formed by real-time computation results. This way, the system delay is reduced while the harmonic sampling rate is still the same.

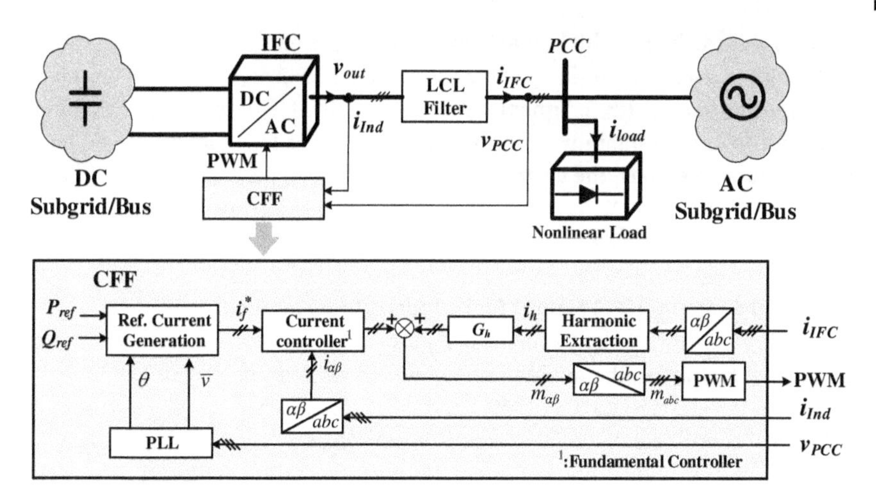

Figure 10.14 An example of harmonic control of low-switching IFCs with current feed-forward strategies.

Multi-rate Sampling Method

The multi-rate method samples the control variables N times (N is an integer, $N > 2$) of the PWM frequency (f_s) as shown in Figure 10.13c. On the other hand, the fundamental frequency controller can still use the regularly sampled variables if necessary. As a result, there are multiple sampling pulses in one PWM period. The system delay is thus reduced, and the sampling rate of harmonics is increased. In addition, the controllers for fundamental frequency control and harmonic control, and the PWM are operated at different frequencies.

Digital Natural Sampling Method

In the digital natural sampling method, both the control sampling and PWM sampling are operated at high frequency, which is N times the PWM frequency. As shown in Figure 10.13d, multiple sampling pulses are generated per PWM period, and the modulation reference will be updated with the same rate. Compared to the multi-rate sampling method, the PWM sampling rate is also increased. However, as the modulation reference will be continuously updated, this sampling method may lead to multiple switching in one PWM period when the modulation reference has a similar slope to carriers due to multiple intersections, which is possible in harmonic control.

10.3.2 Control of Low-switching IFCs for Harmonics Compensation with Feed-forward Strategy

With improved sampling methods, the harmonics information can be accurately extracted. However, the harmonic control loops still need to be improved to have sufficient control bandwidth for harmonic control. It is hard to achieve high bandwidth with a cascaded multiloop control system in low-switching-frequency converters, as the bandwidths of inner loops are constrained with respect to switching frequency to ensure stability. In this case, two different feed-forward-based harmonic control methods, the current feed-forward (CFF) method and the voltage feed-forward (VFF) method, can be used, which will be illustrated in the following sections.

Current Feed-forward (CFF) Method for Harmonics Compensation

In the CFF control method shown in Figure 10.14, the harmonic currents are directly fed forward to the modulation references with designed gain G_h. To realize the control strategies, the output currents are transformed into the α–β or d–q orthogonal coordinate, in which it is possible to construct the complex coefficient G_h. Harmonic extraction modules obtain the harmonics in currents. The extracted harmonics are amplified by the complex gain G_h and transformed back to the a–b–c stationary frame. Finally, they are added to modulation references. As the regulated signals are directly added into the modulation references, neither extension of the controller bandwidth nor extra high bandwidth controller is necessary for the feedback loop. As a result, the feed-forward control is not limited by the feedback control bandwidth.

The equivalent circuits of the IFC, load, and grid system are shown in Figure 10.15. Since IFCs typically employ the CCM as the inner control loop, the Norton equivalent method is adopted in the analysis. The non-linear load is modeled as a harmonic current source in parallel with an impedance that accounts for fundamental frequency power.

It is worth noting that, in a low-switching-frequency IFC system, the system delay would be a significant factor that influences the current harmonic phase angle. Although the gain of feedback fundamental-frequency control loop at harmonic frequency is small, the coupling to the output harmonic still exists. Therefore, the output of the IFC at harmonic frequencies, V_{out_h}, can be

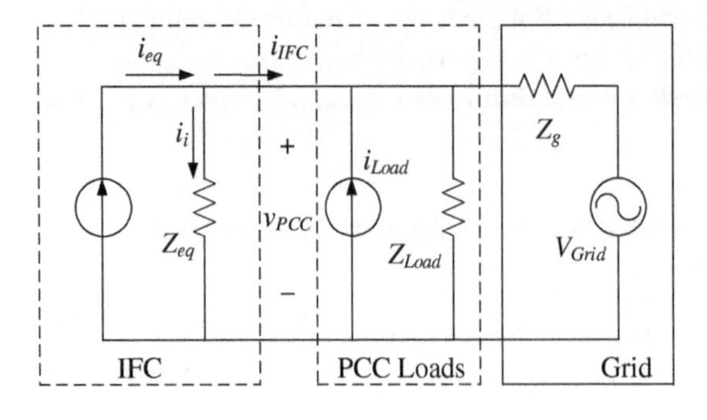

Figure 10.15 Equivalent model of a microgrid with IFC and non-linear loads.

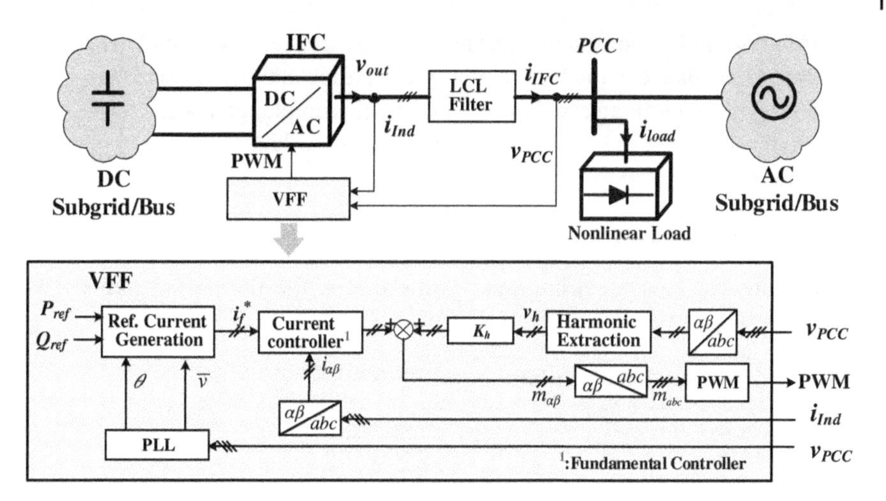

Figure 10.16 An example of harmonic control of low-switching IFCs with voltage feed-forward strategies.

expressed as:

$$v_{\text{out\_h}}(s) = -G_{\text{h}} \cdot i_{\text{IFC\_h}}(s) \cdot \mathrm{e}^{-T_{\text{d}}s} - G_{PR}(s) \cdot \mathrm{e}^{-T_{\text{d}}s} i_{\text{IFC\_h}}(s) \tag{10.6}$$

where T_{d} denotes the system delay.

And the virtual impedance of the CFF control can be expressed as:

$$Z_{\text{V}}(s) = -\frac{v_{\text{PCC\_h}}(s)}{i_{\text{IFC\_h}}(s)} = \left(1 + \frac{G_{\text{h}} \cdot Z_{\text{C}}(s) \cdot \mathrm{e}^{-T_{\text{d}}s} + G_{PR}(s) \cdot Z_{\text{C}}(s) \cdot \mathrm{e}^{-T_{\text{d}}s}}{Z_{L1}(s)Z_{L2}(s) + Z_{L1}(s)Z_{\text{C}}(s) + Z_{L2}(s)Z_{\text{C}}(s)}\right) Z_{\text{eq}}(s) \tag{10.7}$$

where $v_{\text{PCC\_h}}$ is the harmonic component at the PCC; $i_{\text{IFC\_h}}$ is the harmonic current in i_{IFC}.

As can be seen from (10.7), the equivalent output impedance Z_{V}, i.e. the virtual impedance of the VSC, can be changed by the feed-forward gain K_{h}. Assuming all the harmonic currents in the system are injected by the non-linear loads when Z_{V} is increased, i.e. the harmonic equivalent impedance is increased on the IFC side, the IFC will resist the grid background harmonics and maintain high IFC output current quality. On the other hand, when Z_{V} is decreased, the harmonic current will be absorbed by the IFC. In this case, the voltage at the PCC will be improved even local non-linear loads are connected.

Voltage Feed-forward (VFF) Method for Harmonics Compensation

Different from the CFF method, the VFF method applies the PCC voltage as the feed-forward variable, as shown in Figure 10.16. This also avoids cascading multiple controllers, and the AC voltage control loop will not suffer from the bandwidth constraints of multiloop systems. The feed-forward gain, K_h, is applied to reshape the output impedance of the IFC. Similarly, the orthogonal frames, such as α–β or d–q orthogonal coordinates, will be used to enable complex gain K_h.

With the complex gain K_h applied, the output voltage of the IFC can be expressed as:

$$v_{\text{out}_h}(s) = -K_h \cdot v_{\text{PCC\_h}}(s) \cdot e^{-T_d s} - G_{\text{PR}}(s) \cdot e^{-T_d s} \cdot i_{\text{IFC}_h}(s). \tag{10.8}$$

Therefore, the equivalent harmonic impedance of the IFC Z_V is expressed as (10.9).

$$Z_V(s) = -\frac{v_{\text{PCC\_h}}(s)}{i_{\text{IFC\_h}}(s)} = \frac{\left(1 + \frac{G_{\text{PR}}(s) \cdot Z_C(s) \cdot e^{-T_d s}}{Z_{L1}(s)Z_{L2}(s) + Z_{L1}(s)Z_C(s) + Z_{L2}(s)Z_C(s)}\right)}{\left(1 + \frac{K_h \cdot Z_C(s) \cdot e^{-T_d s}}{Z_{L1}(s) + Z_C(s)}\right)} Z_{\text{eq}}(s). \tag{10.9}$$

When the CFF path is added, the harmonic control gain plays the role of a virtual impedance in parallel with the original IFC impedance. By contrast, the VFF paths do not have physical meanings, while the feed-forward gain can reshape the output impedance. Although the different control variables are used, the harmonic impedance can be scaled up and scaled down by the feed-forward gain. Similar control performance can be obtained: when the virtual impedance is increased, the IFC can do harmonic rejection; when the virtual impedance is decreased, the IFC can compensate the voltage harmonics on the PCC.

It is worth noting that with the improved multi-rate sampling method, the control system contains several sampling rates, and as a result, the traditional discrete-time model based on the Z-transform is hard to implement. The continuous-time model can still be built, but the accuracy is not satisfying when the harmonic frequency is closed to the system's Nyquist frequency. Alternatively, the lifting model [9] can be adopted here, building the discrete-time model without compromising the modeling accuracy. This is particularly important when low-switching-frequency converters are used for compensating a relatively high order harmonics, e.g. using 2 kHz switching frequency to compensate 13th order harmonics (780 Hz). Inaccurate modeling and the resulting improper parameter design can lead to instability and amplification of the harmonics. Therefore, it is recommended to use the methods developed in [9] to perform modeling and analysis for converter systems when: (i) the control system contains multiple sampling rates; and (ii) the control system needs to control a signal that is relatively close to its Nyquist frequency.

Case Study Results of CFF and VFF

Figure 10.17 shows the harmonic compensation operation using the VFF method. The system parameters can be found in Table C.11 of Appendix C. When the virtual

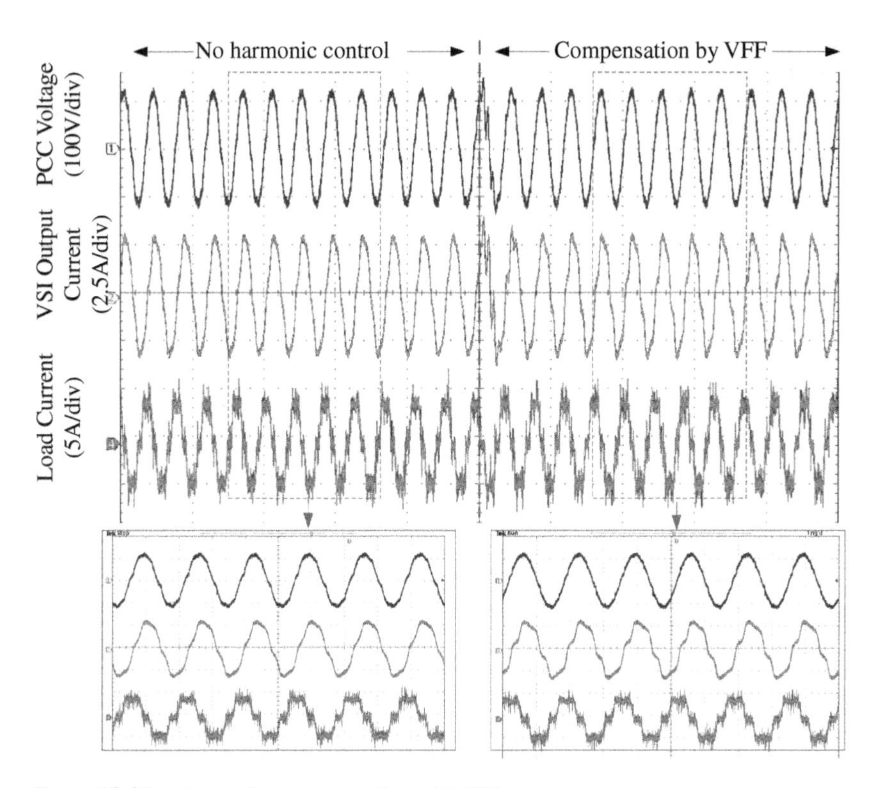

Figure 10.17 Harmonics compensation with VFF.

impedance is set to be a small value, the VFF control system realizes harmonic compensation, i.e. improves the PCC voltage. As shown in Figure 10.17, the distortion of the PCC voltage is reduced significantly after a short transient process. The zoomed-in view shows more distorted output currents are absorbed by the IFC so that the PCC voltages become less distorted. The FFT analysis in Figure 10.18 shows that the THD of PCC voltage is reduced from 7.06 to 4.86% because the fifth order and seventh order harmonics are significantly compensated. Both are reduced from nearly 4% to no more than 1%.

The CFF method has similar performance under different virtual impedance configurations. When the virtual impedance is configured to be small, the CFF can also compensate the PCC voltage harmonics, as shown in Figure 10.19. A smooth transient can be seen when the compensation starts. Meanwhile, the voltage quality improvement can be seen from the zoomed-in view: a sinusoidal PCC voltage appears after the CFF is applied. According to the FFT analysis in Figure 10.20, the fifth and seventh order harmonics are attenuated, decreasing from around 4% to approximately 0.5%.

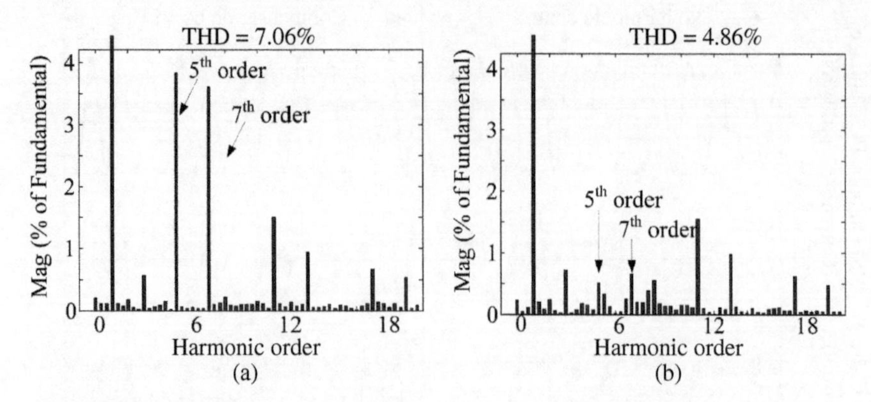

Figure 10.18 FFT analysis of the PCC voltage: (a) no harmonic control; (b) harmonic compensation by VFF.

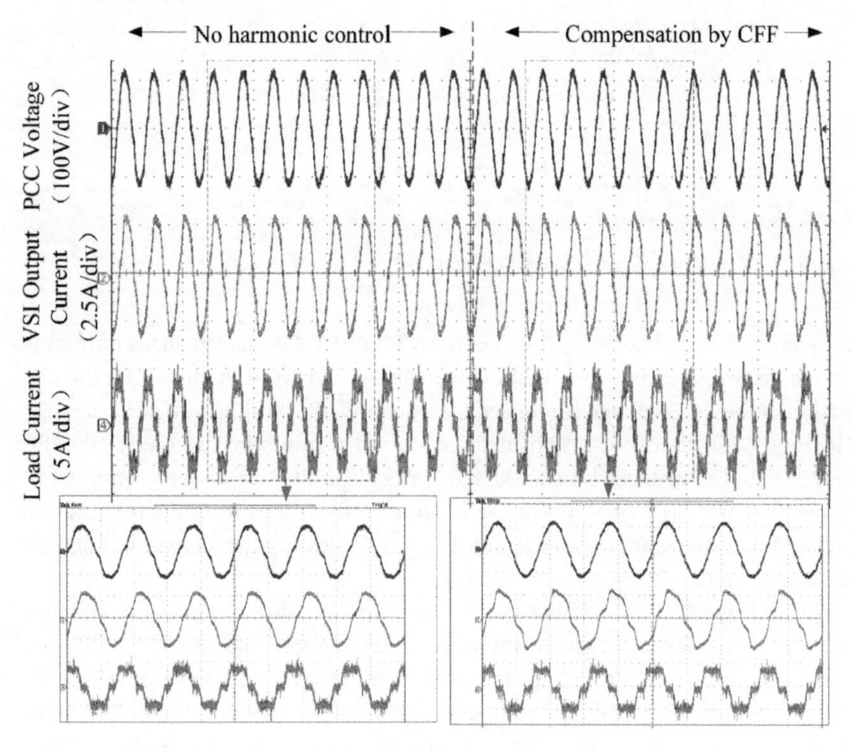

Figure 10.19 Harmonics compensation with CFF.

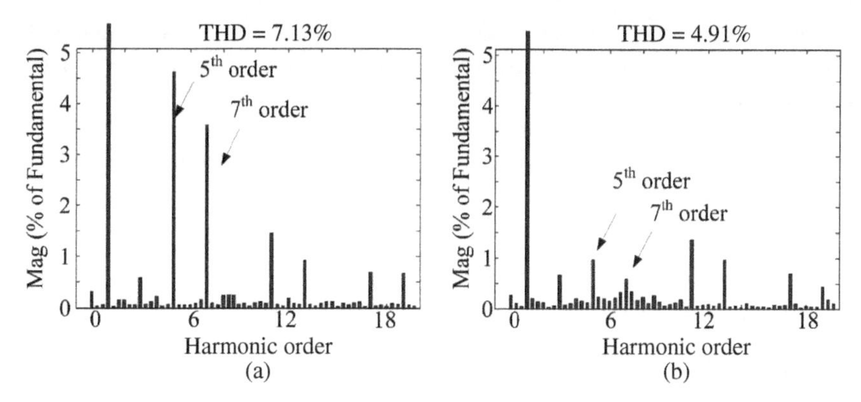

Figure 10.20 FFT analysis of the PCC voltage: (a) no harmonic control; (b) harmonics compensation by CFF.

10.4 Control of Interfacing Power Converters for Harmonics Compensation in a DC Subgrid

In hybrid AC/DC microgrids, the AC and DC subgrids have harmonic interactions due to the frequency coupling. The AC-side harmonics will be transferred to the DC side while the DC side ripple and harmonics will be transferred to the AC side through the IFC PWM process. This can be explained as follows.

A three-phase PWM voltage source converter produces the voltage as:

$$\begin{cases} v_{an}(t) = S_a(t) \times v_{\text{dc}}(t) \\ v_{bn}(t) = S_b(t) \times v_{\text{dc}}(t) \\ v_{cn}(t) = S_c(t) \times v_{\text{dc}}(t) \end{cases} \tag{10.10}$$

where $v_{an}(t)$, $v_{bn}(t)$, and $v_{cn}(t)$ are output voltages in the time domain of each converter's phase. $v_{\text{dc}}(t)$ is the DC-link voltage in the time domain. $S_a(t)$, $S_b(t)$, and $S_c(t)$ are modulation functions of the converters, which are expressed as:

$$\begin{cases} S_a(t) = \sum\limits_{s=1,0,-1} \sum\limits_{h=1}^{\infty} m_{s,h} \cos(\omega_{s,h} t + \varphi_{s,h}) \\ S_b(t) = \sum\limits_{s=1,0,-1} \sum\limits_{h=1}^{\infty} m_{s,h} \cos\left(\omega_{s,h} t + \varphi_{s,h} - \dfrac{2s}{3}\pi\right) \\ S_c(t) = \sum\limits_{s=1,0,-1} \sum\limits_{h=1}^{\infty} m_{s,h} \cos\left(\omega_{s,h} t + \varphi_{s,h} + \dfrac{2s}{3}\pi\right) \end{cases} \tag{10.11}$$

where subscript s represents the sequence (1 for positive sequence, 0 for zero sequence, −1 for negative sequence); subscript h represents harmonic order. Then

$m_{s,h}$, $\omega_{s,h}$, and $\varphi_{s,h}$ represent the modulation index, angular frequency, and initial phase angle for the hth order harmonic in positive ($s = 1$), negative ($s = -1$), or zero ($s = 0$) sequence.

Since the multiplication of two signals in the time domain implies convolution of their Fourier transforms in the frequency domain, (10.12) can be expressed in the frequency domain as:

$$\begin{cases} v_{an}(j\omega) = \frac{1}{2\pi} S_a(j\omega) \otimes v_{dc}(j\omega) \\ v_{bn}(j\omega) = \frac{1}{2\pi} S_b(j\omega) \otimes v_{dc}(j\omega). \\ v_{cn}(j\omega) = \frac{1}{2\pi} S_c(j\omega) \otimes v_{dc}(j\omega) \end{cases} \tag{10.12}$$

As indicated by (10.12), if the DC voltage contains ripples at nth order and the modulation function contains hth harmonics, the AC side will contain $(n \pm h)$th order harmonics.

On the other hand, the AC side harmonics can also transfer back to the DC side. The AC side current injected by the DC/AC converter can be expressed as:

$$\begin{cases} i_a(t) = \sum_{s=1,0,-1} \sum_{k=1}^{\infty} i_{as,k} \cos(\omega_{s,k} t + \varphi_{s,k}) \\ i_b(t) = \sum_{s=1,0,-1} \sum_{k=1}^{\infty} i_{bs,k} \cos\left(\omega_{s,k} t + \varphi_{s,k} - \frac{2s}{3}\pi\right) \\ i_c(t) = \sum_{s=1,0,-1} \sum_{k=1}^{\infty} i_{cs,k} \cos\left(\omega_{s,k} t + \varphi_{s,k} + \frac{2s}{3}\pi\right) \end{cases} \tag{10.13}$$

where subscript k represents the harmonic order contained in the currents; subscript s still represents the sequences; $i_a(t)$, $i_b(t)$, and $i_c(t)$ are currents in the time domain for each phase. It is worth noting that zero-sequence current will not exist in three-phase three-wire AC systems but can be present in three-phase four-wire systems.

Then the DC side current in the time domain will be synthesized as:

$$i_{dc}(t) = S_a(t) \times i_a(t) + S_b(t) \times i_b(t) + S_c(t) \times i_c(t). \tag{10.14}$$

Again, (10.14) can be transformed to the frequency domain as:

$$i_{dc}(j\omega) = \frac{1}{2\pi}[S_a(j\omega) \otimes i_a(j\omega) + S_b(j\omega) \otimes i_b(j\omega) + S_c(j\omega) \otimes i_c(j\omega)]. \tag{10.15}$$

If the AC current contains kth harmonics and the modulation functions contain hth harmonics, the DC link will see a current with $(k \pm h)$th harmonics. This derivation can also be applied to single-phase systems by only considering variables in one phase.

Considering (10.12) and (10.15), the DC/AC harmonic interactions will create a wide spectrum of harmonics. However, the amplitude of the harmonics will descend with increasing distance to the center frequencies in the frequency domain.

In hybrid AC/DC microgrids, the DC subgrid harmonics, particularly the low-order harmonics, are mainly introduced by the AC-side harmonics or unbalanced three-phase AC voltage. Also, the DC harmonics can be caused by DC/DC converter control on the DC side, such as maximum power point tracking control of PV systems, or LC oscillation due to parasitic DC line impedance.

In the DC subgrid, since the DC-bus capacitors are usually large enough to ensure the reliable operation of a DC subgrid under normal conditions, high-frequency voltage ripples and harmonics can be easily filtered. In practice, the second order harmonic caused by the unbalanced AC voltage is the most important harmonic component to be addressed. Also, the sixth order harmonic may be observed due to the AC side fifth and seventh harmonics. For such low-order harmonics compensation, larger capacitors or APFs are required. Alternatively, they can be compensated for using DC/AC IFCs that link DC and AC subgrids (individual or parallel IFCs) or DC/DC IFCs on the DC side integrating distributed generation, energy storage, and DC subgrids.

10.4.1 Harmonics Compensation in a DC Subgrid Using DC/AC Interlinking Power Converters

The DC subgrid harmonics can be compensated by DC/AC IFCs. As discussed earlier, compensating second order DC ripples means adding harmonic signals to the modulation function, leading to third order harmonics in the AC side. Also, the methods focusing on improving the AC current quality under DC ripples while leaving the DC ripple unattenuated, such as directing the DC ripples to modulation references, are not ideal. Therefore, the major challenge is to simultaneously eliminate the second order DC voltage ripples while reducing third order harmonics at the AC side.

Since the unbalanced AC voltage usually causes the DC second order ripples, for unbalanced AC voltage compensation the DC/AC IFC output current positive and negative sequence components can be controlled in a smart way, which has been discussed in detail in Chapter 9. These methods can reduce/mitigate active power oscillation in the AC side and reduce second order ripples at the DC side with good AC current quality (see Section 9.2 for some control examples). For parallel IFCs, the DC/AC IFCs can be coordinated to cancel the active power oscillation cancelation in the AC side and therefore mitigate the second order harmonics in the DC side according to their power rating (see Section 9.3). This way, both the AC subgrid current quality and DC subgrid voltage quality can be improved simultaneously. More details on DC/AC IFC unbalanced voltage compensation strategies can be found in Chapter 9.

10.4.2 Harmonics Compensation in a DC Subgrid Using DC/DC Interfacing Power Converters

In hybrid AC/DC microgrids, harmonics on the DC side can also be directly compensated for by the DC/DC IFCs in the DC subgrid. The DC/DC IFCs, which can be considered to have DC-APF function in this condition, can compensate the low-frequency harmonics caused by both the AC-side and DC-side distortions. These converters provide a low impedance path for the harmonics (usually a virtual resistor proportional to the power rating), directing the common DC link ripples and power/current oscillations to their own DC buses. Energy storage components, which can be capacitors, super-capacitors, batteries, etc. are required to absorb such oscillations. Moreover, to properly act as DC-APFs, they must use bidirectional DC/DC (BDC) converters, such as a non-isolated half-bridge BDC, and an isolated dual active bridge (DAB) converter. In the hybrid AC/DC microgrid, distributed DC-APFs can be coordinated to share the harmonic compensation.

Case Study: Control and Coordination of Two DC/DC IFCs for Harmonics Compensation in a DC Subgrid

In Figure 10.21, an example of DC-subgrid harmonics compensation using two DC-APFs with different power ratings is provided. The system parameters are provided in Table C.12 of Appendix C. In the control strategy [10], the DC-APF capacitor voltage control and virtual resistor-based harmonic control are paralleled. The capacitor voltage control loop can ensure the proper operation of the DC-APF, maintaining the average voltage on its low-voltage side capacitor. The compensation efforts are predefined according to the power rating of the DC-APFs. The DC-APFs can control their input current to compensate the ripple by sensing the voltage ripple on the DC bus. As a result, the second order ripples are effectively mitigated. Also, thanks to the virtual resistor control, the ripple power is shared by the paralleled DC-APFs. As seen from the results in Figure 10.22a, the DC link oscillation is shared between the two DC-APFs with a ratio of 1:2, which is proportional to their power ratings. Also, the DC-APFs capacitor voltages are kept stable.

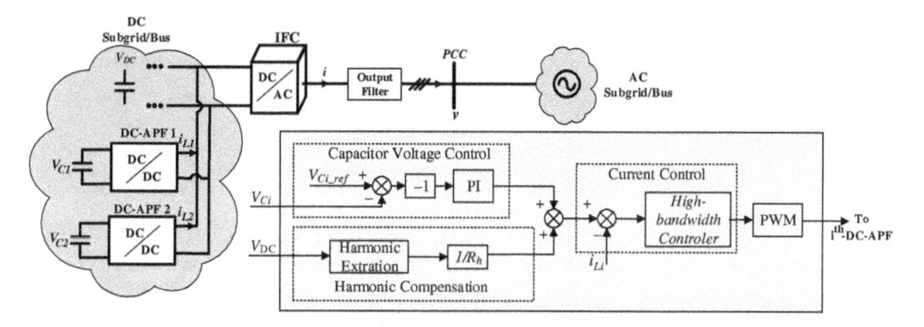

Figure 10.21 Two DC/DC IFC control for DC-subgrid harmonic compensation.

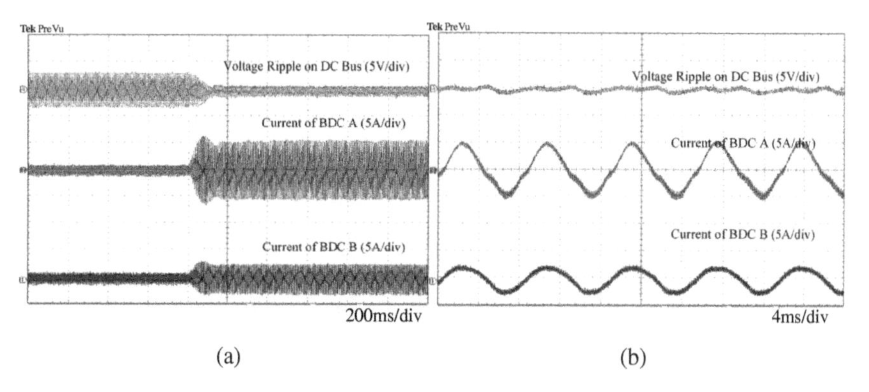

Figure 10.22 Performance of the two DC/DC IFC control for DC-subgrid harmonic compensation (common DC bus voltage and current of DC-APFs): (a) transient waveforms when the compensation function is enabled; (b) zoom-in view of steady-state waveforms [10].

10.5 Coordinated Control of Multiple Interfacing Power Converters for Harmonics Compensation

In hybrid AC/DC microgrids, IFCs can be distributed throughout the system, requiring coordinated control for harmonic compensation. Specifically, the harmonic compensation can be shared among the IFCs of the AC subgrid or DC subgrid based on their available ratings. For practical applications, the sharing of harmonics compensation effort among the IFC units also includes many additional factors, such as power losses on the feeders, economic dispatch, the location of sensitive loads, existing active/passive harmonic filters, and voltage regulators. For a given distribution system and set of load characteristics, it is possible to optimize the virtual impedance of each IFC to achieve the best overall performance considering different factors. The optimal harmonic load flow can be an approach for this optimization. In general, coordinate harmonic control strategies can be classified as autonomous harmonic control and supervisory harmonic control.

10.5.1 Autonomous Harmonic Control

Autonomous control enables the IFCs to compensate harmonics without relying on physical communications collaboratively. Similar to power-sharing control, harmonic sharing can be achieved by adopting droop control (where the harmonic compensation efforts can be automatically determined by the harmonic droop gain proportional to the power ratings [11]) by shaping the IFC output impedance, etc. However, similar to reactive power-sharing, harmonic sharing can also be affected by the mismatched feeder impedance in the weak microgrid and differences in the converter dynamic performance, requiring some supervisory control to correct the sharing errors. Also, optimizing the virtual impedances of IFCs considering different harmonic sharing factors is not possible.

10.5.2 Supervisory Harmonic Control

Supervisory harmonic control is usually realized in a centralized harmonic controller, which distributes the harmonic compensation reference signals to the distributed IFCs with a low-bandwidth communication system [12]. Thus, the harmonic compensation efforts can be further optimized considering different factors rather than simply determined by the IFC power ratings. In the following, an example of supervisory harmonic control is provided.

Case Study: Supervisory Control Strategy Example of Multiple IFC Coordination for Harmonics Compensation

Multiple IFCs at different locations can be coordinated to improve harmonics compensation performance, and the harmonics compensation priorities can be assigned for them. In this example, modal analysis is used to prioritize the IFCs in the AC subgrids [13]. Modal analysis is a frequency-domain method to analyze dynamic performances and is very suitable to identify resonances in a system. Due to the line impedances and power factor correction capacitors, the microgrid nodes can have different sensitivities to harmonics, which show low modal impedance at resonance frequencies.

In Figure 10.23, a typical AC residential community microgrid with multiple distributed IFCs is shown. For the microgrid shown in Figure 10.23a, each paralleled

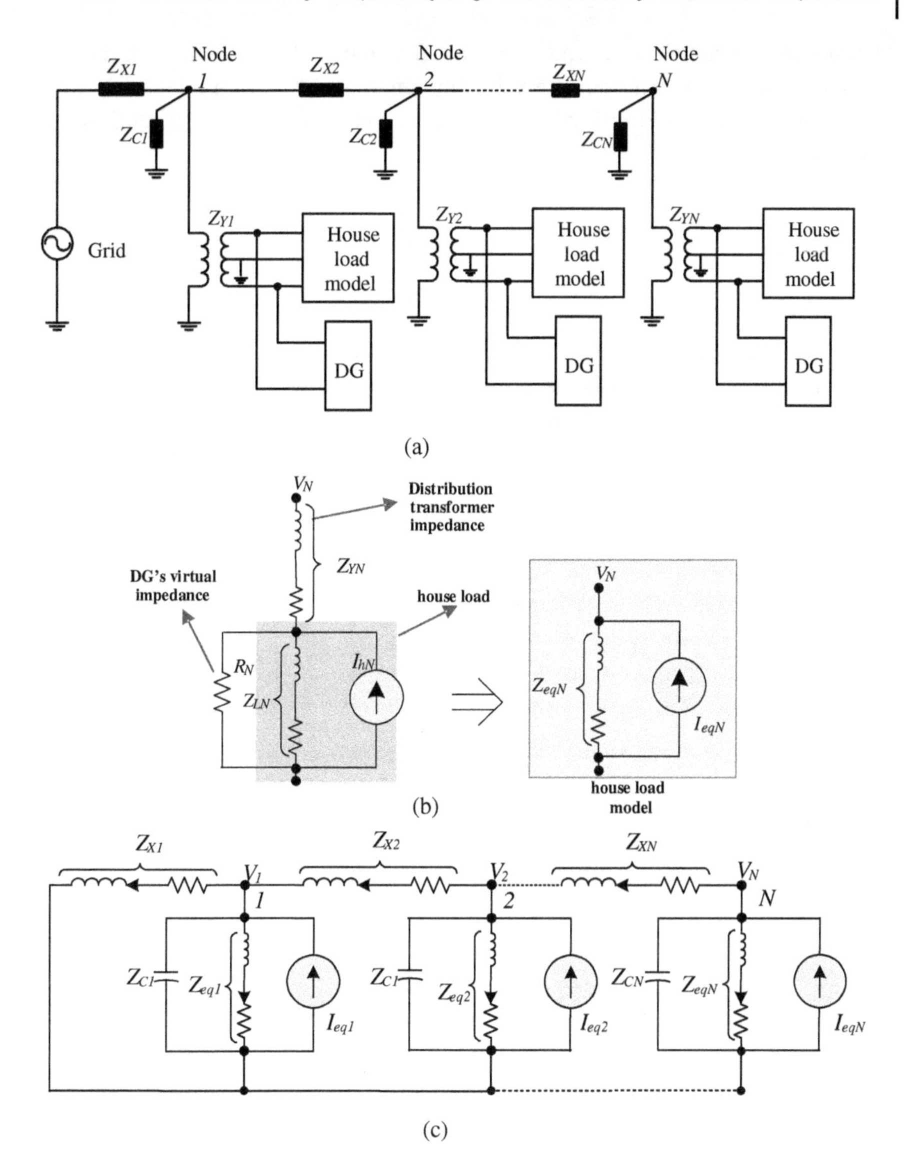

Figure 10.23 AC microgrid structure and the equivalent model: (a) microgrid with distributed IFCs; (b) equivalent circuit of a single DG system; (c) equivalent model of the microgrid.

house can be modeled as a harmonic source and an impedance at the fundamental frequency, which only accounts for the fundamental power. From the equivalent model of the microgrid in Figure 10.23c, the nodal equation can be obtained as follows:

$$[V]_f = [Y]_f^{-1}[I]_f; \quad [Y]_f^{-1} = [Z_N]_f = \begin{bmatrix} Z_{11} & Z_{12} & \vdots & Z_{1n} \\ Z_{21} & Z_{22} & \vdots & Z_{2n} \\ \vdots & \vdots & \vdots & Z_{3n} \\ Z_{N1} & Z_{n2} & Z_{n3} & Z_{nn} \end{bmatrix}_f \quad (10.16)$$

where the subscript f denotes the frequency, the $[V]_f$ and $[I]_f$ are the nodal voltage and current vectors, and the $[Z_N]_f$ is the nodal impedance matrix. For every frequency, the node with resonance will have a high modal impedance value. Thus, resonances and resonance frequencies can be identified by decoupling the coupled nodal impedance matrix $[Z_N]_f$ into decoupled modal impedance matrix $[Z_M]_f$ in the modal domain. The modal equation can be achieved as follows:

$$[v_M]_f = [\Lambda]_f^{-1}[i_M]_f \Rightarrow \begin{bmatrix} V_{m,1} \\ V_{m,2} \\ \vdots \\ V_{m,N} \end{bmatrix}_f = \begin{bmatrix} \lambda_{11}^{-1} & 0 & \vdots & 0 \\ 0 & \lambda_{22}^{-1} & \vdots & 0 \\ \vdots & \vdots & \vdots & 0 \\ 0 & 0 & 0 & \lambda_{NN}^{-1} \end{bmatrix}_f \begin{bmatrix} i_{m,1} \\ i_{m,2} \\ \vdots \\ i_{m,N} \end{bmatrix}_f \quad (10.17)$$

where $[v_M]_f$, $[i_M]_f$, and $[\Lambda]_f$ are the modal voltage vector, the modal current vector, and the diagonal modal impedance matrix, that are obtained using the following expressions:

$$\begin{aligned} [v_M]_f &= [T]_f[V]_f \\ [i_M]_f &= [T]_f[I]_f \\ [Y]_f &= [L]_f[\Lambda]_f[T]_f. \end{aligned} \quad (10.18)$$

In (10.18), $[T]_f$ and $[L]_f$ are the right and left eigenvector matrices, which are related by:

$$[L]_f^{-1} = [T]_f. \quad (10.19)$$

Since the critical modal impedance (highest modal impedance) is much higher than other modal impedances (assume that λ_{11f}^{-1} is the critical modal impedance), the nodal equation can be simplified as in (10.20).

From (10.20), it can be concluded that the harmonic voltage is related to the modal impedance priority, participation factor, and load harmonic currents. In Figure 10.24, the implementation of the selective harmonic compensation scheme with virtual harmonic impedance is shown. As can be seen, the detected harmonics are regulated by the different priorities from participation factor ($PF_{k,h}$), modal

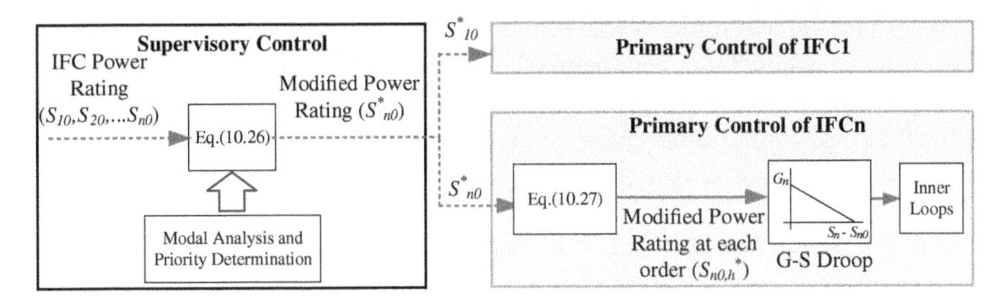

Figure 10.24 Selective harmonics compensation scheme with G–S droop and priorities.

impedance, and the load harmonic factor ($H_{n,h}$), to ensure that the critical harmonics can be compensated with high priority.

$$[Z_M]_f = [\Lambda]_f^{-1} = \underbrace{\begin{bmatrix} \lambda_{11}^{-1} & 0 & \vdots & 0 \\ 0 & \lambda_{22}^{-1} & \vdots & 0 \\ \vdots & \vdots & \vdots & 0 \\ 0 & 0 & 0 & \lambda_{NN}^{-1} \end{bmatrix}_f}_{\text{Modal impedance}} \approx \underbrace{\begin{bmatrix} \lambda_{11}^{-1} & 0 & \vdots & 0 \\ 0 & 0 & \vdots & 0 \\ \vdots & \vdots & \vdots & 0 \\ 0 & 0 & 0 & 0 \end{bmatrix}_f}_{\text{Modal impedance}}$$

$$\Rightarrow \begin{bmatrix} V_1 \\ V_2 \\ \vdots \\ V_N \end{bmatrix}_f = \lambda_{11,f}^{-1} \underbrace{\begin{bmatrix} L_{11}T_{11} & L_{12}T_{12} & \vdots & \vdots \\ L_{21}T_{11} & L_{22}T_{12} & \vdots & \vdots \\ \vdots & \vdots & \vdots & \vdots \\ \vdots & \vdots & \vdots & L_{N1}T_{1N} \end{bmatrix}_f}_{\text{Particapation Factor}} \begin{bmatrix} I_1 \\ I_2 \\ \vdots \\ I_N \end{bmatrix}_f . \tag{10.20}$$

These priorities work as gains to amplify the harmonic signals, which result in different harmonic compensation efforts for different IFCs and harmonic orders. The compensation priority (CP) at node k and h-order harmonics can be expressed as:

$$CP_{k,h} = \frac{PF_{k,h}Z_{k,h}H_{k,h}}{\max\left[\bigcup_{n,h}^{n=1,2,..N;h=3,5,..,h\max}(PF_{n,h}Z_{n,h}H_{n,h})\right]}. \tag{10.21}$$

To apply the priorities in droop control, the conductance–power (G–S) droop can be adopted [14]. In the conductance–power (G–S) based droop algorithm the harmonic conductance command is obtained from the actual output apparent power at harmonic orders (S_n), available capacity of converters (S_{n0}), droop offset (G_{n0}). The G–S droop equation can be expressed as:

$$G_n = G_{n0} + b_n(S_n - S_{n0}) \tag{10.22}$$

where b_n is the droop slope.

As a result, the harmonic voltage content at PCC (V_{G_h}) is modified according to the harmonic conductance command (G_n) to generate the harmonic reference current. Since the main component of the RMS voltage at the PCC is the fundamental voltage, the output harmonic apparent power of the DG (S_n) can be expressed as

$$S_n = |V_n|G_n|V_{n,h}| \approx |V_{n,f}|G_n|V_{n,h}|. \tag{10.23}$$

Combining (10.22) and (10.23), it is easy to obtain:

$$S_n = \frac{|V_{n,f}|(G_{n0} - b_n S_{n0})|V_{n,h}|}{1 - b_n|V_{n,f}||V_{n,h}|} \tag{10.24}$$

where $V_{n,f}$ is the fundamental voltage at node n and $V_{n,h}$ is the harmonic component for node n.

Considering that IFCs only participate in harmonic compensation as ancillary functions, the available power rating for harmonic compensation is generally low, which requires a low droop slope for harmonic sharing. Also, the stability concerns in the residential distribution grid ask for a low droop slope. As a result, in G–S droop controllers for DG, the droop slope b_n can be very small, resulting in $b_n|V_{n,f}||V_{n,h}| \approx 1$. In this case, to ensure proper sharing among different IFCs with small droop slope, it is recommended by [14] that the ratio between the output harmonic power and the available power rating has the same ratio in different IFCs, which can be expressed as:

$$S_n = \text{const} \cdot S_{n0}. \tag{10.25}$$

Then the priority shall be implemented by changing the available power rating for harmonic compensation at each harmonic order and each node. If the available IFC ratings of different IFCs are S_{n0} then the reference IFC ratings (S_{n0}^*) can be identified. Here, the IFC with the highest priority will use most of its rating for harmonic compensation. While an IFC with less priority will only use a small portion of its rating for harmonics compensation. IFCs will effectively utilize their power ratings and will not be overloaded. In practice, the priority will be determined by the supervisory controller. Once the reference harmonic power ratings (S_{n0}^*) are identified, the modified power rating for different harmonic orders will be distributed to the IFCs at different nodes. This is shown in Figure 10.24 and Eqs. (10.26) and (10.27).

$$S_{n0}^* = S_{n0}\frac{\sum_{h=3,5,7\ldots}^{h_{\max}} CP_{k,h}}{\sum_{n=1,2,3\ldots}^{N}\sum_{h=3,5,7\ldots}^{h_{\max}} CP_{n,h}}. \tag{10.26}$$

For individual harmonic order, the available power rating can be calculated by (10.27)

$$S_{n0,h}^* = S_{n0}^*\frac{CP_{k,h}}{\sum_{h=3,5,7\ldots}^{h_{\max}} CP_{n,h}}. \tag{10.27}$$

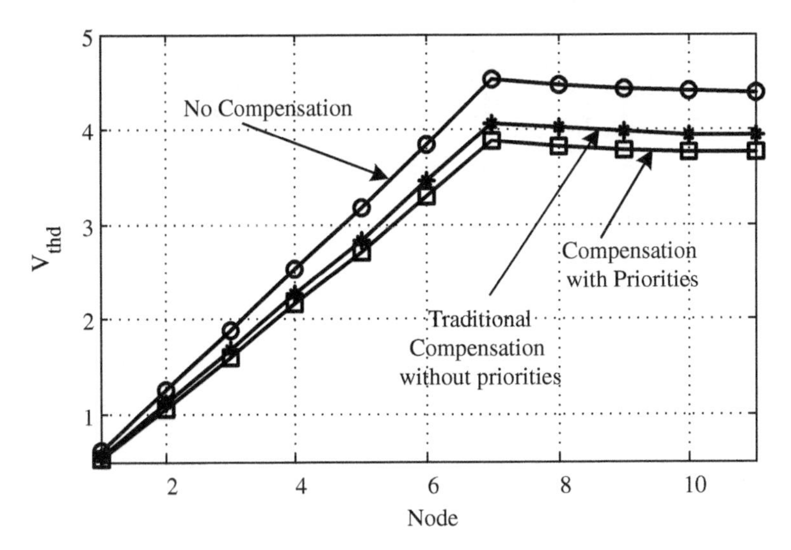

Figure 10.25 THD at each node under different compensation priority schemes.

As shown in Figure 10.24, the priorities are generated by supervisory control, which has complete knowledge of the whole system as modal analysis requires information about the distribution grid's impedances and structure. Considering the development of the smart grid and internet-of-things, this information is getting easier to obtain in the modern distribution grid. Therefore, the priority-based compensation targeted on improving overall THD shows great potential in the foreseeable future. With modal analysis performed, the priorities are distributed to each IFC by low-bandwidth communication links. The priority values only need to be updated when the system has changed in impedance and new installations of IFCs. The systems can have better compensation performance by applying different available power ratings for different IFCs at different harmonic orders.

This control strategy is applied to the AC-coupled microgrid with 11 nodes to validate the scheme, which contains IFCs in each node. The simulated system parameters are provided in Table C.13 of Appendix C. In this system, capacitor banks for power factor correction are installed at node 7. As a result, the capacitor and line impedances cause different modal impedances at each node, requiring different harmonic compensation efforts. The results are shown in Figure 10.25. For a fair comparison, harmonic compensation is carried out using a traditional compensation scheme without priorities and the droop control scheme with priorities while keeping the total capacity the same. As expected, when given different priorities to coordinate the harmonic compensation efforts, the THD is lower than the case when all the IFCs do the harmonic compensation with the same priority.

Table 10.1 Summary of harmonics compensation in hybrid AC/DC microgrids using interfacing power electronics converters.

Challenges	Converter type		Control method	Features
AC harmonics compensation	Individual IFCs	Medium or low power IFCs	Current control method (CCM):	• Direct control of IFC output current • Adjustable compensation • Not recommended for islanded mode
			Voltage control method (VCM):	• Direct control of capacitor voltage • Adjustable compensation • Can operate at both grid-connected and islanded condition
			Hybrid control method (HCM):	• Parallel control of both output current and capacitor voltage • Adjustable compensation • Can operate at both grid-connected and islanded modes
		High power IFCs	Feed forward control method (FCM)	• Harmonic control signal is directly fed into the modulation index • Adjustable compensation • Good dynamics in low switching frequency converters
	Multiple IFC coordination		Autonomous harmonic sharing:	• Communications are not required • Sharing errors cannot be corrected
			Supervisory harmonic sharing:	• Communications are required • Autonomous harmonic sharing capability can be added to increase robustness • Coordination can be made to optimize the compensation performance
DC ripple mitigation	DC/AC IFCs:		Coordinated ripple cancelation	• No need to add extra circuits • May affect the AC-side control system • Typical centralized compensation
	DC/DC IFCs:		Autonomous ripple sharing:	• Dedicated DC ripple mitigation devices or DC/DC IFCs for energy storage systems • No impact on the AC side • Typically distributed compensation

10.6 Summary

In this chapter, harmonics compensation in hybrid AC/DC microgrids by using smart IFCs was addressed. The virtual impedance concept of IFCs for harmonics compensations was presented. Harmonics compensation in AC subgrids was discussed, in which the control of interfacing power converters with CCM, VCM, and HCM were presented. Methods for harmonics compensation with low-switching interfacing power converters in the AC subgrid were introduced. Moreover, harmonics compensation in DC subgrids with DC/AC subgrids interlinking power converters and DC/DC converters in DC subgrids were studied. Finally, coordinated control of multiple IFCs for harmonics compensation was discussed. A summary of all discussed control strategies of IFCs for harmonic compensation is shown in Table 10.1.

References

1. Li, Y.W. and He, J. (2014). Distribution system harmonic compensation methods: an overview of DG-interfacing inverters. *IEEE Industrial Electronics Magazine* 8 (4): 18–31.
2. He, J. and Li, Y.W. (2011). Analysis, design and implementation of virtual impedance for power electronics interfaced distributed generation. *IEEE Transactions on Industry Applications* 47 (6): 2525–2538.
3. Lee, T.T., Cheng, P.T., Akagi, H., and Fujita, H. (2008). A dynamic tuning method for distributed active filter systems. *IEEE Transactions on Industry Applications* 44 (2): 612–623.
4. Li, Y.W. and Kao, C.N. (2009). An accurate power control strategy for power-electronics-interfaced distributed generation units operating in a low-voltage multibus microgrid. *IEEE Transactions on Power Electronics* 24 (12): 2977–2988.
5. He, J., Li, Y.W., and Munir, M.S. (2012). A flexible harmonic control approach through voltage-controlled DG–grid interfacing converters. *IEEE Transactions on Industrial Electronics* 59 (1): 444–455.
6. He, J. and Li, Y. (2013). Hybrid voltage and current control approach for DG-grid interfacing converters with LCL filters. *IEEE Transactions on Industrial Electronics* 60 (5): 1797–1809.
7. Li, W., Ruan, X., Pan, D., and Wang, X. (2013). Full-feedforward schemes of grid voltages for a three-phase LCL-type grid-connected inverter. *IEEE Transactions on Industrial Electronics* 60 (6): 2237–2250.

8 Tian, H., Li, Y.W., and Wang, P. (2018). Hybrid AC/DC system harmonics control through grid interfacing converters with low switching frequency. *IEEE Transactions on Industrial Electronics* 65 (3): 2256–2267.

9 Tian, H., Li, Y.W., and Zhao, Q. (2020). Multirate harmonic compensation control for low switching frequency converters: scheme, modeling, and analysis. *IEEE Transactions on Power Electronics* 35 (4): 4143–4156.

10 Tian, H. and Li, Y. (2021). Virtual resistor based second-order ripple sharing control for distributed bidirectional DC–DC converters in hybrid AC–DC microgrid. *IEEE Transactions on Power Electronics* 36 (2): 2258–2269.

11 Jintakosonwit, P., Akagi, H., Fujita, H., and Ogasawara, S. (2002). Implementation and performance of automatic gain adjustment in a shunt-active filter for harmonic damping throughout a power distribution system. *IEEE Transactions on Power Electronics* 17 (3): 438–447.

12 He, J., Li, Y.W., and Blaabjerg, F. (2015). An enhanced islanding microgrid reactive power, imbalance power, and harmonic power sharing scheme. *IEEE Transactions on Power Electronics* 30 (6): 3389–3401.

13 Munir, M.S., Li, Y.W., and Tian, H. (2016). Improved residential distribution system harmonic compensation scheme using power electronics interfaced DGs. *IEEE Transactions on Smart Grid* 7 (3): 1191–1203.

14 Munir, M.S., Li, Y.W., and Tian, H. (2020). Residential distribution system harmonic compensation using priority driven droop controller. *CPSS Transactions on Power Electronics and Applications* 5 (3): 213–223.

A

Instantaneous Power Theory from Three-phase and Single-phase System Perspectives

A.1 Introduction

This appendix provides an instantaneous power analysis of the three-phase power system from the three-phase and single-phase perspectives. Then, different reference currents to be tracked by the interfacing power converters to achieve specific performances for unbalanced voltage compensation are provided. The reference currents are used in the control strategies in Chapter 9. Finally, in this appendix, discussions and comparisons of the instantaneous power analysis from the three-phase and single-phase system perspectives are provided.

A.2 Principles of Instantaneous Power Theory

From Figure A.1 and according to instantaneous power theory [1, 2], the instantaneous active and reactive powers injected into the grid at the point of common coupling (PCC) can be described as in (A.1) and (A.2).

$$p = v \cdot i = v_a\, i_a + v_b\, i_b + v_c\, i_c \tag{A.1}$$

$$q = v_{\perp} \cdot i = v_{a_{\perp}} i_a + v_{b_{\perp}} i_b + v_{c_{\perp}} i_c \tag{A.2}$$

where $v = [\, v_a, v_b, v_c]$ and $i = [i_a, i_b, i_c]$ are the three-phase PCC voltage vector and the current vector, and $v_{\perp}$ lags v by 90°. The operator "$\cdot$" denotes the dot product of vectors. Considering symmetric-sequence component of the PCC voltage vector and the current vector, (A.1) and (A.2) can be described as:

$$\begin{aligned} p = (v^{+} + v^{-} + v^{0}) \cdot (i^{+} + i^{-} + i^{0}) = (v^{+} \cdot i^{+} + v^{-} \cdot i^{-}) + \\ (v^{+} \cdot i^{-} + v^{-} \cdot i^{+}) + (v^{0} \cdot i^{0}) = P + \Delta P + (P_0 + \Delta P_0) \end{aligned} \tag{A.3}$$

Smart Hybrid AC/DC Microgrids: Power Management, Energy Management, and Power Quality Control, First Edition. Yunwei Ryan Li, Farzam Nejabatkhah, and Hao Tian.

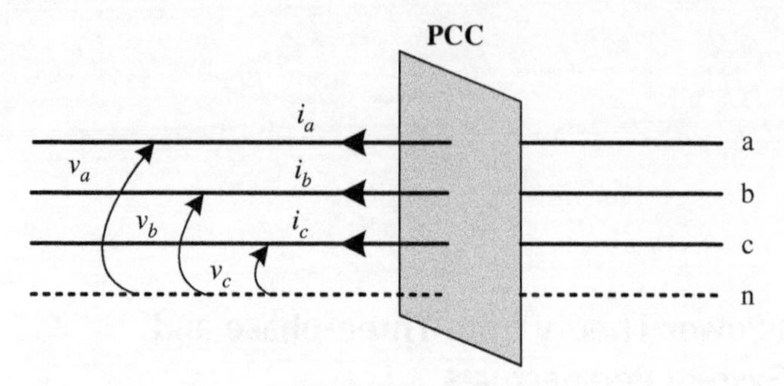

Figure A.1 Three-phase power system at the PCC.

$$\begin{aligned} q &= \left(v_\perp^+ + v_\perp^- + v_\perp^0\right) \cdot (i^+ + i^- + i^0) \\ &= \left(v_\perp^+ \cdot i^+ + v_\perp^- \cdot i^-\right) + \left(v_\perp^+ \cdot i^- + v_\perp^- \cdot i^+\right) \\ &= Q + \Delta Q \end{aligned} \tag{A.4}$$

where v^+, v^-, v^0 and i^+, i^-, i^0 are positive, negative, and zero sequence vectors of the three-phase PCC voltage vector and current vector, P and ΔP are average and oscillatory terms of instantaneous active power, P_0 and ΔP_0 are zero sequence average and oscillatory terms of instantaneous active power, and Q and ΔQ are average and oscillatory terms of instantaneous reactive power. It should be mentioned that in (A.3) and (A.4), $v^+ \centerdot i^0 = v^- \centerdot i^0 = 0$. Moreover, orthogonal vectors for positive, negative, and zero sequence vectors could be achieved using the transformation matrix in (A.5), if the positive direction of phasors rotation is assumed clockwise. From (A.5), it is clear that $v_\perp^+ . i^0 = v_\perp^- . i^0 = 0$ in (A.4), and $v_\perp^0 = 0$.

$$v_\perp^{+,-,0} = \frac{1}{\sqrt{3}} \begin{bmatrix} 0 & 1 & -1 \\ -1 & 0 & 1 \\ 1 & -1 & 0 \end{bmatrix} v^{+,-,0}. \tag{A.5}$$

In (A.3) and (A.4), the average active and reactive powers (P and Q) can be decomposed into their positive and negative sequence components as follows:

$$P = P^+ + P^- = (v^+ \cdot i^+) + (v^- \cdot i^-) \tag{A.6}$$

$$Q = Q^+ + Q^- = \left(v_\perp^+ \cdot i^+\right) + \left(v_\perp^- \cdot i^-\right) \tag{A.7}$$

where P^+ and P^- are the positive and negative sequences of average active power, respectively, and Q^+ and Q^- are the positive and negative sequences of average reactive power, respectively.

A.3 Power Control Using Instantaneous Power Theory from a Three-phase System Perspective

Here, proper reference current calculations to be applied to the three-phase interfacing converters to achieve specific performances are provided. Thus, the instantaneous power analysis from the three-phase system perspective is used in the calculations. In this study, it is assumed that a three-phase interfacing converter (IFC) is connected to the grid at the PCC, thus the IFC output voltage and current are the same as the PCC voltage and current (v, i). Moreover, the power system has three wires, so there is no active power contribution from the zero-sequence current. Therefore, the zero-sequence voltage will be neglected. In Figure A.2, a typical grid-connected three-phase IFC with its control block diagram is shown. Here, the focus is on reference current generation for the IFC.

Two different reference currents calculations are provided in Sections A.3.1 and A.3.2. In the first one, injection of both positive and negative sequences current into the grid at the fundamental frequency is desired, suitable for an adjustable unbalanced compensation [3]. The second reference current is derived from providing an easy way to reduce the adverse effects of an unbalanced voltage on IFC operation in terms of output powers oscillation cancelation [4].

A.3.1 Reference Current Focusing on Unbalanced Condition Compensation

In this strategy, the reference current of a three-phase IFC is calculated to flexibly inject both positive and negative sequences of active and reactive currents into

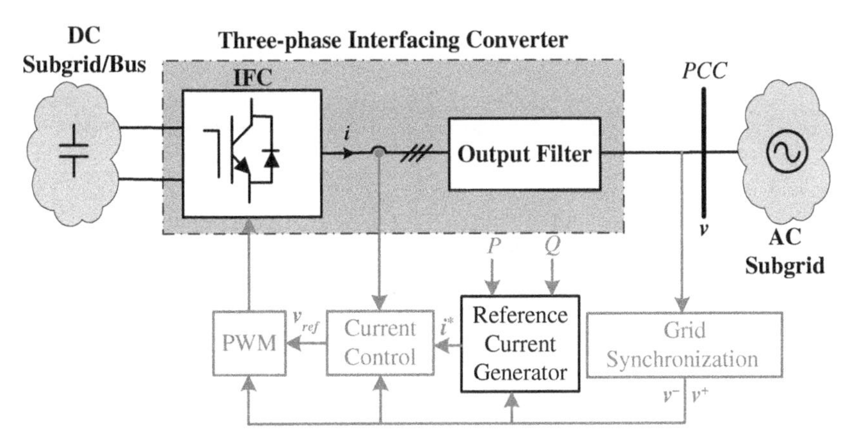

Figure A.2 Typical grid-connected three-phase interfacing converter with its control strategy.

the grid to compensate for unbalanced conditions. In other words, this method is useful if the adjustable unbalanced compensation is a preferential issue.

According to (A.3) and (A.4), the current vector aligned with the voltage vector will generate active power while the current vector aligned with the orthogonal voltage vector will generate reactive power. Therefore, the total IFC reference current vector to inject positive and negative sequences current could be expressed as:

$$i^* = a^+v^+ + a^-v^- + b^+v_\perp^+ + b^-v_\perp^- \tag{A.8}$$

where a^+, a^-, b^+, b^- are positive and negative sequence instantaneous conductances and susceptances. If only either positive or negative sequence current had to be injected to the exchange of a certain amount of power with the grid (P and Q), the values of conductance and susceptance will be as follows:

$$a^+ = \frac{P}{|v^+|^2} \; ; \; a^- = \frac{P}{|v^-|^2} \; ; \; b^+ = \frac{Q}{|v^+|^2} \; ; \; b^- = \frac{Q}{|v^-|^2}. \tag{A.9}$$

In the condition that injection of both positive and negative sequences of active and reactive powers into the grid is desired, both sequences should be regulated to keep the reference active and reactive powers constant. As a result, weighting factors k_1 and k_2 are defined as $k_1 = \frac{P^+}{P}$ and $k_2 = \frac{Q^+}{Q}$ to regulate the proportions of each sequence power delivered into the grid as follows:

$$i^* = \frac{Pk_1}{|v^+|^2}v^+ + \frac{P(1-k_1)}{|v^-|^2}v^- + \frac{Qk_2}{|v^+|^2}v_\perp^+ + \frac{Q(1-k_2)}{|v^-|^2}v_\perp^-. \tag{A.10}$$

As seen from (A.10), k_1 and k_2 can be controlled to adjust positive and negative sequences of active and reactive currents to compensate for the unbalanced condition. From (A.10), the positive and genitive sequence currents can be derived as follows:

$$i^{+*} = \frac{Pk_1}{|v^+|^2}v^+ + \frac{Qk_2}{|v^+|^2}v_\perp^+ \tag{A.11}$$

$$i^{-*} = \frac{P(1-k_1)}{|v^-|^2}v^- + \frac{Q(1-k_2)}{|v^-|^2}v_\perp^-. \tag{A.12}$$

From (A.3), (A.4), (A.11), and (A.12), the active and reactive power oscillations at the IFC output connected to the PCC can be achieved as:

$$\Delta P = \left(\frac{Pk_1}{|v^+|^2} + \frac{P\,(1-k_1)}{|v^-|^2}\right)v^+ \cdot v^- + \left(\frac{Qk_2}{|v^+|^2} - \frac{Q\,(1-k_2)}{|v^-|^2}\right)v_\perp^+ \cdot v^- \tag{A.13}$$

$$\Delta Q = \left(\frac{Pk_1}{|v^+|^2} - \frac{P(1-k_1)}{|v^-|^2}\right)v^+ \cdot v_\perp^- + \left(\frac{Qk_2}{|v^+|^2} + \frac{Q\,(1-k_2)}{|v^-|^2}\right)v_\perp^+ \cdot v_\perp^-. \tag{A.14}$$

Considering (A.13) and (A.14), the active and reactive powers oscillations can be canceled out considering the following values of k_1 and k_2:

$$\Delta P = 0 \Rightarrow k_1 = \frac{|v^+|^2}{|v^+|^2 - |v^-|^2} \ \& \ k_2 = \frac{|v^+|^2}{|v^+|^2 + |v^-|^2} \tag{A.15}$$

$$\Delta Q = 0 \Rightarrow k_1 = \frac{|v^+|^2}{|v^+|^2 + |v^-|^2} \ \& \ k_2 = \frac{|v^+|^2}{|v^+|^2 - |v^-|^2}. \tag{A.16}$$

A.3.2 Reference Current Focusing on Active and Reactive Power Oscillation Cancelation

In this strategy, the reference current is calculated to easily control/cancel out the three-phase IFC output active and reactive powers oscillations due to unbalanced conditions. Although the power oscillations can be canceled out considering (A.15) and (A.16) in the reference current (A.10), this control strategy provides a much easier way for power oscillation cancelation, especially when parallel IFCs are used.

Considering (A.3) and (A.4), in unity PF where $Q = 0$, since $v^+ . i^-$ and $v^- \cdot i^+$ and also $v_\perp^+ \cdot i^-$ and $v_\perp^- \cdot i^+$ are in-phase quantities (dot product gives the scalar answer), the active or reactive power oscillation of an IFC can be compensated for using the scalar coefficient k_p as follows:

$$\Delta P = 0 \Rightarrow v^+ \cdot i_p^- = -k_p \, v^- \cdot i_p^+ \qquad k_p \geq 0 \tag{A.17}$$

$$\Delta Q = 0 \Rightarrow v_\perp^+ \cdot i_p^- = -k_p \, v_\perp^- \cdot i_p^+ \quad k_p \geq 0 \tag{A.18}$$

where subscript "p" is related to unity PF operation. From (A.3), (A.4), (A.17), and (A.18), the IFC reference current vector under unity PF can be obtained:

$$i_p^* = \frac{P}{|v^+|^2 + k_p|v^-|^2} v^+ + \frac{P\,k_p}{|v^+|^2 + k_p|v^-|^2} v^-. \tag{A.19}$$

Similarly, under zero PF operation mode ($P = 0$), the active or reactive power oscillation can be compensated using scalar coefficient k_q as follows:

$$\Delta P = 0 \Rightarrow v^+ \cdot i_q^- = -k_q \, v^- \cdot i_q^+ \qquad k_q \geq 0 \tag{A.20}$$

$$\Delta Q = 0 \Rightarrow v_\perp^+ \cdot i_q^- = -k_q \, v_\perp^- \cdot i_q^+ \quad k_q \geq 0 \tag{A.21}$$

where subscript "q" is related to reactive power control. From (A.3), (A.4), (A.20), and (A.21), the IFC reference current vector under zero PF can be derived as:

$$i_q^* = \frac{Q}{|v^+|^2 + k_q|v^-|^2} v_\perp^+ + \frac{Q\,k_q}{|v^+|^2 + k_q|v^-|^2} v_\perp^-. \tag{A.22}$$

Combining (A.19) and (A.22), the IFC reference current vector is obtained:

$$i^* = i_p^* + i_q^* = \left(\frac{P}{|v^+|^2 + k_p|v^-|^2} v^+ + \frac{P\,k_p}{|v^+|^2 + k_p|v^-|^2} v^- \right) + \left(\frac{Q}{|v^+|^2 + k_q|v^-|^2} v_\perp^+ + \frac{Q\,k_q}{|v^+|^2 + k_q|v^-|^2} v_\perp^- \right). \tag{A.23}$$

From (A.3), (A.4), and (A.23), the IFC output active and reactive powers oscillations at the PCC can be derived as:

$$\Delta P = \frac{P(1 + k_p)(v^+ \cdot v^-)}{|v^+|^2 + k_p|v^-|^2} + \frac{Q(1 - k_q)\left(v_\perp^+ \cdot v^-\right)}{|v^+|^2 + k_q|v^-|^2} \tag{A.24}$$

$$\Delta Q = \frac{P(1 - k_p)\left(v^+ \cdot v_\perp^-\right)}{|v^+|^2 + k_p|v^-|^2} + \frac{Q(1 + k_q)\left(v_\perp^+ \cdot v_\perp^-\right)}{|v^+|^2 + k_q|v^-|^2}. \tag{A.25}$$

Considering (A.24) and (A.25), $k_p = -1$ and $k_q = 1$, and $k_p = 1$ and $k_q = -1$ result in IFC active and reactive power oscillation cancelation, respectively. As is clear, this reference current provides a much easier way for IFC output power oscillations cancelation (see (A.15) and (A.16)).

A.4 Power Control Using Instantaneous Power Theory from a Single-phase System Perspective

In this section, appropriate reference current calculations to be applied to single-phase IFCs to achieve specific performance is provided. Due to the control of single-phase IFCs, the instantaneous power theory from the single-phase system perspective is developed. In this method, it is assumed that one single-phase IFC is connected to each phase of the three-phase grid at the PCC; therefore, the IFC output voltages and currents are similar to the PCC voltages and currents in that phase (similar to Figure A.2 but one single-phase IFC is connected to each phase). In addition, the power system has four wires, which results in zero-sequence current and voltage.

In this strategy, reference currents for single-phase IFCs are calculated to inject the desired active and reactive powers into the unbalanced grid and compensate for the unbalanced condition. In other words, this method is useful if the unbalanced compensation of a power system by single-phase IFCs is a preferential issue.

As mentioned, the current vector that is aligned with the voltage vector will generate active power, while the current vector that is aligned with the orthogonal voltage vector will generate reactive power (see (A.1) and (A.2)). Thus, from a single-phase perspective, total reference currents for single-phase IFCs in phases a, b, and c can be defined as:

$$i_a^* = \underbrace{A_{a_p} v_a}_{i_{ap}^*} + \underbrace{B_{a_q} v_{a_\perp}}_{i_{aq}^*} \tag{A.26}$$

$$i_b^* = \underbrace{A_{b_p} v_b}_{i_{bp}^*} + \underbrace{B_{b_q} v_{b_\perp}}_{i_{bq}^*} \tag{A.27}$$

$$i_c^* = \underbrace{A_{c_p} v_c}_{i_{cp}^*} + \underbrace{B_{c_q} v_{c_\perp}}_{i_{cq}^*} \tag{A.28}$$

where i_a^*, i_b^*, and i_c^* are the total reference currents for single-phase IFCs connected to the phases a, b, and c, (A_{a_p}, B_{a_q}), (A_{b_p}, B_{b_q}), and (A_{c_p}, B_{c_q}) are instantaneous conductances and susceptances in each phase, i_{ap}, i_{bp}, and i_{cp} are active currents of the IFCs in phases a, b, and c, and i_{aq}, i_{bq}, and i_{cq} are reactive currents of the IFCs in phases a, b, and c. The active and reactive currents generate the active and reactive powers in three phases. In (A.26)–(A.28), the values of conductance and susceptance that give rise to the exchange of a certain amount of power with the grid in each phase, (P_a, Q_a), (P_b, Q_b), and (P_c, Q_c), under given voltage conditions can be obtained as:

$$A_{a_p} = \frac{P_a}{|v_a|^2}\ ;\ B_{a_q} = \frac{Q_a}{|v_a|^2}\ ;\ A_{b_p} = \frac{P_b}{|v_b|^2}\ ;\ B_{b_q} = \frac{Q_b}{|v_b|^2}$$
$$A_{c_p} = \frac{P_c}{|v_c|^2}\ ;\ B_{a_q} = \frac{Q_c}{|v_c|^2}. \tag{A.29}$$

Therefore, the total reference currents for single-phase IFCs in each phase will be:

$$\begin{cases} i_a^* = i_{ap}^* + i_{aq}^* = \frac{P_a}{|v_a|^2} v_a + \frac{Q_a}{|v_a|^2} v_{a_\perp} \\ i_b^* = i_{bp}^* + i_{bq}^* = \frac{P_b}{|v_b|^2} v_b + \frac{Q_b}{|v_b|^2} v_{b_\perp} \\ i_c^* = i_{cp}^* + i_{cq}^* = \frac{P_c}{|v_c|^2} v_c + \frac{Q_c}{|v_c|^2} v_{c_\perp} \end{cases}. \tag{A.30}$$

Finally, sequential analysis of three-phase active and reactive currents in (A.30) results in the following positive, negative, and zero sequences current:

$$\begin{aligned} i^{+^*} &= i_p^{+^*} + i_q^{+^*} \\ &= \frac{1}{3}\left(\frac{P_a}{|v_a|^2} v_a + \frac{P_b}{|v_b|^2} e^{j\frac{2\pi}{3}} v_b + \frac{P_c}{|v_c|^2} e^{j\frac{4\pi}{3}} v_c\right) + \\ &\frac{1}{3}\left(\frac{Q_a}{|v_a|^2} v_{a_\perp} + \frac{Q_b}{|v_b|^2} e^{j\frac{2\pi}{3}} v_{b_\perp} + \frac{Q_c}{|v_c|^2} e^{j\frac{4\pi}{3}} v_{c_\perp}\right) \end{aligned} \tag{A.31}$$

$$\begin{aligned} i^{-*} &= i_p^{-*} + i_q^{-*} \\ &= \frac{1}{3}\left(\frac{P_a}{|v_a|^2}v_a + \frac{P_b}{|v_b|^2}e^{j\frac{4\pi}{3}}v_b + \frac{P_c}{|v_c|^2}e^{j\frac{2\pi}{3}}v_c\right) + \\ &\frac{1}{3}\left(\frac{Q_a}{|v_a|^2}v_{a_\perp} + \frac{Q_b}{|v_b|^2}e^{j\frac{4\pi}{3}}v_{b_\perp} + \frac{Q_c}{|v_c|^2}e^{j\frac{2\pi}{3}}v_{c_\perp}\right) \end{aligned} \tag{A.32}$$

$$\begin{aligned} i^{0*} &= i_p^{0*} + i_q^{0*} \\ &= \frac{1}{3}\left(\frac{P_a}{|v_a|^2}v_a + \frac{P_b}{|v_b|^2}v_b + \frac{P_c}{|v_c|^2}v_c\right) + \\ &\frac{1}{3}\left(\frac{Q_a}{|v_a|^2}v_{a_\perp} + \frac{Q_b}{|v_b|^2}v_{b_\perp} + \frac{Q_c}{|v_c|^2}v_{c_\perp}\right). \end{aligned} \tag{A.33}$$

Considering (A.31)–(A.33), it is clear that three-phase active and reactive powers can be controlled to adjust the positive, negative, and zero sequence currents, which leads to adjustable unbalanced compensation.

A.5 Discussion

In Sections A.3 and A.4, the instantaneous power analysis from the three-phase and single-phase system perspectives are used to generate reference currents for three-phase and single-phase IFCs to adjustably compensate the unbalanced condition. Since in both strategies, a certain amount of power is exchanged with the grid (P and Q), the generated reference currents need to be equal under similar operating conditions. Here, two simple examples are provided to compare the generated reference currents from the two perspectives.

A.5.1 Example 1: Only Positive Sequence Active Current Injection

In Section A.3.1, the instantaneous power theory from the three-phase system perspective is used to generate a reference current to flexibly inject both positive and negative sequences current into the unbalanced grid. From (A.10), if just positive sequence current has to be injected to the exchange of a certain amount of active power with the grid (P), (A.10) can be written as follows:

$$i^{+*} = \frac{P}{|v^+|^2}v^+. \tag{A.34}$$

Replacing v^+ with phase voltages, the following relation will be achieved:

$$i^{+*} = \frac{P}{3|v^+|^2}\left(v_a + e^{j\frac{2\pi}{3}}v_b + e^{j\frac{4\pi}{3}}v_c\right). \tag{A.35}$$

On the other hand, in Section A.4, the instantaneous power theory from the single-phase system perspective is used to generate reference currents for single-phase IFCs (connected to the PCC in each phase) to inject desired active and reactive currents into the unbalanced grid. From (A.31), the positive sequence of the PCC current under just active power injection can be obtained as follows:

$$i^{+^*} = \frac{P}{3}\left(\frac{u_1}{|v_a|^2}v_a + \frac{u_2}{|v_b|^2}e^{j\frac{2\pi}{3}}v_b + \frac{u_3}{|v_c|^2}e^{j\frac{4\pi}{3}}v_c\right) \tag{A.36}$$

where $u_1 + u_2 + u_3 = 1$. It is worth mentioning that the average of three-phase instantaneous active power is equal to the summation of average active powers of three phases ($P = P_a + P_b + P_c$).

Comparing (A.35) with (A.36) clarifies that these two positive sequence reference currents look similar; however, they are not equal due to different operating conditions. In other words, in (A.35), just positive sequence current is injected while (A.36) is the positive sequence component of the reference current in (A.30), which also contains negative and zero sequence currents. To compare these two positive sequence currents ((A.35) and (A.36)), the negative and zero sequence currents of (A.30), which are presented in (A.32) and (A.33), should be set to zero as follows:

$$i^{-^*} = \frac{P}{3}\left(\frac{u_1}{|v_a|^2}v_a + \frac{u_2}{|v_b|^2}e^{j\frac{4\pi}{3}}v_b + \frac{u_3}{|v_c|^2}e^{j\frac{2\pi}{3}}v_c\right) = 0 \tag{A.37}$$

$$i^{0^*} = \frac{P}{3}\left(\frac{u_1}{|v_a|^2}v_a + \frac{u_2}{|v_b|^2}v_b + \frac{u_3}{|v_c|^2}v_c\right) = 0. \tag{A.38}$$

Solving (A.37) and (A.38), the following relations are achieved:

$$\begin{cases} \dfrac{u_1}{|v_a|} = \dfrac{u_2}{|v_b|} = \dfrac{u_3}{|v_c|} \\ \angle\theta_{v_a} = \angle\theta_{v_b} + \dfrac{2\pi}{3} = \angle\theta_{v_c} + \dfrac{4\pi}{3} \end{cases}. \tag{A.39}$$

Applying the conditions in (A.39) into (A.35) and (A.36), and assuming that the average active powers in the three phases are equal ($u_1 = u_2 = u_3 = 1/3$), following relations will result:

$$i^{+^*} = \frac{P}{3|v^+|^2}\left(v_a + e^{j\frac{2\pi}{3}}v_b + e^{j\frac{4\pi}{3}}v_c\right) = \frac{P}{3|v^+|^2}v_a \tag{A.40}$$

$$i^{+^*} = \frac{P}{3}\left(\frac{u_1}{|v_a|^2}v_a + \frac{u_2}{|v_b|^2}e^{j\frac{2\pi}{3}}v_b + \frac{u_3}{|v_c|^2}e^{j\frac{4\pi}{3}}v_c\right) = \frac{P}{3|v^+|^2}v_a. \tag{A.41}$$

Thus, from (A.40) and (A.41), it is seen that the reference currents from the three-phase perspective and the single-phase perspective are equal under similar operating conditions.

A.5.2 Example 2: Only Negative Sequence Active Current Injection

As another example, negative sequence active current injection to the grid is considered, and reference currents generated in Sections A.3.1 and A.4 are compared. From the three-phase system perspective, (A.10) should be written as follows:

$$i^{-*} = \frac{P}{3|v^-|^2}\left(v_a + e^{j\frac{4\pi}{3}} v_b + e^{j\frac{2\pi}{3}} v_c\right). \tag{A.42}$$

From (A.32), the negative sequence current under active power injection from the single-phase system perspective will be as follows:

$$i^{-*} = \frac{P}{3}\left(\frac{u_1}{|v_a|^2} v_a + \frac{u_2}{|v_b|^2} e^{j\frac{4\pi}{3}} v_b + \frac{u_3}{|v_c|^2} e^{j\frac{2\pi}{3}} v_c\right). \tag{A.43}$$

However, for only negative sequence current injection from the single-phase system perspective, the positive and zero sequences current of (A.31) and (A.33) should be set to zero as follows:

$$i^{+*} = \frac{P}{3}\left(\frac{u_1}{|v_a|^2} v_a + \frac{u_2}{|v_b|^2} e^{j\frac{2\pi}{3}} v_b + \frac{u_3}{|v_c|^2} e^{j\frac{4\pi}{3}} v_c\right) = 0 \tag{A.44}$$

$$i^{0*} = \frac{P}{3}\left(\frac{u_1}{|v_a|^2} v_a + \frac{u_2}{|v_b|^2} v_b + \frac{u_3}{|v_c|^2} v_c\right) = 0. \tag{A.45}$$

Solving (A.44) and (A.45), the single-phase system parameters should have the following relations:

$$\begin{cases} \dfrac{u_1}{|v_a|} = \dfrac{u_2}{|v_b|} = \dfrac{u_3}{|v_c|} \\ \angle\theta_{v_a} = \angle\theta_{v_b} + \dfrac{4\pi}{3} = \angle\theta_{v_c} + \dfrac{2\pi}{3} \end{cases}. \tag{A.46}$$

Applying the relations in (A.46) in (A.42) and (A.43) to unify the operating conditions and assuming equal average active powers in three phases, the following results will be achieved:

$$i^{-*} = \frac{P}{3|v^-|^2}\left(v_a + e^{j\frac{4\pi}{3}} v_b + e^{j\frac{2\pi}{3}} v_c\right) = \frac{P}{3|v^-|^2} v_a \tag{A.47}$$

$$i^{-*} = \frac{P}{3}\left(\frac{u_1}{|v_a|^2} v_a + \frac{u_2}{|v_b|^2} e^{j\frac{4\pi}{3}} v_b + \frac{u_3}{|v_c|^2} e^{j\frac{2\pi}{3}} v_c\right) = \frac{P}{3|v^-|^2} v_a. \tag{A.48}$$

From (A.47) and (A.48), it is clear that under similar operating conditions, the reference currents from the single-phase system perspective and the three-phase system perspective are equal.

A.6 Summary

This appendix reviewed and derived an instantaneous power analysis from three-phase and single-phase system perspectives to calculate the reference currents for interfacing converters to achieve specific goals. Three methods to generate the reference currents were provided. In the first reference current, which was derived for a three-phase IFC to adjustably compensate for the unbalanced condition, the positive and negative sequence active and reactive currents were directly controlled by the two coefficients k_1 and k_2 (control of positive and negative sequence virtual impedances). The second reference current was derived to easily reduce/minimize the adverse effects of unbalanced voltage on three-phase IFC operation by controlling the two coefficients k_p and k_q. It should be mentioned that adverse effect reduction such as power oscillation cancelation can be done by the first generated reference current (using k_1 and k_2), but the second one provides an easier way to achieve this goal, especially when parallel IFCs are used. The third reference current was developed for single-phase IFCs to compensate for the unbalanced condition adjustably. Three-phase active and reactive powers were controlled in this reference current to adjust the positive, negative, and zero sequence currents.

References

1 Teodorescu, R., Liserre, M., and Rodrıguez, P. (2011). *Grid Converters for Photovoltaic and Wind Power Systems.* New York, NY, USA: Wiley.

2 Akagi, H., Watanabe, E., and Aredes, M. (2007). *Instantaneous Power Theory and Applications to Power Conditioning.* Wiley–IEEE Press ISBN 978-0-470-10761-4.

3 Nejabatkhah, F., Li, Y.W., and Wu, B. (2016). Control strategies of three-phase distributed generation inverters for grid unbalanced voltage compensation. *IEEE Transactions on Power Electronics* 31 (7): 5228–5241.

4 Nejabatkhah, F., Li, Y.W., and Sun, K. (2018). Parallel three-phase interfacing converters operation under unbalanced voltage in hybrid AC/DC microgrid. *IEEE Transactions on Smart Grid* 9 (2): 1310–1322.

References

B

Peak Current of Interfacing Power Converters Under Unbalanced Voltage

B.1 Introduction

Under unbalanced voltage, individual and parallel interfacing converters (IFCs) peak currents can increase, which reduces their power/current transferring capability. In some cases, converters reach their rating limits. Thus, the individual and parallel IFCs peak currents under unbalanced voltage should be controlled to maximize their power/current transferring capability and avoid reaching their rating limits. In this Appendix, two main topics are covered:

(1) A thorough study is conducted on peak currents of individual and parallel IFCs peak currents under unbalanced voltage (Section B.2).
(2) Maximizing the current/power transferring capability of IFCs under unbalanced conditions is investigated in detail (Section B.3).

B.2 Peak Currents of Interfacing Converters

B.2.1 Individual Interfacing Converters

The n-parallel IFCs connecting AC and DC subgrids/buses are shown in Figure B.1. Individual ith IFC peak current is studied in detail (a parallel IFCs configuration is used, Figure B.1, to easily extend individual IFC peak current discussions here to parallel IFCs peak current discussions; see Section B.2.2). Here, ith IFC reference current is considered as in (B.1) for simplicity (for a detailed discussion on the reference current, please see Appendix A).

$$i_i^* = i_{pi}^* + i_{qi}^* = \left(\frac{P_i}{|v^+|^2 + k_{pi}|v^-|^2} v^+ + \frac{P_i k_{pi}}{|v^+|^2 + k_{pi}|v^-|^2} v^- \right) + \left(\frac{Q_i}{|v^+|^2 + k_{qi}|v^-|^2} v_\perp^+ + \frac{Q_i k_{qi}}{|v^+|^2 + k_{qi}|v^-|^2} v_\perp^- \right). \tag{B.1}$$

Smart Hybrid AC/DC Microgrids: Power Management, Energy Management, and Power Quality Control, First Edition. Yunwei Ryan Li, Farzam Nejabatkhah, and Hao Tian.

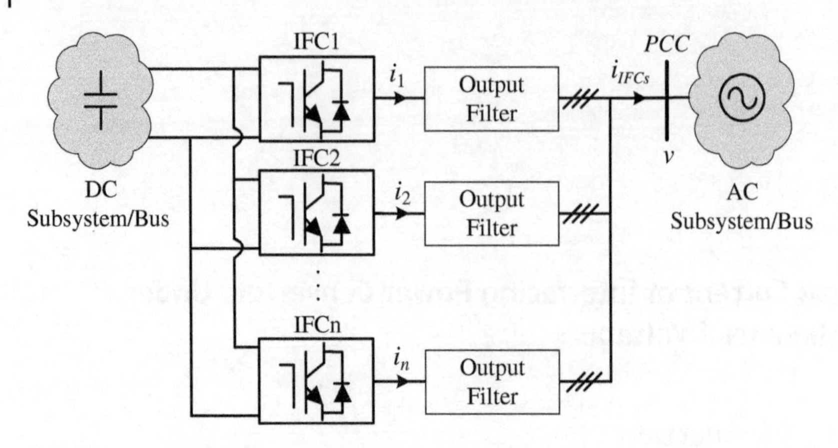

Figure B.1 *n*-parallel three-phase IFCs.

From (B.1), the ith IFC active and reactive reference current vectors can be rewritten as follows:

$$i^*_{pi} = i^+_{pi} + i^-_{pi} = \frac{P_i}{|v^+|^2 + k_{pi}|v^-|^2} v^+ + \frac{P_i k_{pi}}{|v^+|^2 + k_{pi}|v^-|^2} v^- \tag{B.2}$$

$$i^*_{qi} = i^+_{qi} + i^-_{qi} = \frac{Q_i}{|v^+|^2 + k_{qi}|v^-|^2} v^+_\perp + \frac{Q_i k_{qi}}{|v^+|^2 + k_{qi}|v^-|^2} v^-_\perp \tag{B.3}$$

where $v^+ = |v^+|e^{j(\omega t + \theta^+)}$ and $v^- = |v^-|e^{j(-\omega t - \theta^-)}$. The locus of the PCC positive and negative sequence voltage vectors and the IFC current vector is shown in Figure B.2, where ρ is defined as $\rho = (\theta^+ - \theta^-)/2$. Considering (B.2) and (B.3), the loci of the ith IFC active and reactive reference current vectors are ellipses in which their semi-major and semi-minor axis lengths can be achieved as:

$$I_{pLi} = \frac{P_i|v^+|}{|v^+|^2 + k_{pi}|v^-|^2} + \frac{P_i k_{pi}|v^-|}{|v^+|^2 + k_{pi}|v^-|^2} \tag{B.4}$$

$$I_{pSi} = \frac{P_i|v^+|}{|v^+|^2 + k_{pi}|v^-|^2} - \frac{P_i k_{pi}|v^-|}{|v^+|^2 + k_{pi}|v^-|^2} \tag{B.5}$$

$$I_{qLi} = \frac{Q_i|v^+|}{|v^+|^2 + k_{qi}|v^-|^2} + \frac{Q_i k_{qi}|v^-|}{|v^+|^2 + k_{qi}|v^-|^2} \tag{B.6}$$

$$I_{qSi} = \frac{Q_i|v^+|}{|v^+|^2 + k_{qi}|v^-|^2} - \frac{Q_i k_{qi}|v^-|}{|v^+|^2 + k_{qi}|v^-|^2}. \tag{B.7}$$

The maximum current at each phase of the ith IFC is the maximum projection of the current ellipse on the abc axis. From (B.2) and (B.7), the projection of the current ellipse on the abc axis can be derived as:

$$i^{*'}_{xi} = (I_{pLi}\cos\gamma - I_{qLi}\sin\gamma)\cos(\omega t) + (-I_{qSi}\cos\gamma - I_{pSi}\sin\gamma)$$
$$\sin(\omega t) \quad x = a, b, c \tag{B.8}$$

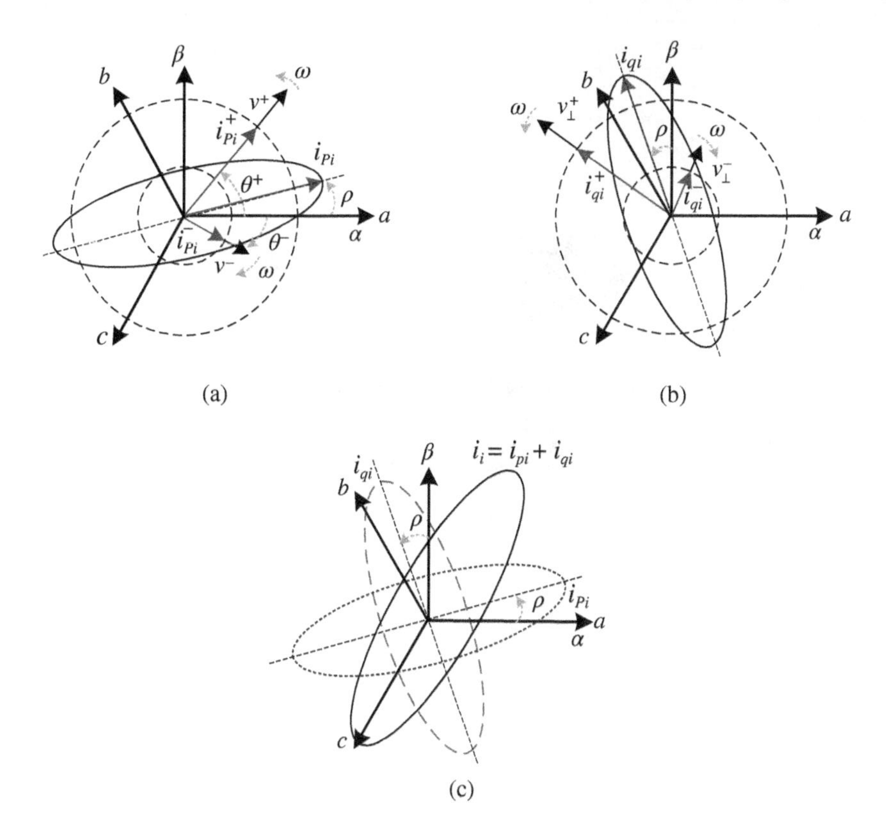

Figure B.2 Locus of the PCC voltage and the ith IFC current: (a) unity power factor operation mode ($Q_i = 0$), (b) reactive power compensation mode ($P_i = 0$), (c) both active and reactive power production.

where γ is the rotation angle which is equal to ρ, $\rho + \pi/3$, and $\rho - \pi/3$ for the a, b, and c axes, respectively, and ρ is defined as $\rho = (\theta^+ - \theta^-)/2$. Using (B.2)–(B.8), the maximum current at each phase of ith-IFC can be expressed as:

$$I_{xi}^{\max} = \sqrt{\begin{array}{l} \frac{P_i^2}{(|v^+|^2+k_{pi}|v^-|^2)^2}\left(|v^+|^2+|v^-|^2k_{pi}^2+2|v^+||v^-|k_{pi}\cos(2\gamma)\right)+ \\ \frac{Q_i^2}{(|v^+|^2+k_{qi}|v^-|^2)^2}\left(|v^+|^2+|v^-|^2k_{qi}^2-2|v^+||v^-|k_{qi}\cos(2\gamma)\right)- \\ \frac{2P_iQ_i|v^+||v^-|}{(|v^+|^2+k_{pi}|v^-|^2)(|v^+|^2+k_{qi}|v^-|^2)}(k_{pi}+k_{qi})\sin(2\gamma) \end{array}} \quad x = \text{a, b, c.} \tag{B.9}$$

From (B.9), the peak current of the individual ith IFC can be achieved as follows:

$$I_i^{\max} = \max\left(I_{ai}^{\max}, I_{bi}^{\max}, I_{ci}^{\max}\right). \tag{B.10}$$

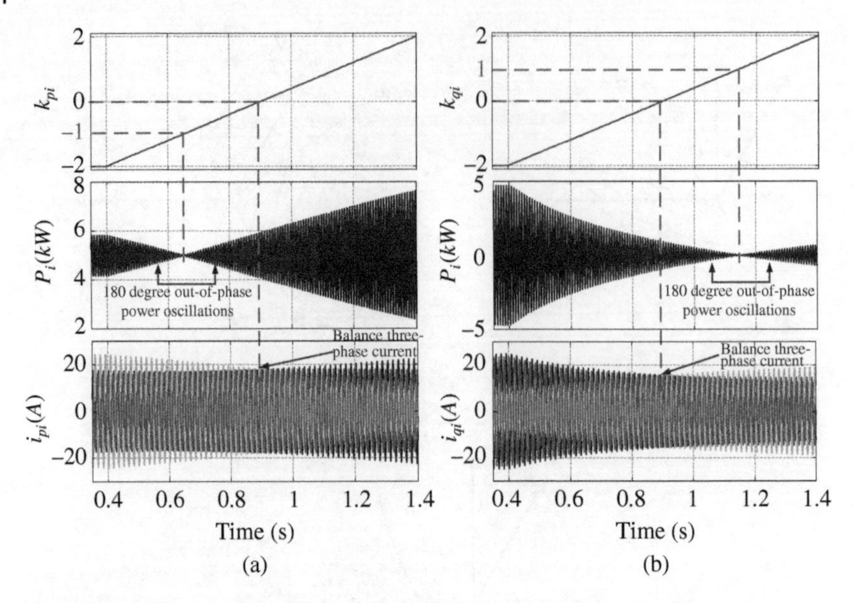

Figure B.3 Individual IFC output active power and three-phase currents under unbalanced voltage condition: (a) different k_{pi} in unity PF operation mode and (b) different k_{qi} in zero PF operation mode.

Moreover, considering (B.4)–(B.8), the phase angle of the ith IFC peak current can be derived as:

$$\delta_i\big|_{I_i^{\max}} = \tan^{-1}\left(\frac{\left(\frac{P_i(|v^+|+k_{Pi}|v^-|)}{|v^+|^2+k_{Pi}|v^-|^2}\right)\cos\gamma + \left(\frac{-Q_i(|v^+|+k_{qi}|v^-|)}{|v^+|^2+k_{qi}|v^-|^2}\right)\sin\gamma}{\left(\frac{-Q_i(|v^+|-k_{qi}|v^-|)}{|v^+|^2+k_{qi}|v^-|^2}\right)\cos\gamma + \left(\frac{-P_i(|v^+|-k_{Pi}|v^-|)}{|v^+|^2+k_{Pi}|v^-|^2}\right)\sin\gamma}\right). \tag{B.11}$$

Considering (B.9)–(B.11), it can be understood that the amplitude and phase angle of the individual IFC peak current depends on the IFC output average active and reactive powers, PCC positive and negative sequence voltages, and k_{Pi} and k_{qi}. As an example, a typical individual IFC output active power and three-phase currents under different k_{pi} and k_{qi} in unity PF and zero PF operations are shown in Figure B.3.

B.2.2 Parallel Interfacing Converters

Considering (B.4)–(B.8), the projection of n-parallel IFCs collective current ellipse on each phase can be expressed as:

$$i^{*'}_{x\text{-IFCs}} = \sum_{i=1}^{n} i^{*'}_x = \left(\sum_{i=1}^{n} \frac{P_i|v^+|}{|v^+|^2 + k_{P_i}|v^-|^2} + \sum_{i=1}^{n} \frac{P_i k_{P_i}|v^-|}{|v^+|^2 + k_{P_i}|v^-|^2}\right)$$

$$
\begin{aligned}
&\cos\gamma\cos\omega t - \left(\sum_{i=1}^{n}\frac{Q_i|v^+|}{|v^+|^2+k_{q_i}|v^-|^2}+\sum_{i=1}^{n}\frac{Q_ik_{q_i}|v^-|}{|v^+|^2+k_{q_i}|v^-|^2}\right)\\
&\sin\gamma\cos\omega t - \left(\sum_{i=1}^{n}\frac{Q_i|v^+|}{|v^+|^2+k_{q_i}|v^-|^2}-\sum_{i=1}^{n}\frac{Q_ik_{q_i}|v^-|}{|v^+|^2+k_{q_i}|v^-|^2}\right)\\
&\cos\gamma\sin\omega t - \left(\sum_{i=1}^{n}\frac{P_i|v^+|}{|v^+|^2+k_{p_i}|v^-|^2}-\sum_{i=1}^{n}\frac{P_ik_{p_i}|v^-|}{|v^+|^2+k_{p_i}|v^-|^2}\right)\\
&\sin\gamma\sin\omega t \qquad x=a,b,c.
\end{aligned} \tag{B.12}
$$

From (B.12), it is clear that the amplitude and phase angle of the IFCs collective current projection on each phase depends on the average active and reactive powers, the positive and negative sequences of the PCC voltage, and k_{Pi} and k_{qi}.

To study n-parallel IFCs peak currents under zero collective active power oscillations, the two following constraints should be applied on (B.12) (please see Chapter 9 for more detail).

$$
\sum_{i=1}^{n}\frac{P_i}{|v^+|^2+k_{p_i}|v^-|^2}=\frac{\sum_{i=1}^{n}P_i}{|v^+|^2-|v^-|^2} \tag{B.13}
$$

$$
\sum_{i=1}^{n}\frac{Q_i}{|v^+|^2+k_{q_i}|v^-|^2}=\frac{\sum_{i=1}^{n}Q_i}{|v^+|^2+|v^-|^2}. \tag{B.14}
$$

Thus, for n-parallel IFCs under zero collective active power oscillations, the maximum collective current amplitude at each phase, and the peak current amplitude and phase angle are:

$$
I_{x-\mathrm{IFCs}}^{\max}\Big|_{\Delta P=0}=\sqrt{\begin{array}{l}\left(\left(\frac{\sum_{i=1}^{n}P_i}{|v^+|^2-|v^-|^2}\right)^2+\left(\frac{\sum_{i=1}^{n}Q_i}{|v^+|^2+|v^-|^2}\right)^2\right)\times\\ \left(|v^+|^2+|v^-|^2-2|v^+||v^-|\cos(2\gamma)\right)\end{array}} \quad x=a,b,c \tag{B.15}
$$

$$
I_{\mathrm{IFCs}}^{\max}\Big|_{\Delta P=0}=\max\left(I_{a-\mathrm{IFCs}}^{\max}\Big|_{\Delta P=0},I_{b-\mathrm{IFCs}}^{\max}\Big|_{\Delta P=0},I_{c-\mathrm{IFCs}}^{\max}\Big|_{\Delta P=0}\right) \tag{B.16}
$$

$$
\delta\Big|_{I_{\mathrm{IFCs}}^{\max}|_{\Delta P=0}}=\tan^{-1}\left(\frac{\left(\frac{(|v^+|-|v^-|)\sum_{i=1}^{n}P_i}{|v^+|^2-|v^-|^2}\right)\cos\gamma+\left(-\frac{(|v^+|+|v^-|)\sum_{i=1}^{n}Q_i}{|v^+|^2+|v^-|^2}\right)\sin\gamma}{\left(-\frac{(|v^+|-|v^-|)\sum_{i=1}^{n}Q_i}{|v^+|^2+|v^-|^2}\right)\cos\gamma-\left(\frac{(|v^+|+|v^-|)\sum_{i=1}^{n}P_i}{|v^+|^2-|v^-|^2}\right)\sin\gamma}\right). \tag{B.17}
$$

From (B.15)–(B.17), it can be seen that under zero active power oscillations of parallel IFCs, the collective peak current amplitude and the phase angle of parallel IFCs are independent of k_{Pi} and k_{qi}, and are constant values under fixed average active and reactive power output.

B.3 Maximizing Power/Current Transfer Capability of Interfacing Converters

Under the unbalanced condition, IFC output peak current increases, and active power oscillation cancelation may worsen this situation (see Figure B.3). As a result, the power/current transfer capability of IFCs is reduced.

From Section B.2.1, a minimum negative sequence current should be injected into the grid (ideally, it should be zero) to maximize individual IFC power/current transfer capability. From Figure B.3, when both k_{Pi} and k_{qi} move toward zero (negative sequence current injection is reduced), the peak current of the individual IFC reduces. At $k_{Pi} = k_{qi} = 0$, no negative sequence current is injected into the grid, and individual IFC power/current transfer capability is maximized. However, it should be noted that k_{Pi} and k_{qi} are usually non-zero values for power oscillation control purposes (see Figure B.3). Thus, both power oscillations and peak currents of individual IFCs should be considered for k_{Pi} and k_{qi} controls.

For parallel IFCs, it will be proved that peak currents of all individual IFCs should follow the following conditions to maximize the power/current transferring capability of parallel IFCs under a given voltage condition [1]:

1. All individual IFCs peak currents should be in the same phase as the collective peak current of parallel IFCs, and
2. All individual IFCs peak currents should be in-phase with the collective peak current of parallel IFCs.

This operation condition can optimize the utilization range of parallel IFCs under unbalanced voltage.

In general, for n-parallel IFCs under various PFs, three phases current phasors based on maximum currents can be represented as follows:

$$\begin{aligned}
I_{a1}^{\max}\angle\delta_{a1}\big|_{I_{a1}^{\max}} + I_{a2}^{\max}\angle\delta_{a2}\big|_{I_{a2}^{\max}} + \ldots + I_{an}^{\max}\angle\delta_{an}\big|_{I_{an}^{\max}} &= I_{a-\mathrm{IFCs}}^{\max}\big|_{\Delta P=0}\angle\delta_a\big|_{I_{a-\mathrm{IFCs}}^{\max}\big|_{\Delta P=0}} \\
I_{b1}^{\max}\angle\delta_{b1}\big|_{I_{b1}^{\max}} + I_{b2}^{\max}\angle\delta_{b2}\big|_{I_{b2}^{\max}} + \ldots + I_{bn}^{\max}\angle\delta_{bn}\big|_{I_{bn}^{\max}} &= I_{b-\mathrm{IFCs}}^{\max}\Big|_{\Delta P=0}\angle\delta_b\big|_{I_{b-\mathrm{IFCs}}^{\max}\big|_{\Delta P=0}} \\
I_{c1}^{\max}\angle\delta_{c1}\big|_{I_{c1}^{\max}} + I_{c2}^{\max}\angle\delta_{c2}\big|_{I_{c2}^{\max}} + \ldots + I_{cn}^{\max}\angle\delta_{cn}\big|_{I_{cn}^{\max}} &= I_{c-\mathrm{IFCs}}^{max}\big|_{\Delta P=0}\angle\delta_c\big|_{I_{c-\mathrm{IFCs}}^{\max}\big|_{\Delta P=0}}.
\end{aligned} \tag{B.18}$$

It is worth mentioning again that $I_{x-\mathrm{IFCs}}^{\max}\Big|_{\Delta P=0}$ (amplitude of collective maximum current of parallel IFCs in phase-x under $\Delta P = 0$) and $\delta_x\big|_{I_{x-\mathrm{IFCs}}^{\max}\big|_{\Delta P=0}}$ (phase angle of collective maximum current of parallel IFCs in phase-x under $\Delta P = 0$) are constant values. Assuming that the phase angles of individual IFCs maximum currents in each phase are equalized with the phase angle of collective maximum currents of parallel IFCs in that phase $\left(\angle\delta_{x1}\big|_{I_{x1}^{\max}} = \angle\delta_{x2}\big|_{I_{x2}^{\max}} = \ldots = \delta_{xn}\big|_{I_{xn}^{\max}} = \angle\delta_x\big|_{I_{x-\mathrm{IFCs}}^{\max}\big|_{\Delta P=0}}; x = a, b, c\right)$, (B.18) can be rewritten as follows:

$$
\begin{aligned}
I_{a1}^{\max'} + I_{a2}^{\max'} + \ldots + I_{an}^{\max'} &= \sum_{i=1}^{n} I_{ai}^{\max'} = I_{a-\text{IFCs}}^{\max}\Big|_{\Delta P=0} \\
I_{b1}^{\max'} + I_{b2}^{\max'} + \ldots + I_{bn}^{\max'} &= \sum_{i=1}^{n} I_{bi}^{\max'} = I_{b-\text{IFCs}}^{\max}\Big|_{\Delta P=0} \\
I_{c1}^{\max'} + I_{c2}^{\max'} + \ldots + I_{cn}^{\max'} &= \sum_{i=1}^{n} I_{ci}^{\max'} = I_{c-\text{IFCs}}^{\max}\Big|_{\Delta P=0}
\end{aligned} \tag{B.19}
$$

where superscript "′" refers to maximum currents values after phase angle equalization. From (B.19), although phase angle equalization can reduce the summation of individual IFCs maximum currents in each phase ($\sum_{i=1}^{n} I_{xi}^{\max'} \leq \sum_{i=1}^{n} I_{xi}^{\max}; x = a, b, c$), it cannot guarantee the minimization of individual IFCs peak currents summation. In other words, it cannot guarantee maximizing the power/current transfer capability of parallel IFCs without considering the phases that the peak currents are in those phases. In more detail, if it is assumed that m number of IFCs peak currents are in phase a, u number are in phase b, and $n-m-u$ number are in phase c ($n \geq m+u$), the following expression will be obtained:

$$
\begin{aligned}
\sum_{i=1}^{m} I_{ai}^{\max'} + \sum_{i=m+1}^{m+u} I_{bi}^{\max'} + \sum_{i=m+u+1}^{n} I_{ci}^{\max'} &> I_{a-\text{IFCs}}^{\max}\Big|_{\Delta P=0} \\
\sum_{i=1}^{m} I_{ai}^{\max'} + \sum_{i=m+1}^{m+u} I_{bi}^{\max'} + \sum_{i=m+u+1}^{n} I_{ci}^{\max'} &> I_{b-\text{IFCs}}^{\max}\Big|_{\Delta P=0} \\
\sum_{i=1}^{m} I_{ai}^{\max'} + \sum_{i=m+1}^{m+u} I_{bi}^{\max'} + \sum_{i=m+u+1}^{n} I_{ci}^{\max'} &> I_{c-\text{IFCs}}^{\max}\Big|_{\Delta P=0}.
\end{aligned} \tag{B.20}
$$

Therefore, considering (B.20), it is concluded that to minimize the summation of individual IFCs peak currents (which is equal to the collective peak current of parallel IFCs; min $\left(\sum_{i=1}^{n} I_{i}^{\max}\right) = I_{\text{IFCs}}^{\max}\Big|_{\Delta P=0}$), or in other words, to maximize the parallel IFC power/current transfer capability, all individual IFCs peak currents should be kept in the same phase and in-phase with parallel IFCs collective peak currents.

It is worth mentioning that this strategy can be used for parallel IFCs operating under different PFs. It should also be highlighted that, here, summation of n-parallel IFCs peak currents are minimized to maximize parallel IFCs power/current transfer capability, and the individual IFCs peak current values are not addressed directly.

B.3.1 Individual IFCs Peak Currents in the Same Phase as the Collective Peak Current of Parallel IFCs

Under zero active power oscillation, the collective peak current of parallel IFCs is independent of k_{pi} and k_{qi}, and it is a constant value under fixed values of active and reactive powers (see (B.15)–(B.17)). If all IFCs peak currents are in the same phase as the collective peak current of parallel IFCs (e.g. all peak currents are in phase b), the summation of their peak current amplitudes will be reduced.

However, considering (B.9) and (B.15), under $\Delta P = 0$, the peak currents of individual IFCs can be in the same phase or different phases with the collective peak current of parallel IFCs, depending on k_{pi} and k_{qi} values with given P_i and Q_i. For example, in two-parallel IFCs with $P_1 = 4$ kW, $Q_1 = 7$ kVar, $P_2 = 5$ kW and $Q_2 = 0.5$ kVar, under $k_{p1} = -0.74$, $k_{q1} = 0.74$, $k_{p2} = -1.20$, and $k_{q2} = 6.27$, the system has $\Delta P = 0$, and the peak currents of the two IFCs and the collective peak current of parallel IFCs are in phase b. In this operating point, the collective peak current is $I_{\text{IFCs}}^{\max}\big|_{\Delta P=0} = 49.02A$ which is shared between the two IFCs as $I_1^{\max} = 30A$ and $I_2^{\max} = 23.79A$ (the difference between $I_1^{\max} + I_2^{\max}$ and $I_{\text{IFCs}}^{\max}\big|_{\Delta P=0}$ is due to phase angle difference between $\delta|_{I_1^{\max}}$ and $\delta|_{I_2^{\max}}$). Under the $k_{p1} = -1.81$, $k_{q1} = -1.39$, $k_{p2} = -0.24$, and $k_{q2} = -15.85$ operating point, the collective active power oscillation is zero again. However, the peak current of the first, second and collective peak current are in phases c, a, and b, respectively. In this operating point, the collective peak current is similar to the previous operating point and equal to $I_{\text{IFCs}}^{\max}\big|_{\Delta P=0} = 49.02$ A (independent of k_{Pi} and k_{qi} under $\Delta P = 0$) while the first and second IFC peak currents are $I_1^{\max} = 42.05$ A and $I_2^{\max} = 33.36$ A. As clear from the example, under $\Delta P = 0$, when all IFC peak currents are in the same phase as the collective peak current of parallel IFCs, the summation of their peak current amplitudes is reduced. For example, in Chapter 9, Section 9.3.2., it has been explained that since all IFCs, except the redundant IFC, are controlled based on their current ratings, keeping all IFC peak currents in the same phase with the collective peak current can reduce the redundant IFC peak current.

Considering the discussions above, k_{pi} and k_{qi} of individual IFCs can be controlled to lead their peak currents in the same phase as the collective peak current of parallel IFCs.

Boundary Conditions

In this section, different conditions in which the peak currents of individual IFCs and the collective peak current of parallel IFCs are in the same phase are studied. Considering these conditions, appropriate boundaries are considered for coefficient factors to keep the peak currents of individual IFCs and collective peak current in the same phase, which leads to a smaller peak current amplitude summation of the parallel IFCs.

Considering (B.15), among three phases, the collective peak current of parallel IFCs is in the phase where $\cos(2\gamma)$ has its minimum value.

For the individual ith IFC, the maximum current expression at each phase in (B.9) can be rewritten as follows:

$$I_{xi}^{\max} = \sqrt{F_{1i} - \sqrt{(F_{2i})^2 + (F_{3i})^2}\cos(2\gamma + \beta_i)} \quad x = a, b, c \tag{B.21}$$

where

$$F_{1i} = \frac{P_i^2}{(|v^+|^2 + k_{pi}|v^-|^2)^2}\left(|v^+|^2 + |v^-|^2 k_{pi}^2\right) + \frac{Q_i^2}{(|v^+|^2 + k_{qi}|v^-|^2)^2}\left(|v^+|^2 + |v^-|^2 k_{qi}^2\right) \tag{B.22}$$

$$F_{2i} = \frac{2|v^+||v^-| \times \left(P_i^2 k_{pi}(|v^+|^2 + k_{qi}|v^-|^2)^2 - Q_i^2 k_{qi}(|v^+|^2 + k_{pi}|v^-|^2)^2\right)}{(|v^+|^2 + k_{pi}|v^-|^2)^2\,(|v^+|^2 + k_{qi}|v^-|^2)^2} \tag{B.23}$$

$$F_{3i} = -\frac{2P_iQ_i|v^+||v^-|(k_{pi} + k_{qi})}{(|v^+|^2 + k_{pi}|v^-|^2)(|v^+|^2 + k_{qi}|v^-|^2)} \tag{B.24}$$

$$\beta_i = \tan^{-1}\left(\frac{P_i^2 k_{pi}(|v^+|^2 + k_{qi}|v^-|^2)^2 - Q_i^2 k_{qi}(|v^+|^2 + k_{pi}|v^-|^2)^2}{-P_iQ_i(k_{pi} + k_{qi})(|v^+|^2 + k_{pi}|v^-|^2)(|v^+|^2 + k_{qi}|v^-|^2)}\right) + \frac{\pi}{2}. \tag{B.25}$$

Considering (B.21)–(B.25), for an individual IFC, the phase with a minimum value of $\cos(2\gamma + \beta_i)$ will have maximum current. Therefore, the phase of the ith IFC peak current depends on γ (or in other words, ρ) and β_i values. Since under zero active power oscillations, $|v^+|$, $|v^-|$, and ρ are constant values under fixed average active and reactive powers, β_i (or in other words, k_{pi} and k_{qi}) can determine the phase of the IFC peak current.

In Figure B.4, conditions in which the peak current of the individual ith IFC and collective peak current of parallel IFCs under $\Delta P = 0$ are in different phases are shown with the shaded area, while in the unshaded area, these peak currents are in the same phase [2]. From Figure B.4, if β_i of the ith IFC is in the boundary of $0 < \beta_i < 2\pi/3$ or $4\pi/3 < \beta_i < 2\pi$ and ρ is in the unshaded areas, the peak current of that ith IFC will be in the same phase as the collective peak current of the parallel IFCs. On the other hand, in the condition that β_i of the ith IFC is in the boundary of $2\pi/3 < \beta_i < 4\pi/3$, the peak current of that individual IFC and the collective peak current of parallel IFCs will always be in different phases, which is not shown in Figure B.4. It is worth mentioning that in Figure B.4, shaded and unshaded areas

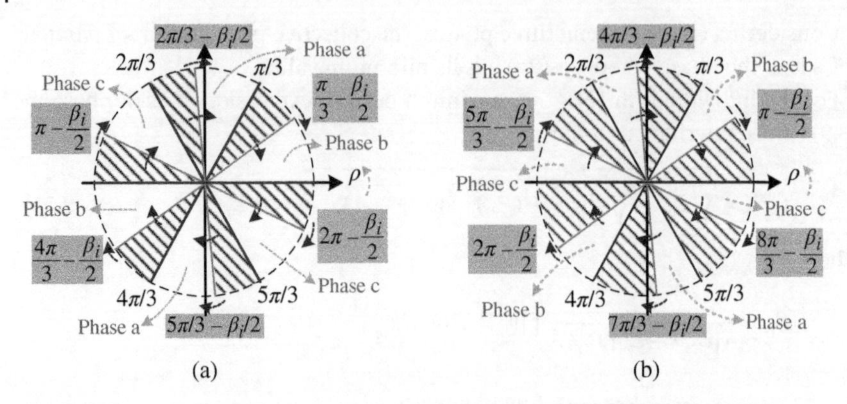

Figure B.4 Relation between the phase of the individual ith IFC peak current and parallel IFC collective peak current (peak currents are in the same phase in unshaded areas and in different phases in shaded areas): (a) $0 < \beta_i < 2\pi/3$, (b) $4\pi/3 < \beta_i < 2\pi$.

are controlled by β_i values (or in other words, by k_{pi} and k_{qi} values). Since ρ is a constant value under fixed values of average active and reactive powers, β_i can be controlled to lead individual IFC peak current to the same phase as the collective peak current of parallel IFCs.

IFCs under Unity Power Factor Operation Mode

Considering (B.25), under unity PF operation mode ($Q_i = 0$), if $k_{Pi} < 0$, β_i is equal to 0, and if $k_{Pi} > 0$, β_i is equal to π. Therefore, when $k_{Pi} < 0$, the peak current of the individual ith IFC will be in the same phase as the collective peak current of the parallel IFCs, regardless of the value of ρ and active power flow direction (under $\beta_i = 0$, the dashed areas do not exist in Figure B.4).

In Figure B.5, the phase in which all IFCs peak currents and the collective peak current of parallel IFCs are in that phase is shown, in which different values of ρ is considered while $k_{pi} < 0$. Also, in a simulated case study in Figure B.6, it is shown that different values of k_{pi} affect the maximum current at each phase and the peak current of individual IFC under different phase angles ρ and average active powers output.

From Figures B.5 and B.6, it is clear that the grid conditions in terms of different values of ρ affect which phase the peak currents are in. In the following, more detail is provided.

Considering (B.9) and Figure B.5, in the grid condition that $\rho = i\pi/6$; $i = 1$, 3,5,7,9,11, (A_i; $i = 1, \ldots, 6$ operation points), individual IFCs peak currents have their maximum possible values ($\cos(2\gamma)|_{I_{pi}^{\max}} = -1$). It is worth mentioning that the range of $\cos(2\gamma)$ variations in the phase that all peak currents are in that phase is $-1 < \cos(2\gamma)|_{I_{pi}^{\max}} < -0.5$ (see (B.9), and Figures B.5 and B.6). When ρ moves toward boundaries in $\rho = i\pi/3$; $i = 1, \ldots, 6$, the peak currents of individual IFCs

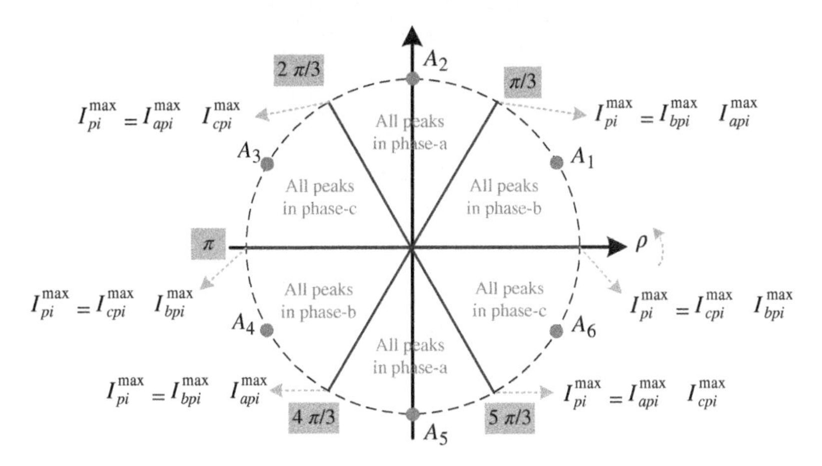

Figure B.5 The phase of IFCs peak currents for different values of ρ when $k_{pi} < 0$ for all IFCs.

will reduce. At $\rho = i\pi/3; i = 1, \ldots, 6$, the peak current of the individual IFC will be equal to one of the other phases' maximum current (see these points in Figure B.5), and increasing ρ will lead the peak current to the other phase. For example, under $\rho = \pi/3$, the peak current, which is in phase b, and the maximum current of phase a are equal, and increasing ρ will lead the peak currents to phase a.

In Chapter 9, Sections 9.3.2 and 9.3.3 in the control strategies under unity PF operation mode, k_{pi} is controlled to be less than zero to keep all individual IFCs peak currents and the collective peak current of parallel IFCs in the same phase, leading to reduced peak currents' summation of parallel IFCs.

IFCs under Non-unity Power Factor Operation Mode

Under non-unity PF operation mode considering (B.25), determination of boundaries in which the individual IFCs peak currents are in the same phase with a collective peak current of parallel IFCs is challenging, and they should be updated under average active–reactive power variations. However, (B.25) can be simplified, and boundaries can be determined under a specific relation between k_{pi} and k_{qi} as:

$$k_{pi} + k_{qi} = 0 \quad i = 1, \ldots, n-1. \tag{B.26}$$

However, this condition cannot be applied to all parallel IFCs. For example, in Chapter 9, Section 9.3.2, since the redundant IFC is utilized for active power oscillation cancelation, (B.26) may not be applicable for the redundant IFC. The relation between k_{pn} and k_{qn} of the redundant IFC will be discussed later.

Applying (B.26), if $k_{pi} < 0; i = 1, \ldots, n-1$ (or $k_{qi} > 0; i = 1, \ldots, n-1$), the peak current of the individual ith IFC and the collective peak current of parallel IFCs will be in the same phase since $\beta_i = 0$, regardless of the value of ρ and average active and reactive powers flow directions (similar to the unity PF operation condition).

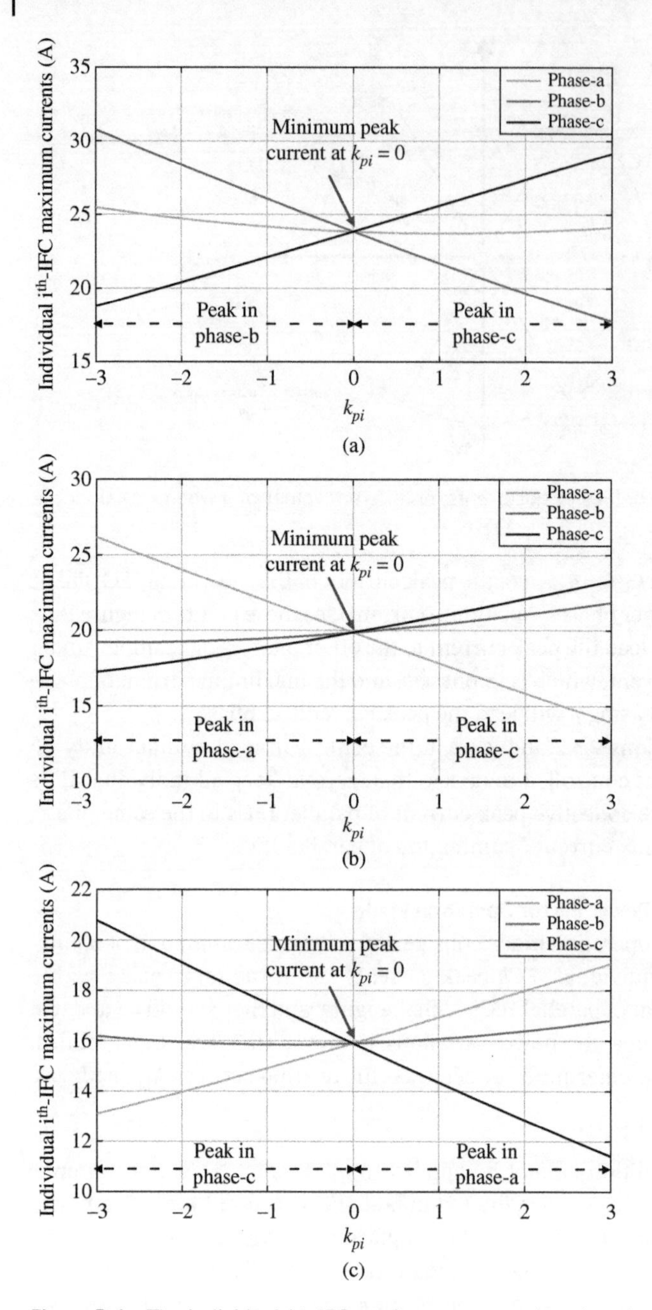

Figure B.6 The individual *i*th IFC maximum currents in *a*, *b*, *c* phases under different k_{pi}, phase angle (ρ), and average active powers ($|v^+| = 168$ V, $|v^-| = 16$ V): (a) $\rho = 45$, $P_i = 6$ kW, (b) $\rho = 85$, $P_i = 5$ kW, and (c) $\rho = 160$, $P_i = 4$ kW.

Moreover, $k_{pi} > 0; i = 1, \ldots, n-1$ (or $k_{qi} < 0; i = 1, \ldots, n-1$) will lead peak currents to different phases since $\beta_i = \pi$.

Considering the individual IFC power oscillations and their peak current, (B.26) could also satisfy the active power oscillation-free operation of individual IFCs under $k_{pi} = -1$ and $k_{qi} = 1$, and provide a minimum peak current of the individual IFC under $k_{pi} = k_{qi} = 0$ (for more information, see Figure B.3).

In the parallel IFCs control strategy of Section 9.3.2, if IFCs are under non-unity PF operation mode, (B.26) is applied to all individual IFCs except the redundant one, and their k_{pi} will be controlled to be less than zero to keep their peak currents in the same phase with the collective peak current of parallel IFCs. The redundant IFC cancels out active power oscillations using (B.13) and (B.14) in that control strategy. As a result, the relation between k_{pn} and k_{qn} of the redundant IFC can be achieved as following:

$$G_n = \frac{k_{qn}}{k_{pn}} = \frac{\left[Q_n - |v^+|^2\left(\frac{\sum_{i=1}^{n} Q_i}{|v^+|^2+|v^-|^2} - \sum_{i=1}^{n-1}\frac{Q_i}{|v^+|^2-k_{p_i}|v^-|^2}\right)\right]}{\left[P_n - |v^+|^2\left(\frac{\sum_{i=1}^{n} P_i}{|v^+|^2-|v^-|^2} - \sum_{i=1}^{n-1}\frac{P_i}{|v^+|^2+k_{p_i}|v^-|^2}\right)\right]} \times \frac{\left[\frac{\sum_{i=1}^{n} P_i}{|v^+|^2-|v^-|^2} - \sum_{i=1}^{n-1}\frac{P_i}{|v^+|^2+k_{p_i}|v^-|^2}\right]}{\left[\frac{\sum_{i=1}^{n} Q_i}{|v^+|^2+|v^-|^2} - \sum_{i=1}^{n-1}\frac{Q_i}{|v^+|^2-k_{p_i}|v^-|^2}\right]}. \tag{B.27}$$

All IFCs will work under the same power factor in this control strategy. Thus, the following relation can be considered:

$$\frac{P_i}{Q_i} = \frac{\sum_{i=1}^{n} P_i}{\sum_{i=1}^{n} Q_i} = \frac{1}{U} \quad i = 1, \ldots, n \tag{B.28}$$

where U is a number. Assuming $|v^-| = M \times |v^+|$ in which M is the unbalanced ratio and $0 \leq M \leq 1$, (B.27) can be rewritten as follows:

$$G_n = \frac{\left[P_n - \left(\frac{\sum_{i=1}^{n} P_i}{1+M^2} - \sum_{i=1}^{n-1}\frac{P_i}{1-k_{p_i}M^2}\right)\right] \times \left[\frac{\sum_{i=1}^{n} P_i}{1-M^2} - \sum_{i=1}^{n-1}\frac{P_i}{1+k_{p_i}M^2}\right]}{\left[P_n - \left(\frac{\sum_{i=1}^{n} P_i}{1-M^2} - \sum_{i=1}^{n-1}\frac{P_i}{1+k_{p_i}M^2}\right)\right] \times \left[\frac{\sum_{i=1}^{n} P_i}{1+M^2} - \sum_{i=1}^{n-1}\frac{P_i}{1-k_{p_i}M^2}\right]}. \tag{B.29}$$

In a practical power system, M is a small value (based on an IEEE Standard, a typical value of M in a three-phase power system under steady-state operation is less than 3%). Assuming that $|v^+|$ and $|v^-|$ are constant values and since $-1 \leq k_{pi} \leq 0; i = 1, \ldots, n-1$, $k_{pi}M^2$ will be small enough to be neglected in (B.29).

Therefore, (B.29) can be simplified as follows:

$$G_n = \frac{k_{qn}}{k_{pn}} = \frac{M^2\left(P_n + M^2\sum_{i=1}^{n-1} P_i\right)\sum_{i=1}^{n} P_i}{-M^2\left(P_n - M^2\sum_{i=1}^{n-1} P_i\right)\sum_{i=1}^{n} P_i} = -\frac{P_n + M^2\sum_{i=1}^{n-1} P_i}{P_n - M^2\sum_{i=1}^{n-1} P_i}. \quad \text{(B.30)}$$

Considering (B.30), G_n will be close to −1, depending on the number of parallel IFCs, the power rating of the redundant IFC in comparison to other IFCs, and the value of M. As a result, considering (B.30), β_n will be a very small value close to zero degrees, which results in the small shaded area in Figure B.4. In this case, even though we fall in this small area (this area is the transition that the peak current is switched from one phase to another phase), the peak current does not change that much. In other words, the peak currents of the two phases are almost the same. Therefore, it does not matter in which phase the peak current is.

As a numerical example, three-parallel IFCs have been simulated under the same PF ($U = 1/15$) and different apparent powers and coefficient factors. In this example, $M = 0.3$, $S_3 = 0.5S_T$ (S_T is the total apparent power), $k_{p1} = -k_{q1} : -0.9 \longrightarrow -0.1$, $k_{p2} = -k_{q2} = -0.9$, and the third IFC (the redundant one) cancels active power oscillations. The variations of β_3 under different operating conditions are shown in Figure B.7. From the figure, β_3 changes within 0.4 degree under different operating conditions, leading to small

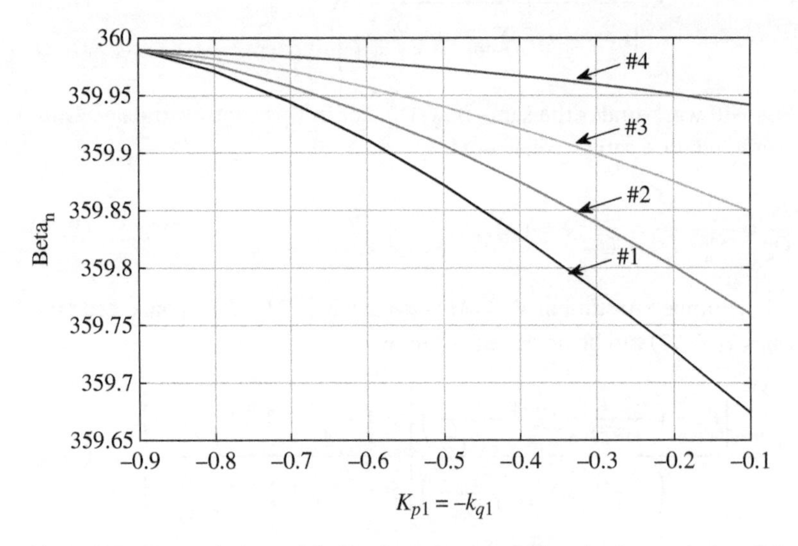

Figure B.7 The variations of B_3 (for the redundant IFC) under $k_{p2} = -k_{q2} = -0.9$, $S_3 = 0.5S_T$, $U = 1/15$, $M = 0.3$, and different IFCs apparent powers: case#1: $P_1 = 7$ kW, $Q_1 = 466.6$ Var, $P_2 = 2.979$ kW, $Q_2 = 198.52$ Var; case#2: $P_1 = 5$ kW, $Q_1 = 333.3$ Var, $P_2 = 4.977$ kW, $Q_2 = 331.85$ Var; case#3: $P_1 = 3$ kW, $Q_1 = 200$ Var, $P_2 = 2.979$ kW, $Q_2 = 198.52$ Var; case#4: $P_1 = 1$ kW, $Q_1 = 66.6$ Var, $P_2 = 8.977$ kW, $Q_2 = 598.52$ Var.

shaded areas. Thus, k_{pi}; $i = 1, \ldots, n$ will be controlled to be less than zero to provide reduced peak currents summation of parallel IFCs (in other words, to reduce the redundant IFC peak current).

B.3.2 Individual IFCs Peak Currents In-phase with the Collective Peak Current of Parallel IFCs

As mentioned, to optimize the utilization range of parallel IFCs and maximize their power/current transfer capability under a given voltage condition (in other words, to minimize the peak current summation of parallel IFCs), the peak currents of all individual IFCs should be kept in the same phase and in-phase with the collective peak current of parallel IFCs. The previous section provided a detailed discussion about how the peak currents can be kept in the same phase. Here, it is studied how individual IFCs peak currents phasors can be kept in-phase with the n-parallel IFCs collective peak current phasor. It should be assumed that all the peak currents are in the same phase.

Under unity PF operation, from (B.11) and (B.17), the phase angles of individual IFCs maximum currents in each phase depend on k_{pi} while in parallel IFCs under $\Delta P = 0$, the phase angles of collective maximum currents in each phase are independent of k_{pi}. However, since in parallel IFCs with common DC and AC links, $|v^+|$, $|v^-|$ and γ are common values in individual and parallel IFCs, and the variations of $\delta_{xi}\big|_{I_{xpi}^{\max}}$ (when k_{pi} deviates from $k_{pi} = -1$) is small enough (under $k_{pi} = -1$, $\delta_{xi}\big|_{I_{xpi}^{\max}} = \delta_x\big|_{I_{xp-\text{IFCs}}^{\max}\big|_{\Delta P=0}}$; $x = a, b, c$), the $\delta_{xi}\big|_{I_{xpi}^{\max}}$ in each phase can be assumed constant under k_{pi} variations and equal to $\delta_x\big|_{I_{xp-\text{IFCs}}^{\max}\big|_{\Delta P=0}}$ with a good approximation. For more investigation, the variation of $\delta_{xi}\big|_{I_{xpi}^{\max}}$ with respect to k_{pi} deviation is derived as in (B.31) using (B.11).

$$\Delta\delta_{xi}\big|_{I_{xpi}^{\max}} = \frac{-2|v^+||v^-|\cot\gamma}{(|v^+| - k_{pio}|v^-|)^2 + (\cot\gamma)^2 \times (|v^+| + k_{pio}|v^-|)^2}\Delta k_{pi}$$

$$x = a, b, c \tag{B.31}$$

where k_{pio} is the initial operation point which is −1 in our case. From (B.31), it can be understood that $\Delta\delta_{xi}\big|_{I_{xpi}^{\max}}$ over Δk_{Pi} is a small value in each phase. As a numerical example, in three-parallel IFCs with $P_1 = 6$ kW, $P_2 = 2$ kW, $P_3 = 3.6$ kW, $k_{pio} = -1, p = 55°$, $|v^+| = 168$ V and $|v^-| = 16$ V, the deviation of $\delta_{xi}\big|_{I_{xpi}^{\max}}$ in the three a, b, c phases from its initial point (where $k_{pi} = -1$ and $\delta_{xi}\big|_{I_{xpi}^{\max}} = \delta_x\big|_{I_{xp-\text{IFCs}}^{\max}\big|_{\Delta P=0}}$; $x = a, b, c$) under $\Delta k_{pi} = 2$ are $\Delta\delta_{ai}\big|_{I_{xpi}^{\max}} \approx 8^\circ$, $\Delta\delta_{bi}\big|_{I_{xpi}^{\max}} \approx 6^\circ$, and $\Delta\delta_{ci}\big|_{I_{xpi}^{\max}} \approx 2^\circ$. Therefore, it can be concluded that $\delta_{xi}\big|_{I_{xpi}^{\max}} \approx \delta_x\big|_{I_{xp-\text{IFCs}}^{\max}\big|_{\Delta P=0}}$; $x = a, b, c$ in the range of k_{pi} operation with a good approximation.

With the discussions above, under $k_{pi} < 0$; $i = 1, \ldots, n$, following the relation among peak currents of individual IFCs and the collective peak current of parallel IFCs with unity PF operation can be derived:

$$I_{p-\text{IFCs}}^{\max} \cong \sum_{i=1}^{n} I_{pi}^{\max}. \tag{B.32}$$

From (B.32), two conclusions can be obtained; first, when $k_{pi} < 0$; $i = 1, \ldots, n$, the maximum power/current transfer capability of n-parallel IFCs under unity PF operation can be achieved, and second, the collective peak current of n-parallel IFCs, which is a constant value under fixed output active powers, can be shared linearly among individual IFCs.

For parallel IFCs operating under non-unity PFs, it is challenging to keep individual IFCs peak current phasors in-phase with the collective peak current phasor since two control variables k_{pi} and k_{qi} are already used to put peak currents in the same phase. However, since all IFCs work under the same power factor most of the time, the phase angle difference between n-parallel IFCs collective peak current and individual IFCs peak currents would be small.

B.4 Summary

In this Appendix, individual and parallel IFCs peak currents under unbalanced voltage were analyzed. The study concluded that for individual IFCs, reducing output active power oscillation increases the output peak current, and both of them should be considered for individual IFCs control under unbalanced voltage. On the other hand, in parallel IFCs under zero active power oscillations, the collective peak current is constant in the fixed average active–reactive powers. Also, it was proven that the power/current transfer capability of parallel IFCs under a given voltage condition is maximized (in other words, the utilization range of parallel IFCs is optimized) when the peak currents of all individual IFCs are kept in the same phase and in-phase with the collective peak current of parallel IFCs.

References

1 Nejabatkhah, F., Li, Y.W., Sun, K., and Zhang, R. (2018). Active power oscillation cancelation with peak current sharing in parallel interfacing converters under unbalanced voltage. *IEEE Transactions on Power Electronics* 33 (12): 10200–10214.

2 Nejabatkhah, F., Li, Y.W., and Sun, K. (2018). Parallel three-phase interfacing converters operation under unbalanced voltage in hybrid AC/DC microgrid. *IEEE Transactions on Smart Grid* 9 (2): 1310–1322.

C

Case Study System Parameters

Table C.1 Parameters used in simulation of microgrid system formed by interfacing converters with enhanced droop control.

Category	Parameter	Value
Parameters of interfacing converters	Filter inductors	5 mH, 0.2 Ω
	Filter capacitor	40 μF
	Sampling frequency	4.5 kHz–9 kHz
Droop coefficient	Frequency droop	0.00143 rad/s/W
	Voltage droop	0.00167 V/Var
	Integration deadband	6 W
	Integral gain	0.0286 V/s/W
	Time constant of low pass filter	0.0159 s
Parameter of microgrid	Rated voltage	208 V/60 Hz
	Impedance of feeder 1–3	3 mH 0.3 Ω
	Impedance of feeders between loads	2 mH 0.2 Ω
	Load 1 power	2350 W, 950 Var
	Load 2 power	1175 W, 475 Var

Smart Hybrid AC/DC Microgrids: Power Management, Energy Management, and Power Quality Control, First Edition. Yunwei Ryan Li, Farzam Nejabatkhah, and Hao Tian.

Table C.2 Parameters used in simulation of virtual impedance based fault ride-through control scheme.

Category	Parameter	Value
Parameters of interfacing converters	AC voltage	110 V 60 Hz
	DC voltage	230 V
	Filter inductors	5 mH, 0.2 Ω
	Filter capacitor	40 μF
	Sampling frequency	4.5 kHz
Control parameters	Proportional gain of outer loop	0.22
	Resonant gain of outer loop	25
	Proportional gain of inter loop	22

Table C.3 System parameters for experiments of adjustable unbalanced voltage compensation with IFC active power oscillation minimization.

Parameter	Value
DC link voltage	150 V
Reference active power	280–70 W
Reference reactive power	70–280 VAR
Three phase unbalanced load	9 Ω, 5 Ω, 3 Ω
Grid phase voltage (rms) and frequency	50 V, 60 Hz
Grid coupling resistance	0.2–1.9 Ω
Grid coupling reactance	1.88–0.94 Ω

Table C.4 System parameters for simulations of parallel three-phase IFC control under unbalanced voltage with redundant IFC for active power oscillation cancelation.

Parameter	Value
DC link voltage	800 V
First IFC power rating	9 kVA
Second IFC power rating	4 kVA
Third IFC power rating	10 kVA
First IFC current rating	30 A
Second IFC current rating	13 A
Third IFC current rating	34 A
Grid voltage (rms) and frequency	240 V, 60 Hz
Grid coupling resistance	0.2 Ω
Grid coupling reactance	1.88 Ω

Table C.5 System parameters for simulations of parallel three-phase IFC control under unbalanced voltage with all IFCs participate in active power oscillation cancelation.

Parameter	Value
DC link voltage	800 V
First IFC power rating	9 kVA
Second IFC power rating	4 kVA
Third IFC power rating	10 kVA
First IFC current rating	30 A
Second IFC current rating	13 A
Third IFC current rating	34 A
Grid voltage (rms) and frequency	240 V, 60 Hz
Grid coupling resistance	0.2 Ω
Grid coupling reactance	1.88 Ω

Table C.6 Values of C_1 to C_{10}; coefficients of the amplitude of the PCC current vector negative sequence.

Parameter	Equation
C_1	$C_1 = \frac{1}{9\vert v_{PCC_a}\vert^2}$
C_2	$C_2 = \frac{1}{9\vert v_{PCC_b}\vert^2}$
C_3	$C_3 = \frac{1}{9\vert v_{PCC_c}\vert^2}$
C_4	$C_4 = \frac{2}{9\vert v_{PCC_a}\vert\vert v_{PCC_b}\vert}\cos\left(\theta_{v_{PCC_a}} - \theta_{v_{PCC_b}} - \frac{4\pi}{3}\right)$
C_5	$C_5 = \frac{2}{9\vert v_{PCC_a}\vert\vert v_{PCC_c}\vert}\cos\left(\theta_{v_{PCC_a}} - \theta_{v_{PCC_c}} - \frac{2\pi}{3}\right)$
C_6	$C_6 = \frac{2}{9\vert v_{PCC_b}\vert\vert v_{PCC_c}\vert}\cos\left(\theta_{v_{PCC_b}} - \theta_{v_{PCC_c}} + \frac{2\pi}{3}\right)$
C_7	$C_7 = \frac{2P_{PCC,b}}{9\vert v_{PCC_a}\vert\vert v_{PCC_b}\vert}\sin\left(\theta_{v_{PCC_a}} - \theta_{v_{PCC_b}} - \frac{4\pi}{3}\right) + \frac{2P_{PCC,c}}{9\vert v_{PCC_a}\vert\vert v_{PCC_c}\vert}\sin\left(\theta_{v_{PCC_a}} - \theta_{v_{PCC_c}} - \frac{2\pi}{3}\right)$
C_8	$C_8 = \frac{2P_{PCC,a}}{9\vert v_{PCC_a}\vert\vert v_{PCC_b}\vert}\sin\left(\theta_{v_{PCC_b}} - \theta_{v_{PCC_a}} + \frac{4\pi}{3}\right) + \frac{2P_{PCC,c}}{9\vert v_{PCC_b}\vert\vert v_{PCC_c}\vert}\sin\left(\theta_{v_{PCC_b}} - \theta_{v_{PCC_c}} + \frac{2\pi}{3}\right)$
C_9	$C_9 = \frac{2P_{PCC,a}}{9\vert v_{PCC_a}\vert\vert v_{PCC_c}\vert}\sin\left(\theta_{v_{PCC_c}} - \theta_{v_{PCC_a}} + \frac{2\pi}{3}\right) + \frac{2P_{PCC,b}}{9\vert v_{PCC_b}\vert\vert v_{PCC_c}\vert}\sin\left(\theta_{v_{PCC_c}} - \theta_{v_{PCC_b}} - \frac{2\pi}{3}\right)$
C_{10}	$C_{10} = \left(\frac{P_{PCC,a}\cos(\theta_{v_{PCC_a}})}{3\vert v_{PCC_a}\vert} + \frac{P_{PCC,b}\cos\left(\theta_{v_{PCC_b}} + \frac{4\pi}{3}\right)}{3\vert v_{PCC_b}\vert} + \frac{P_{PCC,c}\cos\left(\theta_{v_{PCC_c}} + \frac{2\pi}{3}\right)}{3\vert v_{PCC_c}\vert}\right)^2 + \left(\frac{P_{PCC,a}\sin(\theta_{v_{PCC_a}})}{3\vert v_{PCC_a}\vert} + \frac{P_{PCC,b}\sin\left(\theta_{v_{PCC_b}} + \frac{4\pi}{3}\right)}{3\vert v_{PCC_b}\vert} + \frac{P_{PCC,c}\sin\left(\theta_{v_{PCC_c}} + \frac{2\pi}{3}\right)}{3\vert v_{PCC_c}\vert}\right)^2$

Table C.7 Values of D_1 to D_{10}; coefficients of amplitude of the PCC current vector zero sequence.

Parameter	Equation
D_1	$D_1 = \frac{1}{9\lvert v_{PCC_a}\rvert^2}$
D_2	$D_2 = \frac{1}{9\lvert v_{PCC_b}\rvert^2}$
D_3	$D_3 = \frac{1}{9\lvert v_{PCC_c}\rvert^2}$
D_4	$D_4 = \frac{2}{9\lvert v_{PCC_a}\rvert\lvert v_{PCC_b}\rvert}\cos(\theta_{v_{PCC_a}} - \theta_{v_{PCC_b}})$
D_5	$D_5 = \frac{2}{9\lvert v_{PCC_a}\rvert\lvert v_{PCC_c}\rvert}\cos(\theta_{v_{PCC_a}} - \theta_{v_{PCC_c}})$
D_6	$D_6 = \frac{2}{9\lvert v_{PCC_b}\rvert\lvert v_{PCC_c}\rvert}\cos(\theta_{v_{PCC_b}} - \theta_{v_{PCC_c}})$
D_7	$D_7 = \frac{2P_{PCC,b}}{9\lvert v_{PCC_a}\rvert\lvert v_{PCC_b}\rvert}\sin(\theta_{v_{PCC_a}} - \theta_{v_{PCC_b}}) + \frac{2P_{PCC,c}}{9\lvert v_{PCC_a}\rvert\lvert v_{PCC_c}\rvert}\sin(\theta_{v_{PCC_a}} - \theta_{v_{PCC_c}})$
D_8	$D_8 = \frac{2P_{PCC,a}}{9\lvert v_{PCC_a}\rvert\lvert v_{PCC_b}\rvert}\sin(\theta_{v_{PCC_b}} - \theta_{v_{PCC_a}}) + \frac{2P_{PCC,c}}{9\lvert v_{PCC_b}\rvert\lvert v_{PCC_c}\rvert}\sin(\theta_{v_{PCC_b}} - \theta_{v_{PCC_c}})$
D_9	$D_9 = \frac{2P_{PCC,a}}{9\lvert v_{PCC_a}\rvert\lvert v_{PCC_c}\rvert}\sin(\theta_{v_{PCC_c}} - \theta_{v_{PCC_a}}) + \frac{2P_{PCC,b}}{9\lvert v_{PCC_b}\rvert\lvert v_{PCC_c}\rvert}\sin(\theta_{v_{PCC_c}} - \theta_{v_{PCC_b}})$
D_{10}	$D_{10} = \left(\frac{P_{PCC,a}\cos(\theta_{v_{PCC_a}})}{3\lvert v_{PCC_a}\rvert} + \frac{P_{PCC,b}\cos(\theta_{v_{PCC_b}})}{3\lvert v_{PCC_b}\rvert} + \frac{P_{PCC,c}\cos(\theta_{v_{PCC_c}})}{3\lvert v_{PCC_c}\rvert}\right)^2 + \left(\frac{P_{PCC,a}\sin(\theta_{v_{PCC_a}})}{3\lvert v_{PCC_a}\rvert} + \frac{P_{PCC,b}\sin(\theta_{v_{PCC_b}})}{3\lvert v_{PCC_b}\rvert} + \frac{P_{PCC,c}\sin(\theta_{v_{PCC_c}})}{3\lvert v_{PCC_c}\rvert}\right)^2$

Table C.8 Distributed load modifications connected to Bus#671 (Model: Y-PQ) in IEEE 13-Node Test System.

	Phase *a*	Phase *b*	Phase *c*
Old values	17 kW–10 kVAr	66 kW–38 kVAr	117 kW–68 kVAr
New values	400 kW–0 kVAr	300 kW–0 kVAr	1040 kW–0 kVAr

Table C.9 Single-phase IFCs of DG connected to IEEE 13-Node Test System.

Phase	IFC number	Bus number	Power rating of DG inverters	Operating active power	Max available reactive power
Phase *a*	IFC1	#633	1 MVA	120 kW	±992.7 kVAr
	IFC2	#684	1 MVA	80 kW	±996.7 kVAr
Phase *b*	IFC1	#692	500 kVA	50 kW	±497.4 KVAr
	IFC2	#680	300 kVA	40 kW	±297.3 kVAr
	IFC3	#632	700 kVA	60 kW	±697.4 kVAr
Phase *c*	IFC1	#611	800 kVA	150 kW	±785.8 kVAr
	IFC2	#645	1 MVA	120 kW	±992.7 kVAr

Table C.10 Parameters used for simulation of CCM, VCM and HCM for harmonic control.

Parameter	Value
Grid voltage	110 V 60 Hz
DC-link voltage	260 V
LC filter	$L = 1.25$ mH, $C = 40$ μF
Switching frequency	12 kHz
Power reference	$P^* = 300$ W, $Q^* = 125$ Var
Feeder impedance	2.5 mH, 1 Ω
Grid impedance	2.5 mH, 1 Ω

Table C.11 Parameters used for experiment of VFF and CFF for harmonic control.

Parameter	Value
Grid voltage	110 V 60 Hz
DC-link voltage	260 V
LCL filter	$L1 = 2.5$ mH, $L2 = 2.5$ mH, $C = 40$ μF
Switching frequency	2 kHz
Grid impedance	2.5 mH, 1 Ω

Table C.12 Parameters used for experiment of virtual impedance based ripple mitigation control in a DC microgrid.

Category	Parameter	Value
DC bus	DC bus voltage	24 V
	DC bus capacitor	1000 μF
BDC A	Power	480 W
	Inductor	0.5 mH
	Low-voltage side capacitor	2200 μF
BDC B	Power rating	240 W
	Inductor	1 mH
	Low-voltage side capacitor	2200 μF

Table C.13 Parameters used in simulation of microgrid system with priority driven harmonic compensation control.

Parameter	Value
Distribution line voltage	7.2 kV, 60 Hz
Customer side voltage	120 V, 60 Hz
Equivalent impedance of distribution transformer	0.015 + 0.03j p.u
Distribution line impedance	0.43 Ω/km, 150 μH/km
Number of nodes	11

Index

Note: Page numbers in *italics* denote figures and page numbers in **bold** denot tables.

a

Smart Hybrid AC/DC Microgrids: Power Management, Energy Management, and Power Quality Control, First Edition. Yunwei Ryan Li, Farzam Nejabatkhah, and Hao Tian.

e

m

r